Gene Targeting

The Practical Approach Series

SERIES EDITOR

B. D. HAMES
Department of Biochemistry and Molecular Biology
University of Leeds, Leeds LS2 9JT, UK

See also the Practical Approach web site at **http://www.oup.co.uk/PAS**

★ indicates new and forthcoming titles

Affinity Chromatography
Affinity Separations
Anaerobic Microbiology
Animal Cell Culture (2nd edition)
Animal Virus Pathogenesis
Antibodies I and II
Antibody Engineering
Antisense Technology
★ Apoptosis
Applied Microbial Physiology
Basic Cell Culture
Behavioural Neuroscience
Bioenergetics
Biological Data Analysis
Biomechanics – Materials
Biomechanics – Structures and Systems
Biosensors
★ Caenorhabditis Elegans
Carbohydrate Analysis (2nd edition)
Cell-Cell Interactions
The Cell Cycle
Cell Growth and Apoptosis
★ Cell Growth, Differentiation and Senescence
★ Cell Separation
Cellular Calcium
Cellular Interactions in Development
Cellular Neurobiology
Chromatin
★ Chromosome Structural Analysis
Clinical Immunology
Complement
★ Crystallization of Nucleic Acids and Proteins (2nd edition)
Cytokines (2nd edition)
The Cytoskeleton
Diagnostic Molecular Pathology I and II
DNA and Protein Sequence Analysis
DNA Cloning 1: Core Techniques (2nd edition)
DNA Cloning 2: Expression Systems (2nd edition)

DNA Cloning 3: Complex Genomes (2nd edition)
DNA Cloning 4: Mammalian Systems (2nd edition)
★ DNA Microarrays
★ DNA Viruses
Drosophila (2nd edition)
Electron Microscopy in Biology
Electron Microscopy in Molecular Biology
Electrophysiology
Enzyme Assays
Epithelial Cell Culture
Essential Developmental Biology
Essential Molecular Biology I and II
★ Eukaryotic DNA Replication
Experimental Neuroanatomy
Extracellular Matrix
Flow Cytometry (2nd edition)
Free Radicals
Gas Chromatography
Gel Electrophoresis of Nucleic Acids (2nd edition)
★ Gel Electrophoresis of Proteins (3rd edition)
★ Gene Probes 1 and 2
★ Gene Targeting (2nd edition)
Gene Transcription
Genome Mapping
Glycobiology
Growth Factors and Receptors
Haemopoiesis
★ High Resolution Chromotography
Histocompatibility Testing
HIV Volumes 1 and 2
★ HPLC of Macromolecules (2nd edition)
Human Cytogenetics I and II (2nd edition)
Human Genetic Disease Analysis
★ Immobilized Biomolecules in Analysis
Immunochemistry 1
Immunochemistry 2
Immunocytochemistry
★ *In Situ* Hybridization (2nd edition)
Iodinated Density Gradient Media
Ion Channels
★ Light Microscopy (2nd edition)
Lipid Modification of Proteins
Lipoprotein Analysis
Liposomes
Mammalian Cell Biotechnology
Medical Parasitology
Medical Virology
MHC Volumes 1 and 2
★ Molecular Genetic Analysis of Populations (2nd edition)
Molecular Genetics of Yeast
Molecular Imaging in Neuroscience
Molecular Neurobiology
Molecular Plant Pathology I and II
Molecular Virology
Monitoring Neuronal Activity

★ Mouse Genetics and Transgenics
Mutagenicity Testing
Mutation Detection
Neural Cell Culture
Neural Transplantation
Neurochemistry (2nd edition)
Neuronal Cell Lines
NMR of Biological Macromolecules
Non-isotopic Methods in Molecular Biology
Nucleic Acid Hybridisation
★ Nuclear Receptors
Oligonucleotides and Analogues
Oligonucleotide Synthesis
PCR 1
PCR 2
★ PCR 3: PCR In Situ Hybridization
Peptide Antigens
Photosynthesis: Energy Transduction
Plant Cell Biology
Plant Cell Culture (2nd edition)
Plant Molecular Biology
Plasmids (2nd edition)
Platelets
Postimplantation Mammalian Embryos
★ Post-translational Processing
Preparative Centrifugation
Protein Blotting
★ Protein Expression
Protein Engineering
Protein Function (2nd edition)
Protein Phosphorylation (2nd edition)
Protein Purification Applications
Protein Purification Methods
Protein Sequencing
Protein Structure (2nd edition)
Protein Structure Prediction
Protein Targeting
Proteolytic Enzymes
Pulsed Field Gel Electrophoresis
RNA Processing I and II
RNA-Protein Interactions
Signalling by Inositides
★ Signal Transduction (2nd edition)
Subcellular Fractionation
Signal Transduction
★ Transcription Factors (2nd edition)
Tumour Immunobiology
★ Virus Culture

Gene Targeting

A Practical Approach

Second Edition

Edited by

ALEXANDRA L. JOYNER
Howard Hughes Medical Institute and Skirball Institute of Biomolecular Medicine, New York University School of Medicine

UNIVERSITY PRESS

OXFORD
UNIVERSITY PRESS

Great Clarendon Street, Oxford OX2 6DP

Oxford University Press is a department of the University of Oxford and furthers the University's aim of excellence in research, scholarship, and education by publishing worldwide in

Oxford New York

Athens Auckland Bangkok Bogotá Buenos Aires Calcutta Cape Town Chennai Dar es Salaam Delhi Florence Hong Kong Istanbul Karachi Kuala Lumpur Madrid Melbourne Mexico City Mumbai Nairobi Paris São Paulo Singapore Taipei Tokyo Toronto Warsaw

and associated companies in Berlin Ibadan

Oxford is a registered trade mark of Oxford University Press

Published in the United States
by Oxford University Press Inc., New York

A catalogue record for this book is available from the British Library

Library of Congress Cataloging in Publication Data
Gene targeting : a practical approach / edited by Alexandra L. Joyner.
—2nd ed.

Includes bibliographical references and index.

1. Gene targeting Laboratory manuals. I. Joyner, Alexandra L.

QH442.3.G456 1999 660.6′5–dc21 99–36660

ISBN 0-19-963793-8 (Hbk)
0-19-963792-X (Pbk)

Typeset by Footnote Graphics,
Warminster, Wilts
Printed in Great Britain by Information Press, Ltd,
Eynsham, Oxon.

Preface

Over the past ten years it has become possible to make essentially any mutation in the germline of mice by utilizing recombination and embryonic stem (ES) cells. Homologous recombination when applied to altering specific endogenous genes, referred to as gene targeting, provides the highest level of control over producing mutations in cloned genes. When this is combined with site specific recombination, a wide range of mutations can be produced. ES cell lines are remarkable since after being established from a blastocyst, they can be cultured and manipulated relatively easily in vitro and still maintain their ability to step back into a normal developmental program when returned to a pre-implantation embryo. With the exponential increase in the number of genes identified by various genome projects and genetic screens, it has become imperative that efficient methods be developed for determining gene function. Gene targeting in ES cells offers a powerful approach to study gene function in a mammalian organism. Gene trap approaches in ES cells, in particular when they are combined with sophisticated prescreens, offer not only a route to gene discovery, but also to gain information on gene sequence, expression and mutant phenotype.

The basic technology necessary for making designer mutations in mice has become widespread and researchers who have traditionally used cell biology or molecular experiments are adding gene targeting techniques to their repertoire of experimtal approaches. A second edition of this book was written for two main reasons. The first was to update previously described techniques and to add new techniques that have greatly expanded the types of mutations that can be made using recombination in ES cells. A chapter in this new edition describes the design and use of site specific recombination for gene targeting approaches and production of conditional mutations. The second reason for the new book was to provide a more in depth discussion of the experimental design considerations that are critical to a successful gene targeting study and to add approaches for analyzing mutant phenotypes, the most interesting part of an experiment. Gene targeting experiments should be designed to go far beyond just making a mutant mouse. The success of a gene targeting experiment no longer lies in the making of the mutation, but depends on the imaginative and insightful analysis of the mutant phenotypes that the mutation provides. A chapter in this edition describes the use of classical genetics in combination with gene targeting to get the most out of a genetic approach to a biological question.

The nature of in vivo gene targeting studies of gene function are such that critical design decisions must be made at every step in the experiment, and each decision can have a major impact on the value of the information obtained. From the start, the type of mutation to be made must be considered carefully. Whereas 10 years ago most mutations were designed to create null mutations and were therefore relatively simple to design, at present, a null mutation is only one of a long list of mutations that can be made, each providing different insight into the function of a gene. Point mutations, large deletions, gene exchanges (knock-ins) and conditional mutations are but a few of the choices one faces at the start of a gene targeting experiment. The next choice is the source of DNA for the targeting experiment and ES cell line to be used for the manipulations. Once the mutant ES cell clone has been obtained, there are then a number of alternative approaches that can be used to make ES cell chimeras that depend on the ES cell line which was used. Finally, and most importantly, is the analysis of any phenotype that arises. This second addition discusses techniques used to analyze mutant mice, ranging from standard descriptive evaluation, to a chimera analysis or complicated breeding experiments that utilize double mutants. If mice are simply considered as a 'bag of cells' or an in vivo source of selected cell types, then the tremendous resource which mice offer as a model organism is not being realized. The life of a mouse represents a continuum of dynamic processes, including pattern formation, organ development, learning, homeostasis and disease. By making genetic alterations in mice using gene targeting and ES cells, the effects of a given change can be studied in the context of the whole organism.

My goal in editing this book was to provide a manual that could take a newcomer to the exciting field of gene targeting and mutant analysis in mice from a cloned gene to a basic understanding of the genetic approaches available using ES cells, and how each technique can be used to design a particular in vivo test of gene function. The book should also provide a valuable bench side resource for anyone carrying out gene targeting or gene trap experiments, a chimera analysis or classical genetic approaches. I would once again like to extend many thanks and my deepest appreciation to all the authors for their great efforts in including detailed protocols and lucid discussions of the various approaches presented. I would also like to thank my family for their strong support and laboratory members past and present for helping to make gene targeting a reality. Finally, since many of the techniques use mice, the experiments should be carried out in accordance with local regulations.

New York, NY

A.L.J.

Contents

Contributors

ALEJANDRO ABUIN
Department of Comparative Genetics, SmithKline Beecham Pharmaceuticals, New Frontiers Science Park, Third Avenue, Harlow, Essex CM19 5AW, UK.

WOJTEK AUERBACH
HHMI and Developmental Genetics Program, Skirball Institute of Biomolecular Medicine, New York University School of Medicine, 540 First Avenue, 4th Floor, New York, NY 10016, USA.

ALLAN BRADLEY
HHMI and Baylor College of Medicine, Department of Molecular and Medical Genetics, One Baylor Plaza, Houston, TX 77030, USA.

SCOTT BULTMAN
Department of Genetics, Case Western Reserve University, 2119 Abington Road, Cleveland, OH 44106, USA.

SUSAN M. DYMECKI
Harvard Medical School, Department of Genetics, 200 Longwood Avenue, Boston, MA 02115, USA.

ACHIM GOSSLER
The Jackson Laboratory, 600 Main Street, Bar Harbor, ME 04609–1500, USA.

PAUL HASTY
Lexicon Genetics, 4000 Research Forest Drive, The Woodlands, TX 77381, USA.

RANDALL JOHNSON
Department of Biology, UCSD, 9500 Gilman Drive, La Jolla, CA 92093–0116, USA.

ALEXANDRA L. JOYNER
HHMI and Developmental Genetics Program, Skirball Institute of Biomolecular Medicine, Departments of Cell Biology and of Physiology and Neuroscience, New York University School of Medicine, 540 First Avenue, 4th Floor, New York, NY 10016, USA.

TERRY MAGNUSON
Department of Genetics, Case Western Reserve University, 2119 Abington Road, Cleveland, OH 44106, USA.

MICHAEL P. MATISE

HHMI and Developmental Genetics Program, Skirball Institute of Biomolecular Medicine, New York University School of Medicine, 540 First Avenue, 4th Floor, New York, NY 10016, USA.

ANDRAS NAGY

Mount Sinai Hospital, Samuel Lunenfeld Research Institute, 600 University Avenue, Toronto, Ontario M5G 1X5, Canada.

VIRGINIA PAPAIOANNOU

Columbia University, Department of Genetics and Development, 701 West 168th Street, HHSC-1402, New York, NY 10032, USA.

JANET ROSSANT

Mount Sinai Hospital, Samuel Lunenfeld Research Institute, 600 University Avenue, Toronto, Ontario M5G 1X5, Canada.

WOLFGANG WURST

Max Planck Institute of Psychiatry, Molecular Neurogenetics, Kraepelinstr. 2-16, 80804 Munich, Germany.

Abbreviations

βGal	β-galactosidase
°C	degree Celsius
dpc	days post-coitus
E	embryonic stage (in dpc)
ENU	*N*-ethyl-*N*-nitrosourea
ES	embryonic stem
EST	expressed sequence tag
FBS	fetal bovine serum
FRT	Flp recognition target
g	gravity
GFP	green fluorescent protein
GPI	glucose phosphate isomerase
GT	gene trap
h	hour(s)
hAP	human alkaline phosphatase
HCG	human chorionic gonadotropin
HPAP	human placental alkaline phosphotase gene
ICM	inner cell mass
IRES	internal ribosomal entry site
IU	international units
lacz	*beta* galactosidase gene
LH	luteinizing hormone
LIF	leukaemia inhibitory factor
LOD	likelihood of the odds
loxP	locus of cross-over (*x*) in P1
Mb	megabase (10^6 bases)
min	minute(s)
ml	milliliters
neo^r	neomycin phosphotransferase-encoding gene
NLS	nuclear localization signal
oligo	oligodeoxyribonucleotide
ORF	open reading frame
pA	polyadenylation sequence
PBS	phosphate-buffered saline
PCR	polymerase chain reaction
PMSG	pregnant mare serum gonadotropin
RT	room temperature
SA	spliceacceptor
sec	second(s)

SLT	specific-locus test
TC	tissue culture
TE	Tris-EDTA
tk	thymidine kinase gene
UTR	untranslated region
μg	microgram(s)
tet	tetracycline
wt	wild-type
X-Gal	5-bromo-4-chloro-3-indolyl-β-D-galactopyranoside
ZP	zona pellucida
::	novel joint (fusion or insertion)

1

Gene targeting, principles, and practice in mammalian cells

PAUL HASTY, ALEJANDRO ABUIN, and ALLAN BRADLEY

1. Introduction

When a fragment of genomic DNA is introduced into a mammalian cell it can locate and recombine with the endogenous homologous sequences. This type of homologous recombination, known as gene targeting, is the subject of this chapter. Gene targeting has been widely used, particularly in mouse embryonic stem (ES) cells, to make a variety of mutations in many different loci so that the phenotypic consequences of specific genetic modifications can be assessed in the organism.

The first experimental evidence for the occurrence of gene targeting in mammalian cells was made using a fibroblast cell line with a selectable artificial locus by Lin *et al.* (1), and was subsequently demonstrated to occur at the endogenous β-globin gene by Smithies *et al.* in erythroleukaemia cells (2). In general, the frequencies of gene targeting in mammalian cells are relatively low compared to yeast cells and this is probably related to, at least in part, a competing pathway: efficient integration of the transfected DNA into a random chromosomal site. The relative ratio of targeted to random integration events will determine the ease with which targeted clones are identified in a gene targeting experiment. This chapter details aspects of vector design which can determine the efficiency of recombination, the type of mutation that may be generated in the target locus, as well as the selection and screening strategies which can be used to identify clones of ES cells with the desired targeted modification. Since the most common experimental strategy is to ablate the function of a target gene (*null allele*) by introducing a selectable marker gene, we initially describe the vectors and the selection schemes which are helpful in the identification of recombinant clones (Sections 2–5). In Section 6, we describe the vectors and additional considerations for generating subtle mutations in a target locus devoid of any exogenous sequences. Finally, Section 7 is dedicated to the use of gene targeting as a method to express exogenous genes from specific endogenous regulatory elements *in vivo*, also known as 'knock-in' strategies.

1.1 Targeting vectors

A targeting vector is designed to recombine with and mutate a specific chromosomal locus. The minimal components of such a vector are sequences which are homologous with the desired chromosomal integration site and a plasmid backbone. Since both the transfection efficiency and targeting frequency of such a vector can be low, it is desirable to include other components in the vector such as positive and negative selection markers which provide strong selection for the targeted recombination product (Sections 4 and 5). The positive selectable marker in a targeting vector may serve two functions. Its primary purpose is as a selection marker to isolate rare transfected cells that have stably integrated DNA (which occurs at a frequency of about one in every 10^4 treated cells). Its other purpose is to serve as a mutagen, for instance if it is inserted into the coding exon of a gene or replaces coding exons.

Two distinct vector designs can be used for targeting in mammalian cells, replacement and insertion vectors (*Figure 1*). These vector types are configured differently so that following homologous recombination they yield different integration products. Replacement vectors are the most widely used type of vector in gene targeting experiments. For the purpose of clarity, aspects of vector design which are relevant to both replacement and insertion vectors, such as the length of homology, selection cassettes, and enrichment schemes, will be discussed in Sections 4 and 5. However, there are a number of unique considerations for both vector types and these are discussed in Sections 2 and 3.

2. Replacement vectors

The fundamental elements of a replacement vector are the homology to the target locus (see Section 4.1), a positive selection marker (see Section 5), bacterial plasmid sequences, and a linearization site outside of the homologous sequences of the vector. In some cases a negative selectable marker may also be used to enrich the transfected cells against random integration events (Sections 4.2.1 and 5). The basic design of a replacement vector is illustrated in *Figure 1A*. The mechanistic details of recombination pathways used by replacement vectors are beyond the scope of this chapter but the final recombinant allele can be effectively described as a consequence of double reciprocal recombination which takes place between the vector and the chromosomal sequences (*Figure 2A*). The final recovered product is equivalent to a replacement of the chromosomal homology with all components of the vector which are flanked on both sides by homologous sequences; any heterologous sequences at the ends of the vector homology are excised from the vector and are not recovered as stable genomic sequences in the recombinant allele following targeting. This latter feature has been used as a basis to

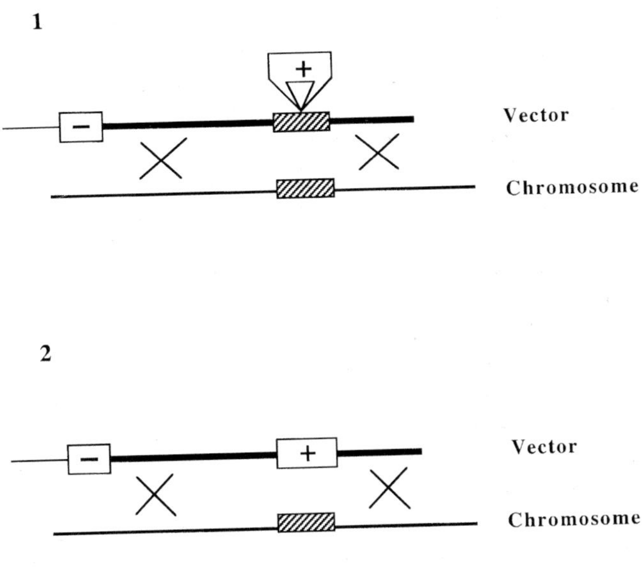

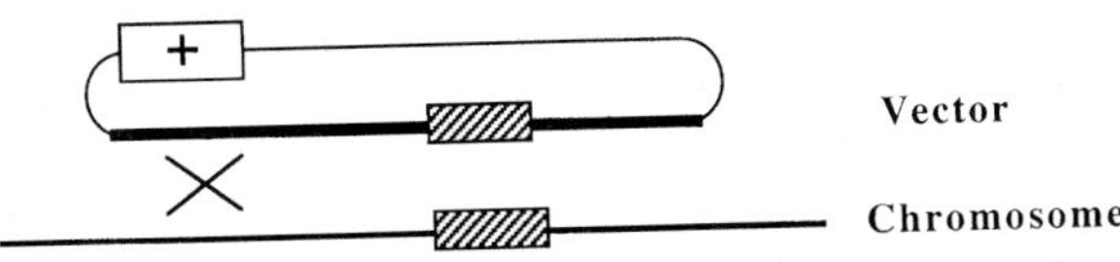

Figure 1. Replacement and insertion vectors. Thick line represents vector homology to target locus, thin line represents bacterial plasmid, line of intermediate thickness represents the target locus in the chromosome. Striped rectangle represents an exon. Positive selection marker, a box that contains a '+'. (A) Replacement vector. Positive selectable marker interrupts (A1) or deletes and replaces (A2) the coding region of the target. Negative selection marker, a box that contains a '–'. The replacement vector is linearized outside the target homology prior to transfection. (B) Insertion vector. A positive selectable marker may be cloned either into the homologous sequences or the vector backbone. A double-strand break is generated in the target homology using a unique restriction enzyme prior to transfection (not shown).

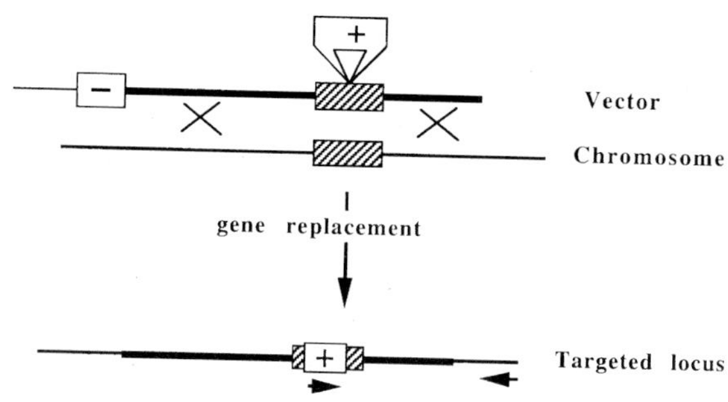

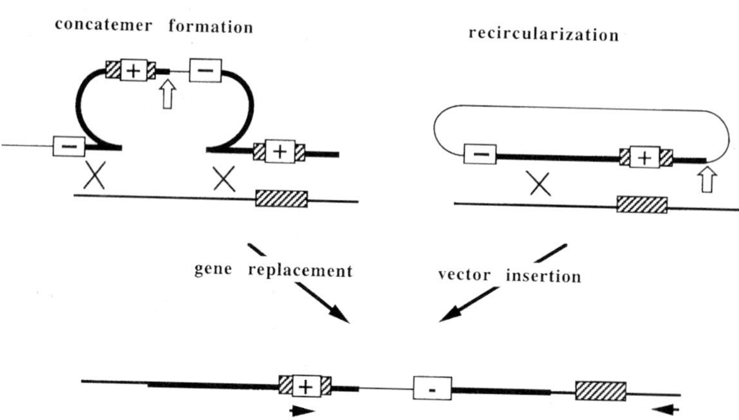

Figure 2. Integration patterns for a replacement vector. Vectors and chromosomal targets are drawn as in *Figure 1*. The 'X's represent cross-over points. The arrow heads represent PCR primers, one in the positive selection marker and the other in the target locus adjacent to the vector homology. (A) Gene targeting that results in a simple gene replacement product and a PCR junction fragment. (B) Gene targeting that results from the integration of all vector components. The open arrows point to the location for extrachromosomal end-to-end joining that concatenates or recircularizes the vector. The PCR junction fragment is probably too large to be amplified.

enrich populations of transfected cells for targeted integration events (Section 4.2.1).

2.1 Design considerations of a replacement vector

The principal consideration in the design of a replacement vector, is the type of mutation generated. Secondary (yet still important) considerations relate to the selection scheme and screening techniques required to isolate the recombinant clones. The recombinant alleles generated by replacement vectors typically have a selection cassette inserted into a coding exon or replacing part of the locus. It is important to consider that, exon interruptions and small deletions will not necessarily ablate the function of the target gene to generate a null allele. Consequently, it is necessary to confirm that the allele which has been generated is null by RNA and/or protein analysis and in many cases transcripts and truncated proteins from such a mutant allele can be detected. Considering that products from the mutated locus may have some function (normal or abnormal) it is important to design a replacement vector so that the targeted allele is null, particularly in the absence of a good assay for the gene product. Disruption or deletion of the coding sequence by the positive selection marker will in most instances ablate a gene's function. However in some situations a truncated protein may be generated which retains some biological activity, thus some knowledge of mutations in a related gene in another organism can be helpful in the determination of the possible function of a targeted allele. Null alleles are more likely to occur by deleting or recombining a selection cassette into more 5′ exons rather than exons that encode the C-terminus of the protein, since under these circumstances minimal portions of the wild-type polypeptide would be made.

There are several considerations to take into account when a positive selection marker is to be inserted into an exon. One critical consideration is that since the length of an exon can influence RNA splicing (3), an artificially large exon caused by the insertion of a selectable marker may not be recognized by the splicing machinery and could be skipped. Thus, transcripts initiated from the endogenous promoter may delete the mutated exon from the mRNA species or even additional exons. If a skipped exon is a coding exon whose nucleotide length is not a multiple of three (codon) the net result will be both a deletion and a frame-shift mutation of the gene, which will often generate a null allele. However, if the disrupted coding exon has a nucleotide length which is a multiple of three, if spliced out, this would result in a protein with a small in-frame deletion which may retain partial or complete function. The same concept applies to gene targeting vectors in which exons are being deleted and replaced by the selectable marker. Deletion of an exon or group of exons with a unit number of codons may also result in a functional protein product with an in-frame deletion. For most purposes it is advisable to delete portions or all of the target gene so that the genetic consequences are not ambiguous. A study of the relationship between deletion size and targeting frequency with replacement vectors at the mouse *Hprt* locus showed that deletions of 19 kb occurred at a similar frequency to small deletions or exon disruptions (4). Although there may be significant variability amongst different loci, it is clearly feasible to delete a significant portion of the coding region for small or compact genes. However, it is also important to be aware that deletions, particularly if they are large, may affect multiple genes in situations where they are located adjacent or internal to the deleted sequences or share regulatory elements with the target locus.

The length and sequence of the homologous sequences in the vector will affect the targeting frequency (see Section 4.1). For most vectors the length of homology should be in the range of 5–8 kb. The homologous sequences in the vector should be derived from genomic libraries isogenic with the specific mammalian cells used in the targeting experiment (see Section 4.1.2). The position of the positive selection marker with respect to the homologous sequences of the vector is important in deciding the type of screen used to find the clones targeted with a gene replacement event. One common screening tool for targeted clones is based on the polymerase chain reaction (PCR) which can be designed to detect the juxtaposition of the vector and the target locus (5, 6). The use of PCR to detect targeted clones is described in detail in Chapter 3. This is accomplished by using one primer which anneals to the positive selection marker in the targeting vector and a second primer which primes from the target chromosomal sequences just beyond the homologous sequences used in the vector (*Figure 2A*).

The efficiency of such a PCR amplification is related to the distance between the unique primer site in the vector (usually in the positive selection marker) and the sequences external to the homologous elements of the vector as well as the specific composition of the DNA sequence to be amplified. The amplified product should be in the 500 base pair to 2 kb range. Thus, replacement vectors configured for screens by PCR require the positive selection marker to be inserted at an asymmetric location near one end of the homologous sequences, while still leaving sufficient homology for the formation of a cross-over. This will give vectors with one long arm and one short arm of homologous sequences (*Figure 2A*). The minimum sequence requirement for recombination is about 500 base pairs, but this is not recommended for routine use given that this will give a very low efficiency of recombination.

Another common screen for clones targeted with gene replacement vectors is by Southern blot analysis. For this it is important to design the vector and identify unique probes flanking the homologous sequences in the vector and restriction sites so that the analysis is both unambiguous and can discriminate the various categories of recombinant clones (Section 2.3).

Since a replacement vector should be linearized before transfection into cells at a site outside the homologous sequences, the cloning steps must incorporate at least one unique restriction enzyme site outside the homologous sequences. There is no advantage in releasing the homologous sequences of the vector from the bacterial plasmid (7).

The following are guidelines for the construction of a replacement vector which generates an easily identifiable null allele.

(a) Use a fragment of isogenic homologous sequences of 5–8 kb.
(b) Insert the positive selectable marker into an upstream exon, delete the entire gene if it is small, or important 5′ exons if it is large.
(c) Interrupt or delete exons without a unit number of codons to avoid generating a protein product with partial or novel function after RNA splicing.
(d) Avoid interrupting or deleting exons which are known to be alternatively spliced *in vivo*.
(e) If PCR is to be used to screen for gene replacement events, clone the positive selection marker so that one arm of homology is 0.5–2 kb.
(f) Linearize the vector outside the homologous sequences.
(g) Design a diagnostic Southern screening strategy. Probes should be tested and fragment sizes for both alleles (endogenous and targeted) with specific enzymes should be known.

2.2 Recombinant alleles generated by replacement vectors

The desired genetic exchange with a replacement vector is one in which vector sequences effectively replace the homologous region in the genome (*Figure 2A*). However, undesirable targeted products occur with replacement vectors (7); primarily with vectors whose ends joined to form concatemers, circles, or both (*Figure 2B*). The final recombination product generated by a concatemer or a circle is the integration of the entire vector including the bacterial plasmid and other associated heterologous sequences. If the vector was intact at the time of the ligation event many of these undesirable products will be eliminated by negative selection because they incorporate the entire vector including the negative selection marker. These undesirable products will also not be scored as positive by most PCR screens for targeted events since these screens are usually specific for the recombination events that have crossed-over on the short arm of the vector while most of the alternative events are the products of recombination events which occur through the long arm of the vector.

2.3 Replacement vectors: screening for targeted events

One of the most important aspects of any gene targeting experiment is to confirm that the desired genetic change has occurred (*Figure 3*). Given that replacement vectors may integrate via a number of different pathways and result in different targeted mutations, it is important to resolve the different

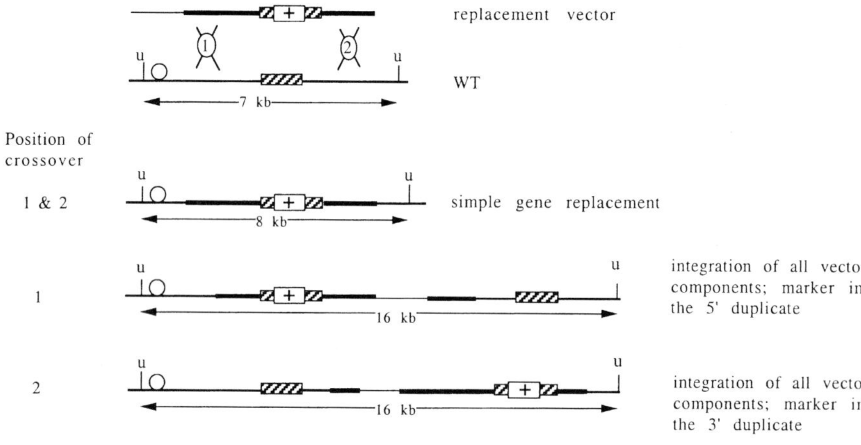

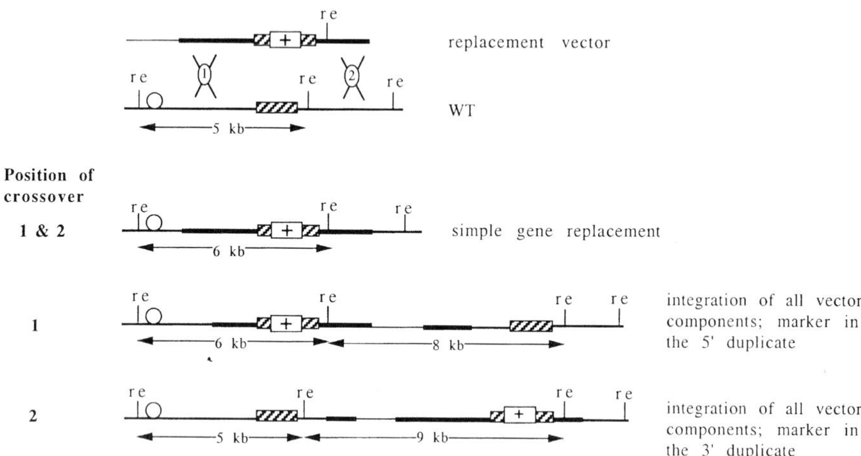

Figure 3. Discriminating different types of targeting events. Southern blot analysis differentiates simple gene replacement events from the integration of all vector components into the target locus. The lines and boxes are the same as in *Figure 1.* The first and second cross-over point are depicted by an 'X' labelled 1 and 2 respectively. The circle represents a probe external to the vector. The bacterial plasmid and the positive selection markers can be used as probes internal to the vector. The panels to

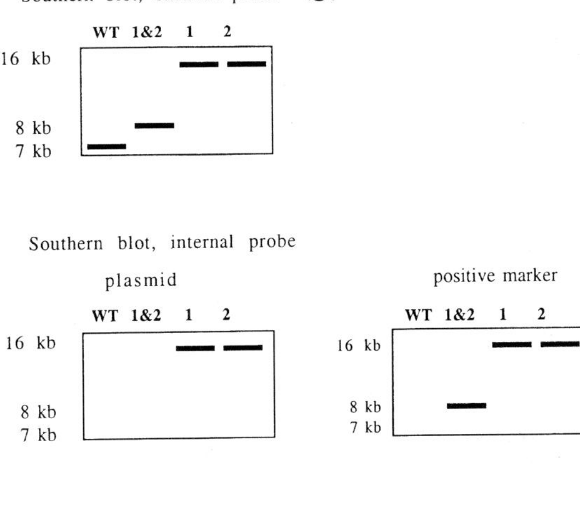

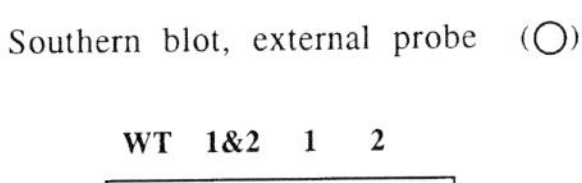

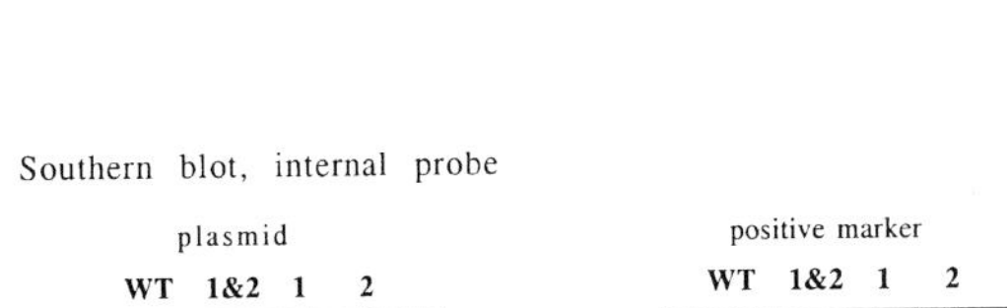

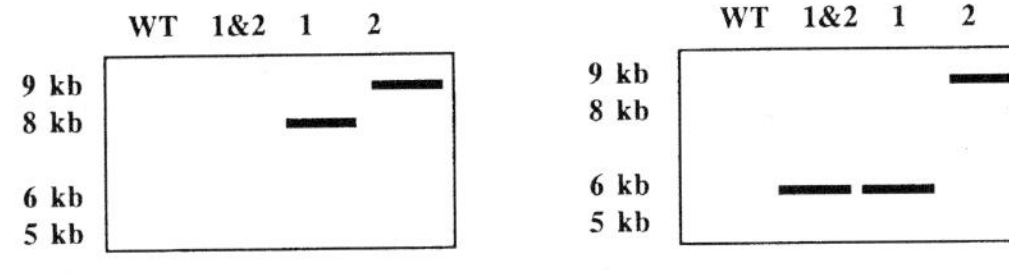

the right are examples of Southern blots. WT, wild-type; 1 and 2, simple gene replacement event: 1, integration of the entire vector with the positive selection marker in the 5′ duplicate; 2, integration of the entire vector with the positive selection marker in the 3′ duplicate. (A) Southern analysis with a unique restriction enzyme, 'u', that does not cut in the vector. (B) Southern analysis with a restriction enzyme, 're', that cuts once in the vector.

classes of integration events, particularly gene replacement events that introduce only a positive selection marker from recombination events which result from integration of the entire vector or concatemers. To clarify the following discussion, targeted clones which have only incorporated the positive selection marker are referred to as simple gene replacement events.

Following the transfection of a replacement vector, colonies which survive positive and negative selection should be clonally isolated and screened by either PCR or Southern blot analysis for a specific recombination allele (see Chapter 3). For PCR-based screens the primers are chosen so that they will amplify a specific junction fragment following homologous recombination (5, 6). If the PCR amplification is efficient the PCR product may be visually detected on an ethidium bromide stained gel. Occasionally it may be necessary to hybridize the PCR product to a labelled oligonucleotide probe (specific for sequences outside the vector) to confirm the identity of the amplification product. Southern blot-based screens are also possible at an early stage in the expansion of the transfected clones. These screens may be performed on the small amounts of DNA obtained from cells grown in 96- or 24-well plates (8) (see Chapter 3 for details); for Southern blot analysis the choice of restriction digest and probe must readily distinguish the wild-type from the predicted targeted allele. Following a primary screen, the recombinant allele present in the putative positive clone must be subjected to extensive restriction analysis to confirm its structure. This analysis is ideally performed with a probe that is not contained in the target vector (external probe) and a restriction digest with an enzyme that does not cut in the vector (*Figure 3A*). Under these circumstances a simple gene replacement event will increase the length of the wild-type fragment by the size of the positive selection marker. The integration of the entire vector will be readily detectable by a very large increase in fragment size. Although this is the ideal situation, it is often difficult to find a restriction enzyme that does not cut in the vector while still generating genomic fragments that can be readily resolved on a gel. Nevertheless, an external probe should always be used, since internal probes will result in many false positives. Internal probes detect the randomly integrated vector and some of these insertions by chance will give the expected restriction fragment size and appear to be targeted insertions.

The allele generated by targeted integration of the entire vector or concatemers of the vector may be difficult to discriminate from simple replacement events if the restriction digest detects a site within the vector and the probe is from outside the vector (*Figure 3B*). This type of Southern blot analysis can only analyse one arm of the target locus, and will score the integration of the entire vector either as a simple gene replacement event or as a wild-type allele depending upon the location of the cross-over. It is therefore important to analyse both the 5′ and 3′ aspects of the target locus by using external probes from sequences flanking both ends of the vector. An internal probe (such as one which hybridizes to the positive selectable marker) should also be used since a simple gene replacement allele should give a different size fragment compared to an allele which is generated by the integration of the entire vector. Plasmid sequences may also be used as probes since they will only detect recombinant alleles generated during vector insertion-like events.

One problem with probes that are from the targeting vector (internal probes) is that they will also detect random integration events. This can result in misdiagnosis of an allele, if for instance a clone with a simple gene replacement allele and a separate random integration event is checked with an internal probe it may appear to have integrated all of the vector components into the target locus. To avoid these false negatives, it is advisable to perform Southern blot analysis with several restriction digests in which at least one of the restriction sites is outside the target vector. If this is done, predicted size fragments would only be consistently seen with simple gene replacement alleles whereas non-predicted fragments would be seen following random integration events.

One class of integration event which may be detected in the primary screens may falsely appear to be targeted (6). Mechanistically this event involves a strand exchange of the vector with the target which results in the vector picking up some sequences which flank the target. This homologous recombination intermediate is not resolved, the vector dissociates from the target and may integrate into a random location to generate a partial third allele. Under these circumstances, the recombinant clones will score as positive by PCR and by Southern analysis (with some restriction digests and some probes). However, if the Southern analysis is extensive these types of clones can be identified because the generation of a novel targeted restriction fragment would not be accompanied by the usual reduction in intensity of the endogenous fragment.

3. Insertion vectors

The basic elements of an insertion vector are the same as those in a replacement vector (*Figure 1B*). The major difference between the two vector types is that the linearization site of an insertion vector is made in the homologous sequences.

An insertion vector undergoes single reciprocal recombination (vector insertion) with its homologous chromosomal target which is stimulated by the double-strand break or gap in the vector (*Figure 4*). Our observations have indicated that an insertion vector can target at a 5- to 20-fold higher frequency than replacement vectors given the same homologous sequences (7). However if an insertion vector is configured so that there are severe topological constraints then this elevation in targeting frequency may not be noted (9). Since the entire insertion vector is integrated into the target site, including the homologous sequences of the vector, the recombinant allele generated by

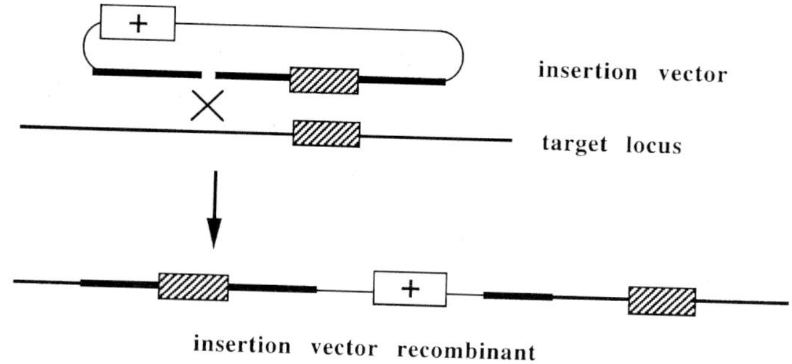

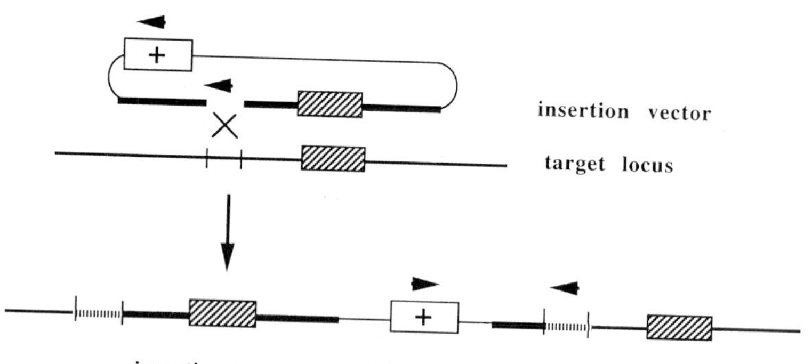

Figure 4. The integration pattern for an insertion vector and gap repair. (A) Integration of an insertion vector. The lines and boxes are as indicated in *Figure 1*. The repaired sequence is represented by the striped line. The insertion vector is linearized inside the target homology prior to transfection. Homologous recombination with the target locus results in the integration of all vector elements and duplication of the homologous sequences present in the vector. (B) PCR screen for insertion events via gap repair. The insertion vector contains a gap at the linearization point, demarcated by two vertical lines in the target locus and the vector insertion recombinant. The arrow heads represent PCR primers; one is specific for the positive selection marker and the other for sequences which correspond to the gap that is repaired during the target event.

such a vector becomes a duplication of the target homology separated by the heterologous sequences in the vector backbone.

3.1 Vector design for insertion vectors

The essential components of an insertion vector are a region of homology to the target locus which has a unique linearization site, a positive selectable marker (which serves to select for transfected cells in the same way as it would in a replacement vector), and a bacterial plasmid backbone. If the homologous sequences in the vector are wild-type (no mutations), then the position of the linearization point within the homologous sequences should not significantly affect either the frequency of the recombination event or the structure of the mutant allele. However, if one of the exons has been mutated, the position of the linearization site will affect the structure of the recombinant allele. These considerations are discussed in more detail in Section 6.3. Because the entire vector integrates into the target locus, the positive selection marker may be positioned in either the vector backbone or the homologous sequences of the vector. Since insertion vectors duplicate the homologous sequences in the vector, the position of a selection cassette may not play a critical role in mutating the gene. However, if the selection cassette is cloned into the only exon in the vector, then after vector insertion, the targeted allele will contain one normal and one artificially large exon. Skipping of this large exon may result in normal transcripts being generated from such a recombinant locus at a low frequency (10).

When designing an insertion vector it is important to consider the potential RNA species and coding possibilities of the recombinant allele. For most recombinant alleles generated with insertion vectors the duplication of exonic sequences is sufficient to mutate the gene, however, this will depend on a number of factors. For instance, a duplication of exonic sequence without a frame-shift mutation may not ablate protein function. To ensure the exon duplication will create a mutation at the protein level it may be necessary to introduce stop codons into a single exon in the targeting vector. A mutation created by frame-shifting in the carboxyl-terminus of a protein may leave a functional N-terminal domain which may retain partial or complete activity. Exon duplications which are 5′ or 3′ to coding sequences or include the first or last coding exon will still have a collinear intact genomic sequence after the targeted integration and may be functionally wild-type.

At the design stage, consideration should be given to the screening procedure for the targeted allele. Because the genomic locus contains a duplication, the screen for an insertion event may be complex. These considerations are discussed in Section 3.2.

The basic design requirements of an insertion vector are as follows:

(a) Use 5–8 kb of isogenic homology.

(b) A unique restriction enzyme site is needed to generate a double-strand break in the homology.

(c) A positive selection marker located either in the homologous sequences or in the plasmid backbone.

(d) If there is only one exon in the vector, do not disrupt it with the positive selection marker.

(e) Do not use a genomic fragment with the first or last coding exon in the vector.

(f) Destroy restriction sites for easy Southern blot analysis.

(g) A Southern blot screening strategy should be designed. Single copy external probes should be identified and restriction fragment sizes that uniquely identify the targeted and wild-type alleles must be known.

3.2 Screening for recombinant alleles generated with insertion vectors

In contrast to the different classes of integration events that may be generated with a replacement vector, an insertion vector usually integrates in a highly predictable way. The first step in screening transfected cells for an allele generated with an insertion vector may be done by PCR or Southern blot analysis on clones selected for the presence of the positive selection marker. There are two ways to configure an insertion vector so that the recombinant locus can generate a specific PCR product. The most reliable PCR screen is based on the repair of DNA deleted from the insertion vector at the position of the linearization site to form a gap as shown in *Figure 4B* (11). Thus, PCR for the recombinant locus will use a primer that corresponds to the deleted sequence and a second primer which is specific for heterologous sequences in the vector. Upon insertion into the target locus, the gap will be repaired using the chromosomal sequences as a template. This will allow amplification by PCR while insertions of the vector into random locations in the genome will only have the single primer site in the vector and thus, will not provide an appropriate template for PCR amplification. To prepare a gapped vector it is necessary to identify two restriction sites 1–4 kb apart in the homologous sequences of the vector. The fragment between these sites is deleted, the gapped vector is purified, ligated, and cloned back into *E. coli* to ensure that the gap sequence never contaminates the transfected vector. If the restriction sites are not compatible oligonucleotide linkers can be used to circularize the vector, however it is essential to ensure that the cloned vector can be easily cut at the gap site prior to transfection. If linkers are used to generate a unique site the few extra nucleotides on the ends of the gapped vector will have a negligible effect on the targeting efficiency.

A second way to design a PCR screen is to use a vector in which the homologous sequences have been identified with a mutation, this mutation can serve as a primer site for PCR coupled with a primer site external to the vector. However, since the mutation in an insertion vector can end up on either side of the duplicated sequence, some targeted clones will not be identified by this assay (for reasons described in Section 6.3.1). Southern blot-based screens using the 96- or 24-well protocol are also possible with insertion vectors (8). Probe(s) used for this preliminary screen should not be included in the homologous sequences of the vector, these probes may be from flanking sequences or if a gapped vector is used, the probe may be from the fragment that constitutes the gap.

After the PCR or Southern blot analysis, positive clones identified in the preliminary screen should be expanded and the integration pattern confirmed by detailed Southern blot analysis preferably using a probe which is external to the target sequences. The best diagnostic digest for Southern blot analysis would be with an enzyme that does not cut within either the homology of the target locus or the vector backbone. A clone which has inserted one unit of the vector into the target locus will show an increase in the restriction fragment by the size of the vector. Concatemers of the vector integrated into the target locus may also be readily identified in the same assay. Since it may not be possible to identify a restriction enzyme that does not cut the vector or target locus to generate fragments which are readily resolvable (less than 20 kb) on an normal agarose gel, it is often necessary to use multiple digests as well as internal and external probes as described for replacement vectors (see Section 2.3). As stated previously, it may be possible to avoid problems in identifying unique restriction sites for Southern blot analysis by mutating sites in the vector or by choosing a region of homology for vector construction which lacks specific enzyme sites. *The Southern blot strategy and single copy probes should be defined before the vector is made.*

4. Maximizing the targeting frequency and selection of targeted clones

A targeting vector transfected into cells is subject to two competing reactions, it can recombine into the desired chromosomal target, or into a random genomic site. Unfortunately, the latter event tends to predominate. It is thus desirable to design the vector to increase the chance of, and selection for, clones with a targeted integration event and against clones with random insertion events. Since gene targeting is designed to precisely modify a chromosomal locus, it is possible to utilize aspects of the recombinant locus and the recombination event itself to provide strong selection and enrichments for clones of cells with targeted integration events. At any given locus there are two major variables which can influence the targeting frequency that may be controlled experimentally. First, the choice of an insertion or replacement vector can significantly affect the targeting frequency (7, 11). Secondly, the homologous sequences in the vector, specifically their length (9, 12) and the degree of polymorphic variation between the vector and the chromosome (13), have been shown to affect targeting frequencies.

4.1 Homology to the target locus

4.1.1 Length of homology

An important factor that affects the recombination frequency in mammalian cells is the length of homology to the target locus. A number of groups have described a relationship between the length of homology and targeting frequency (9, 12). As a general rule the greater the length of homology the higher the targeting frequency. For most vectors we recommend the ideal length of homologous sequences in the vector to be in the range of 5–10 kb. Although the use of even greater lengths of homology will probably increase the targeting frequency, the vectors will become unwieldy due to both their physical size and the limited choice of unique restriction enzyme sites which are required to linearize the vector prior to transfection and later for detailed unambiguous Southern blot analysis of transfected clones.

As discussed previously, the positive selection cassette/mutation present in a replacement vector may divide the homology into a long and short arm. The experimental desire to distribute the homology asymmetrically is associated with the use of PCR to screen for specific junction fragments. An asymmetry can have an effect on the targeting frequency since the length of the short arm could be suboptimal. Although cross-overs have been observed to occur with less than 500 base pairs on the short arm (12), it is desirable to have at least 2 kb for efficient pairing and cross-over formation (14). If PCR screens are not a part of the screening strategy, then it is advantageous to evenly distribute the homology on both sides of the positive selection marker.

Insertion vectors can also be described as having a long and short arm if a mutation disrupts the homologous sequence. If the double-strand break is made too close to a large interruption of homology (such as a selectable marker) this can decrease the targeting frequency. The vector should be designed so that the double-strand break can be made at least 1.5 kb away from the selection cassette and backbone. The location of the double-strand break with respect to subtle mutations in the homologous sequences does not appear to greatly affect the targeting frequency, although this is an important consideration when using 'hit-and-run' vectors to make subtle mutations (see Section 6.3).

4.1.2 Degree of the homology

Studies on extrachromosomal and intrachromosomal recombination, as well as gene targeting experiments (13) have shown that a significant variation in sequence homology between the two elements involved in the genetic exchange can reduce the homologous recombination/targeting frequency. DNA mismatch repair, a biochemical pathway conserved from bacteria to humans, is the major mechanism by which sequences that are not completely homologous are hampered in their ability to recombine. The DNA mismatch repair pathway recognizes and repairs base pair mismatches and heterologies that arise as errors of DNA replication (for review, see refs 15 and 16). Mutations in components of the DNA mismatch repair system lead to an increase in spontaneous mutation rates as well as increased levels of recombination between diverged sequences in *E. coli* (17), *Saccharomyces cerevisiae* (18), and mammalian cells (19). These findings suggest that the DNA mismatch repair machinery recognizes mismatches and heterologies that arise in heteroduplex intermediates of recombination between diverged sequences, blocking their recombination. Notably, targeting frequencies with non-isogenic vectors are elevated to the levels of isogenic constructs in mouse ES cells which are mutant for DNA mismatch repair genes (19).

The number and extent of polymorphic variation between two mouse strains in any given locus is usually unknown and may vary widely from gene to gene. In particular, introns are likely to be more divergent than exonic sequences. It is now clear that the existence of sequence mismatches between the homology in the vector and the target locus will reduce the targeting frequency, particularly if a considerable length of uninterrupted perfect homology doesn't exist (20). Therefore, the DNA used to construct the targeting vector should ideally be isolated from genomic libraries isogenic with the cells used in the targeting experiments.

4.2 Enrichment schemes for targeted clones in culture

When a targeting vector is transfected into cells it can either integrate into its target locus or into a random chromosomal location. The relative ratio of these two pathways depends upon a number of factors which can not be experimentally controlled, such as the location of the target gene in the genome. In most cases, the frequency of random integration is far greater than for targeted recombination. While factors such as increasing the amount of sequence homology or using an insertion vector can increase the representation of targeted clones within a transfected population, it is also possible to significantly reduce the number of clones with random integration events in a population by several methods. Replacement vectors may include a negative selection cassette outside the region of homology. Another method involves trapping the promoter or polyA site of the endogenous gene.

4.2.1 Positive-negative selection for targeted clones

It is possible to use the components of the vector to select against clones that have integrated the vector into random genomic sites. This technique is known as positive-negative selection (24) and may be used with standard replacement vectors (*Figures 5A* and *5B*). In this selection scheme the positive selection works exactly as previously described to isolate all of the clones which stably incorporate the vector DNA, irrespective of the integration site (targeted or random). The vector also contains a negative selection cassette at one or both ends of the vector. During gene replacement events this cassette is lost and degraded whereas clones in which the vector has integrated at random will, at some frequency, incorporate this entire cassette. Thus, selection against this cassette will kill clones of cells which have integrated the vector at a random location, while targeted clones will survive. Those clones which have integrated all of the vector components into the

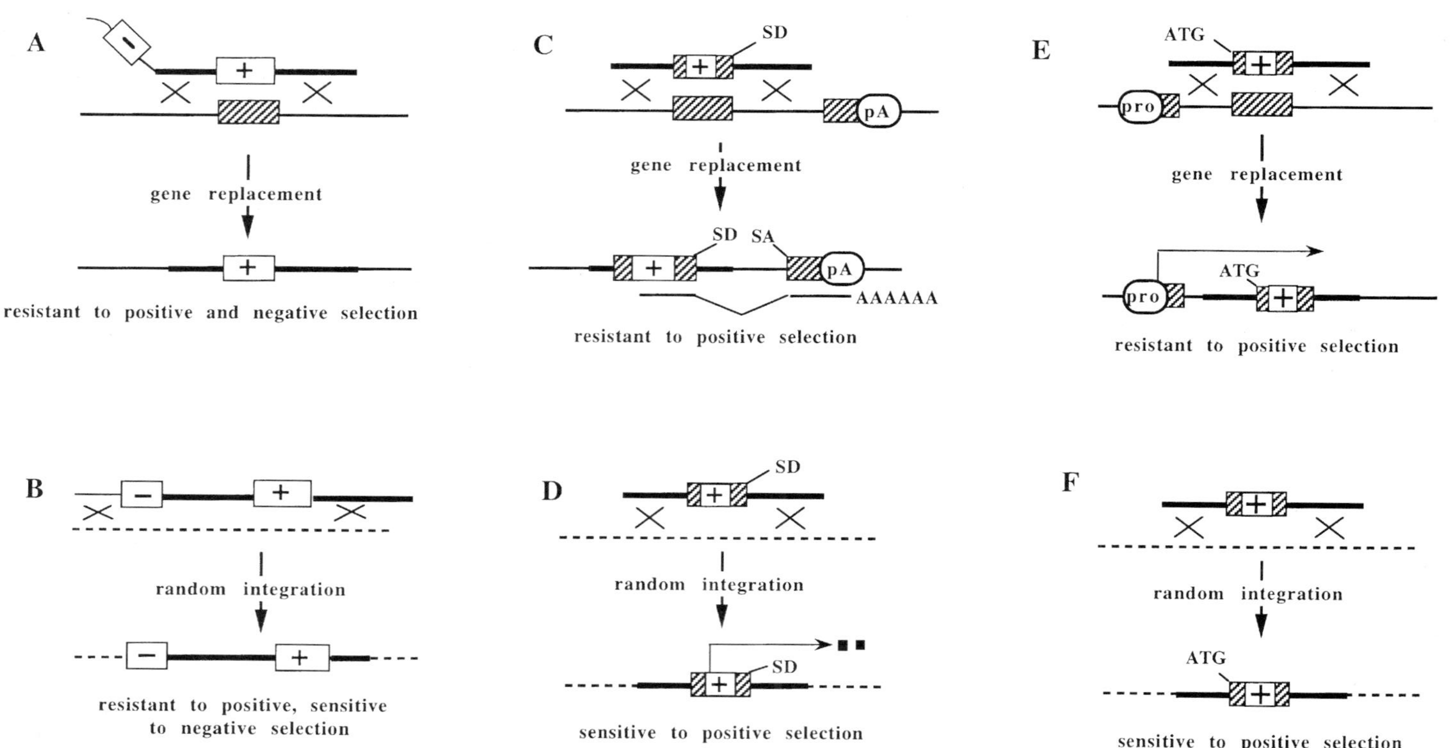

A
gene replacement
resistant to positive and negative selection
C
SD
gene replacement
SD SA
pA
AAAAAA
resistant to positive selection
E
ATG
pro
gene replacement
resistant to positive selection
B
random integration
resistant to positive, sensitive to negative selection
D
SD
random integration
sensitive to positive selection
F
random integration
ATG
sensitive to positive selection

Figure 5. Selection for gene targeting events (A and B), positive-negative selection. The replacement vector is linearized outside the target homology prior to transfection and the positive and negative selection cassettes contain a promoter and polyadenylation signal. The lines and boxes are as indicated in *Figure 1*. The dashed line represents non-homologous chromosomal DNA. SA, splice acceptor; SD, splice donor; pA, transcription termination and polyadenylation site; pro, promoter; ATG, initiation codon; the black square boxes represent read-through transcription. A targeted integration event removes the negative selection marker (A) whereas the integration of the vector into a random chromosomal site includes the negative selection maker (B). These clones survive or are killed by negative selection, respectively. (C and D) Transcription termination (3′) positive selection. (C) Targeted integration enables the positive selectable marker without a poly-adenylation signal, but with a promoter, to splice into and utilize the 3′ transcription processing signals from the target locus. This chimeric transcript is spliced, poly-adenylated, exported from the nucleus, and translated. These targeted clones survive positive selection. (D) random integration of the vector does not usually provide the appropriate transcription termination and polyadenylation signals. These clones do not make stable message and die under positive selection. (E and F) Promoter trap (5′) selection. (E) Targeted integration of the positive marker without a promoter, but with a polyadenylation signal, into an actively transcribed locus in ES cells results in expression of the positive selectable marker. These clones survive positive selection. (F) Random insertion of the vector does not usually result in expression of the marker, these clones are sensitive to positive selection.

target locus by homologous recombination will also die under negative selection. The enrichment achieved with negative selection is highly variable, the modal range in the literature is between 2- and 20-fold. Since the non-targeted clones which survive selection have usually mutated the negative selection cassette, it is possible to increase the efficiency by cloning an additional negative selection marker into the targeting vector. One virtue of positive-negative selection is that it does not depend upon elements of the targeted locus for its application. It is thus generally applicable to most replacement vectors, at any genetic locus irrespective of the gene structure or expression level. In contrast to the predictable structure of a targeted allele, the structure of the targeting vector integrated at random sites in the genome is highly variable. Thus, negative selection against such sites is usually much less impressive than achieved with positive selection using promoter-less or polyadenylation-less vectors, as described below. A family of vectors configured for rapidly generating vectors for positive-negative selection is available from Lexicon Genetics Inc.

4.2.2 Positive selection for targeted clones: promoter, enhancer, and polyadenylation trap targeting vectors

Promoter and enhancer trap targeting vectors are designed to use the transcriptional activity of the endogenous target gene to drive the positive selection cassette cloned in the targeting vector. Vectors of this type are thus restricted to genes which are transcriptionally active in the target cells. For promoter trap vectors, the positive selection cassette is either cloned in-frame with the endogenous translated product, or the cassette may encode its own translation initiation site and is positioned upstream or in place of the nominal translational start site. In either case the homologous sequences used in the vector must be such that they do not retain any promoter activity, thus, the majority of random integrations will be transcriptionally silent and not survive positive selection (*Figures 5E* and *5F*). Typically, promoter trap selection will yield a 100-fold enrichment for targeted clones and will work for both replacement and insertion vectors (21).

There is some risk with this strategy in that the fusion gene product with the positive selection marker may not always be functional, thus it is important to test for activity of the positive selection cassette fusion with a control vector (21). As discussed in Section 7, the use of internal ribosome entry site (IRES) elements has greatly simplified the design of such vectors. Good candidate genes for this selection scheme are those with high expression levels in the target cells, although genes with low expression levels have been successfully targeted with promoter trap vectors (22).

An enhancer trap targeting vector is conceptually similar to a promoter targeting vector. Such vectors use a selection marker with a weak position-dependent promoter which can be activated if integration occurs in the proximity of a transcriptional enhancer element. The enrichment factors attained with such vectors will be less significant than with promoter traps, although vector design will be simpler because a fusion transcript/gene product is not required.

Polyadenylation trap targeting vectors are designed to use the transcription termination/polyadenylation signals of the target gene to generate a stable hybrid transcript consisting of elements from both the target gene and the positive selection cassette (*Figures 5C* and *5D*). To make a targeting vector which relies on polyadenylation selection, the positive selection cassette would have its own promoter and no polyadenylation signal. In contrast to promoter trap selection schemes, this type of positive selection should work for most genes irrespective of whether they are expressed in the transfected cells. The best location for the positive selection cassette is in an exon of the target gene so that the splice donor sequence of the disrupted exon can allow normal splicing to occur downstream of the expression cassette. The homologous sequences in the vector should be truncated to remove the endogenous transcription termination/polyadenylation signals. Typical enrichment factors obtained with such vectors have been in the range of 5- to 50-fold (e.g. ref. 23). In principle, this selection scheme should work with both insertion and replacement vectors although the hybrid mRNA species generated with some fusions may be unstable. Consequently it is worthwhile to test the selection with an appropriate vector before initiating the gene targeting experiments.

5. Selection markers

As described previously, positive selection markers are a necessary component of a targeting vector because they facilitate the isolation of rare transfected cells from the majority of treated cells in a population. Negative selection markers are also a key component in that they facilitate the elimination of various subpopulations of transfected cells. A variety of selection markers have been described which act in either a dominant or recessive context. The dominant markers are more commonly used with ES cells since there are very few endogenous mutations in ES cells that have been identified and are used for selection purposes, the one exception is the X-linked *Hprt* gene (25). The selection cassettes summarized below are mainly derived from bacteria, fungi, or viruses. They are generally used as short intron-less transcription units cloned adjacent to mammalian promoter/enhancer sequences and polyadenylation elements. A list of the commonly used cassettes and selections is detailed in *Table 1*.

5.1 Promoters and polyadenylation sites used for selection markers

The uses of appropriate promoter and transcription processing signals is very important to obtain adequate expression of the resistance markers in the target construct (6). For the positive selection marker it is imperative that expression occurs at all integration sites (random and targeted) generated in a transfection experiment. Thus, for most purposes, it is desirable to use promoters which express in the greatest number of genomic locations. The phosphoglycerate kinase I and RNA polymerase II promoters appear to be relatively position-independent in ES cells (6) and are widely used. A number of position-dependent weak promoters have been successfully used in gene targeting experiments, such as the synthetic mutant polyoma enhanced *HSVtk* promoter, MC1 (9). In some contexts weak promoters can offer a selective advantage in the isolation of targeted clones provided there is adequate expression of the selection cassette at the target locus to give resistance to the selective drug. This type of enrichment is analogous to an enhancer trap. For negative selection purposes the MC1 promoter (9) appears to be adequate, this may be a function of the minimal expression levels required for toxicity and/or the possibility that the MC1 promoter may be up-regulated by the promoter/enhancers associated with the positive selectable markers present in the targeting vector.

The 'quality' of RNA processing signals may also affect the efficiency of each selectable marker (6). For most purposes the PGK and bovine growth hormone polyadenylation sites have been shown to function as efficient terminators.

Table 1. Commonly used selection cassettes

Dominant selectable markers

Selection cassette	Origin[a]	Positive selection drug	Negative selection drug[b]
Neomycin phosphotransferase (*neo*)	B	G418	–
Puromycin-*N*-acetyltransferase (*puro*)	B	Puromycin	–
Hygromycin B phosphotransferase (*hph*)	B	Hygromycin B	–
Blasticidin S deaminase (*bsr*)	B	Blasticidin S	–
Xanthine/guanine phosphoribosyl transferase (*gpt*)	B	Mycophenolic acid	6-Thioxanthine
Herpes simplex thymidine kinase (*HSVtk*)	V	–	GANC, FIAU
Dipheria toxin A fragment (*DTA*)	B	–	None required

Recessive selectable markers

Selection cassette	Origin[a]	Positive selection drug[c]	Host genotype[d]	Negative selection drug[b]	Host genotype[d,e]
Hypoxanthine phosphoribosyl transferase (*hprt**)[f]	M	HAT	*hprt*$^-$	6TG	*hprt*$^-$
Xanthine/guanine phosphoribosyl transferase (*gpt*)	B	HAT	*hprt*$^-$	6TG	*hprt*$^-$
				6TX	wt or *hprt*$^-$
Thymidine kinase (*tk*)	M	HAT	*TK*$^-$	5BdU	*TK*$^-$
Herpes simplex thymidine kinase (*HSVtk*)	V	HAT	*TK*$^-$	GANC, FIAU	wt

[a] B = bacterial, V = viral, M = mammalian.
[b] GANC = gancyclovir, FIAU = 1(1-2-deoxy-2-fluoro-β-D-arabinofuranosyl)-5-iodouracil, 6TG = 6-thioguanine, 5BdU = 5-bromodeoxyuridine.
[c] HAT = aminopterin, hypoxanthine, thymidine.
[d] 'Genotype' refers to the mutant background of the cell in which selection can be performed.
[e] wt = wild-type.
[f] * This is a minigene (25).

5.2 Effects of selection markers on phenotypes

The selectable markers associated with targeted alleles carry promoter and enhancer sequences which can interfere with other genes linked to the targeted locus (26–29). Long-range disruptions of gene expression at distances greater than 100 kb have been reported (30). These long-range effects *in cis* confound the analysis of the mutant phenotype since effectively one can be studying a double mutant. It is therefore extremely desirable to generate mutations without the complications of interpreting the effects of selectable markers on the phenotype. In many cases it is possible to remove selectable markers after targeting to eliminate these undesirable effects. Marker removal is now readily accomplished with the use of the Cre-*loxP* site-specific recombinase system (Chapter 2). Alternatively this can also be accomplished with the use of homologous recombination techniques (see below) with appropriate vector design. One final consideration is that the bacterial *neo* gene has been found to have cryptic splice sites (see Chapter 7, Section 2.2.2 for discussion).

6. Generating subtle mutations with gene targeting techniques

The strategies described in the previous sections have been used to establish disrupted alleles of numerous genes in tissue culture cells and the mouse germline. These methods of gene inactivation have been widely used, but they require the introduction of a positive selectable marker in order to select for the rare transfected cell, including targeted integrations. The positive selection marker also serves as the mutagen if inserted into the coding sequence since it generates a major disruption of the gene. Although major disruptions of a locus represent a valuable starting point for genetic analysis, a series of changes at the nucleotide level in both coding and control regions of the gene are important for the full understanding of a gene's function. Four techniques to introduce a small mutation into a gene are described which have been designed to avoid including a selectable marker in the recombinant locus. These techniques can be used to generate subtle mutations, such as single base substitutions, in mammalian cells. The more recently adopted combination of homologous and site-specific recombination in mammalian cells is versatile and efficient and allows for the removal of selectable markers to generate small mutations. The nature of site-specific recombination, however, does not allow for the generation of true subtle mutations such as single base pair changes. Techniques which facilitate this approach are discussed in Chapter 2.

Small mutations can be introduced into the genome in a single step, without including a positive selection marker, for instance with a replacement vector.

However, if such a vector is used selection for transfected cells is impossible with such a vector, which makes identification of targeted clones very difficult since non-transfected cells would also be present and represent the vast majority of the population to be screened. Therefore, the vectors must be introduced into cells under conditions where the transfected cells can be identified. In practice this requirement can be met by an efficient DNA delivery system such as microinjection (where each injected cell may be identified by location) or the co-introduction of a positive selection marker with the replacement vector where transfected cells can be selected. Alternatively, two-step procedures have been devised to create subtle mutations.

6.1 Subtle mutations generated by microinjection

The introduction of DNA into cells by direct microinjection is very efficient, up to 20% of cells integrate the injected DNA. Thus, if each injected cell is identified, non-selectable genetic modifications can be efficiently introduced into their target locus by homologous recombination (31). After microinjection of the vector into ES cells, the cells should be clonally expanded so that gene replacement events can be directly detected by Southern blot or PCR analysis. Although successful targeting has been reported in both fibroblast (32, 33) and ES (31) cells using microinjection, these results have not been repeated. Moreover, there are severe technical constraints to this method, particularly when applied to ES cells. As a consequence, this technique is not used for ES cells. Readers interested in the practical application of this method should consult a specialist description of microinjection techniques (34).

6.2 Non-selectable mutations generated by co-electroporation

The co-introduction of two DNA molecules into a cell, where one is a positive selection marker and the other is a non-selectable vector (35, 36), can efficiently identify cells that have taken up DNA by positive selection. In fact, under a variety of different transfection protocols the co-introduction of DNA is highly efficient (up to 75% of the cells which integrate DNA will include at least two copies). Although the majority of the transfected cells will contain concatemers of the targeting vector and the positive selection marker integrated into a single locus, some of the clones will have integrated the transfected DNA into different chromosomal sites. Thus, the co-introduction of two DNA molecules with a selection for one component and a screen of the selected clones for a targeted recombination event will generate three categories of clones; non-targeted clones, clones which have integrated concatemers of the targeting vector and the selection marker into the target locus, as well as clones targeted by simple gene replacement in which the positive selection marker has integrated in another location. This latter class of clone are the desired recombinants.

Co-electroporation vectors are configured as replacement vectors and they should be designed so that the recombinant clones can be readily identified by PCR and/or Southern blot analysis (see Section 3). In situations where the desired genetic modification would be minor, such as a single amino acid change, the ability to screen for targeted clones by either PCR or Southern blot analysis is best achieved by changing the wobble base pairs of a number of codons to create a unique PCR primer site or a novel restriction enzyme site.

The desired clone should have a recombinant allele with the designed mutation and the positive selection marker integrated elsewhere in the genome. Since the targeted allele should be the product of a simple gene replacement event many of the considerations for analysis have been previously described in Section 2.3. Analysis of transfected clones should be done in two steps. First the clones should be screened for the mutation at the target locus by PCR and/or Southern blot analysis. Putative positive clones should be further analysed to ensure that concatemers of the selection cassette did not integrate into the target locus. The simplest diagnostic procedure for the absence of concatemers is to use a restriction enzyme that does not cut in either target vector, the positive selection marker, or the bacterial plasmid, and hybridize the Southern blot with an external probe. If a concatemer has integrated into the target locus, the recombinant allele will not be the predicted size.

6.3 Subtle mutations generated with a hit-and-run vector

The 'hit-and-run' (37) or 'in-and-out' (38) targeting procedure utilizes two steps of homologous recombination to introduce mutations as small as a single base change into the gene of interest (*Figure 6*). Hit-and-run vectors are modified insertion vectors containing both positive and negative selectable markers outside the region of homology and the desired mutation in the region of homology. In the first step, homologous recombination and positive selection are used to generate a duplication at the target locus via the insertion of the targeting vector (see Section 3.2). The second step relies on spontaneous intrachromosomal recombination (pop-out) which occurs between the duplicated homologous sequences. The pop-out event can be stimulated 100-fold by incorporating a rare cutting endonuclease site in the plasmid backbone. *1SceI* has been commonly used. Expression of *1SceI* endonuclease from an expression plasmid will generate a double-strand break which will stimulate recombination or single-strand annealing. A pop-out event (either by recombination or single-strand annealing) results in the removal of the plasmid sequences, selection markers, and one complement of the duplication from the target locus. Since the negative selection marker is excised, clones which undergo a pop-out event can be selected with drugs used against the negative selection cassette.

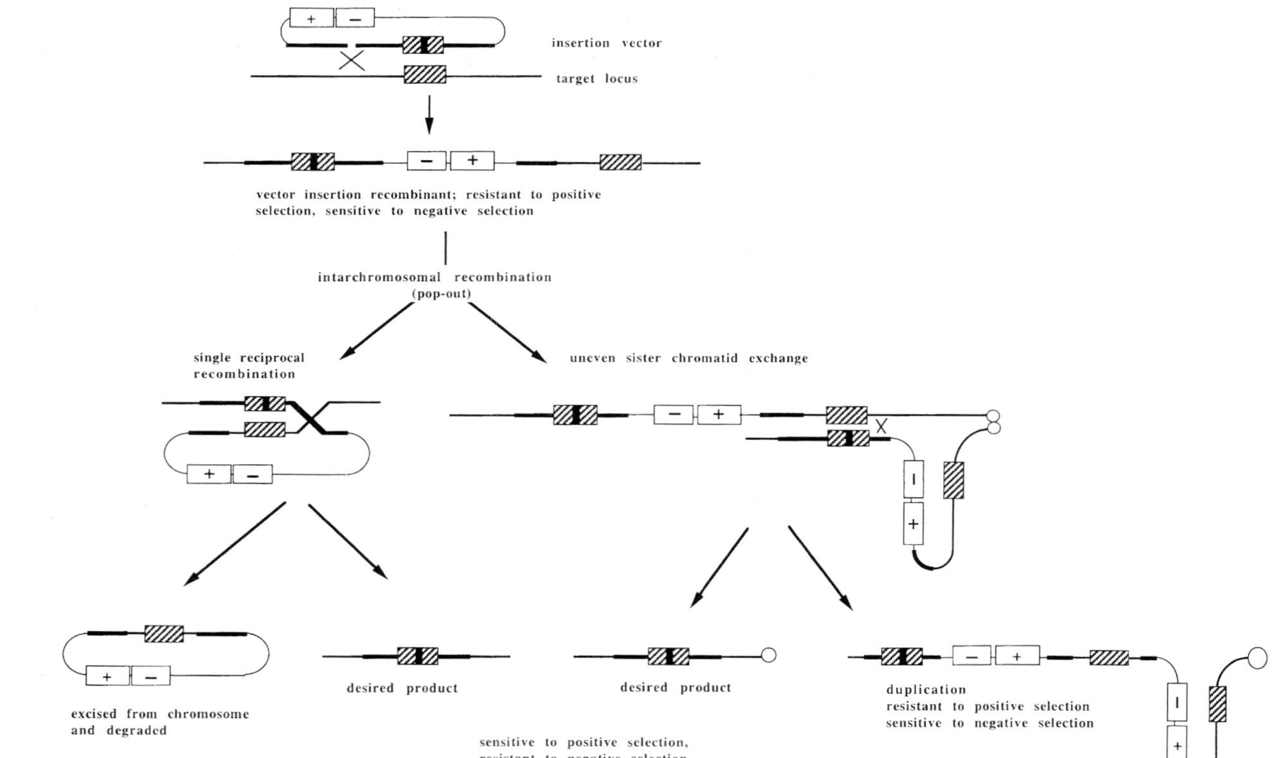

Figure 6. Hit-and-run targeting. The insertion vector is linearized inside the target homology prior to transfection. The lines and boxes are as indicated in *Figure 1*. The small, non-selectable mutation is represented by a thick line in the exon. The centromeres are represented by the empty circles. An insertion event is followed by intrachromosomal recombination; either single reciprocal recombination (*left*), uneven sister chromatid exchange (*right*), or followed by single-strand annealing.

There are a number of considerations in the vector design, screening, and selection procedures used in a hit-and-run experiment. Hit-and-run insertion vectors, which contain the desired mutation in the homologous sequences, will form a variety of different integration products upon targeting. This is associated with gap formation and repair, mismatch heteroduplex repair, and the branch migration of the targeted allele. For instance, the mutation may be removed and replaced with wild-type sequences by gap formation and repair or mismatch heteroduplex repair, during integration into the target locus. The frequency with which the mutation is removed is related to the distance between the mutation and the double-strand break in the homology region; the closer the mutation is to the break the more frequently it is removed by gap enlargement or mismatch heteroduplex repair and repaired using the chromosomal sequences as a template. The Holliday junctions (39) formed during the integration event often migrate. Thus, if the mutation is close to the double-strand break, the Holliday junctions will commonly migrate across the mutation and resolve on the other side of the homologous sequences. This will result in the mutation appearing in the opposite duplication than initially planned. Mismatch heteroduplex repair may result in the duplication or removal of the mutation in some of the integrated clones.

The duplicated homologous sequences generated by a hit-and-run vector will recombine by intrachromosomal homologous recombination (single reciprocal recombination) or more frequently by uneven sister chromatic exchange (40) (*Figure 6*) or possibly by single-strand annealing. The frequency of these excision events is dependent on both the length and the degree of homology between the duplicates (41). Because negative selection is employed in the second step with the hit-and-run procedure, there is a background caused by the loss of the negative selectable marker due to events other than the desired pop-out event. In practice, the frequency of the desired event is at least comparable, if not greater than the background frequency so that clones with the desired alteration are easily identified. However, background should be much less a problem if the pop-out is stimulated by generation of a double-strand break with a rare cutting endonuclease.

The screen for vector insertion is done in two stages. First, targeted clones are identified from clones which survived positive selection after electroporation. After the clones with an insertion event have been isolated, it is necessary to verify that the mutation is present. These correctly targeted clones bearing the desired mutation are then expanded without selection to generate a sufficient number of cells so that by chance some will have undergone the 'pop-out' event. Subsequently these cells are then subjected to negative selection in order to recover pop-out clones. A screen of clones resistant to the negative selection drug is then carried out to identify pop-out clones carrying the desired mutation. See Chapter 3 for details of tissue culture procedures.

There are a number of considerations and parameters that can optimize and increase the chance of obtaining the desired mutation in hit-and-run experiments, both in the vector integration step and in the pop-out step. For a more detailed description of vector design, screening, and selection procedures in hit-and-run experiments, see refs 41 and 42.

6.4 Subtle mutations generated by double replacement

The 'double replacement' technique (43, 44), can be used to generate subtle mutations in the desired locus. As with hit-and-run, double replacement relies on two rounds of homologous recombination to generate the desired small mutation, although unlike hit-and-run, the second round of homologous recombination in double replacement is also a gene targeting event (vector–chromosome recombination), rather than a pop-out event.

Double replacement vectors are, as the name indicates, replacement vectors containing both positive and negative selectable markers inside the region of homology (*Figure 7*). Alternatively, a single marker, such as *Hprt*, which provides the opportunity for both positive and negative selection in the appropriate genetic background, can be used (43). In the first step, replacement targeted clones are generated. In the second step, a different targeting vector homologous to the same chromosomal target which is devoid of any selectable markers and which is carrying the desired mutation is electroporated into targeted clones identified in the first screen. Selection against the negative selectable marker is then applied and resistant clones are screened for the presence of the desired mutation. As with the hit-and-run method, a limitation of this procedure is the spontaneous loss of sensitivity to the negative selection agent due to events other than homologous recombination, such as mutation of the negative selectable marker. In practice the background loss of the negative selectable marker is a much more significant problem in this technique compared with hit-and-run, because the absolute frequency of intrachromosomal recombination can be orders of magnitude higher than the frequency of vector–chromosome recombination. Since the generation of the desired mutation at the target locus depends on two rounds of gene targeting, loci displaying low targeting frequencies may be difficult to target by double replacement if the background in the negative selection screen overshadows the frequency of targeted events in the second step. An attractive aspect of the double replacement method is that after the first targeting step is carried out, a number of different vectors can be used in the second step to generate an array of mutations at the same locus. Incorporation of a rare cutting endonuclease should increase the frequency of the second targeting event. Do not linearize the second targeting vector if using a rare cutting endonuclease (for detailed description refer to ref. 45). However, many laboratories have experienced severe difficulties implementing this technique, even after generating a double-strand break in the target locus.

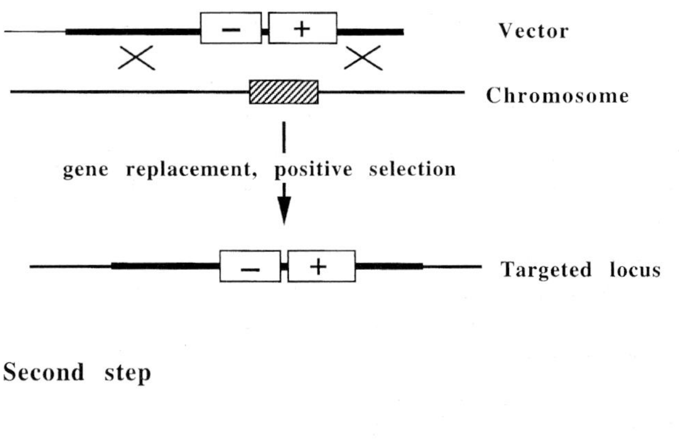

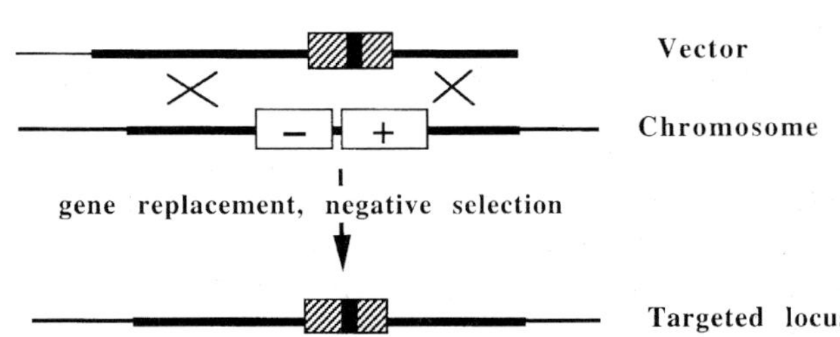

Figure 7. Double replacement targeting. The replacement vectors are linearized outside the target homology prior to transfection. The lines and boxes are as indicated in *Figure 1*. The small, non-selectable mutation is represented by a thick line in the exon. First targeting event with positive selection (A) is followed by a second targeting event with negative selection (B).

7. 'Knock-in' targeting vectors: simultaneous study of gene function and expression

The use of gene targeting to replace an endogenous gene with another, such as a homologue or a marker for gene expression, has been termed 'knock-in' (46). Knock-in experiments, particularly in mouse ES cells, are used to place a transgene (cDNA) contained in a targeting vector, under the transcriptional control of an endogenous gene. When a reporter gene, such as LacZ or GFP, is used in a knock-in experiment, the spatial and temporal expression pattern of the endogenous gene can be monitored during development and in the adult mouse. When the endogenous gene is replaced with a homologue, it can be used to assess whether members of the same gene family have identical biological function when expressed in the same spatial and temporal pattern

(46). The effect of knocking-in a reporter or homologue typically results in the loss-of-function of the endogenous gene.

The ultimate goal of a knock-in targeting experiment is to place the gene of interest under the transcriptional control of the endogenous locus. The first knock-in studies utilized the double replacement technique described in Section 6.4 to replace the mouse alpha-lactalbumin and p53 genes with the human alpha-lactalbumin and LacZ expression marker, respectively, in mouse ES cells (43, 47). Subsequently knock-in experiments have been performed using conventional gene targeting approaches, usually in combination with the Cre-*loxP* system to remove the selection cassette (46). These knock-in vectors are replacement targeting constructs containing a positive selectable marker and the transgene to be expressed. This transgene is arranged in such a way that it falls under the transcriptional control of the endogenous locus upon homologous recombination. Optimally knock-in constructs should be designed so that no endogenous sequences are deleted and so that regulatory sequences associated with positive selectable markers can be deleted after the targeting event.

Figure 8 shows three possible vector designs for knock-in experiments. *Figure 8A* shows a knock-in construct in which the promoter-less fusion protein beta-geo (48), (lacZ–neo fusion) is fused in-frame to the endogenous gene. Targeting with this construct knocks-out the endogenous gene and generates a fusion protein which can be used to monitor the expression of this transcript in the mouse. *Figure 8B* shows a targeting construct in which the transgene (a reporter or homologue) is fused in-frame to the endogenous gene. Removal of the positive selectable marker using the Cre-*loxP* system (see Chapter 2), assures that the strong promoter driving the expression of the selectable marker does not interfere with the normal regulation of the knocked-in allele. *Figure 8C* shows a construct in which the transgene is placed downstream of a splice acceptor sequence in an intron of the endogenous locus. Targeting by the vector, followed by in-frame splicing leads to the production of the desired fusion protein, analogous to a gene trap product

Figure 8. Knock-in strategies. The positive selectable marker is represented by a +. *loxP* sites are depicted as open arrow heads. Exons of the endogenous gene are represented by grey boxes. The black arrow pointing right represents the transcriptional start site, the black line in the grey box represents the translation start codon. The striped box represents the transgene being knocked-in, such as a homologue or reporter. pA, polyadenylation signal; SA, splice acceptor. (A) Knock-in of beta-geo into the target locus in-frame with the target. (B) Knock-in of a reporter or homologue into the target locus, followed by Cre-mediated removal of the selectable marker. The transgene is in-frame with the target. An alternative is to place the cDNA with an ATG in 5′ untranslated sequences and/or without a polyadenylation signal to allow 3′ processing and polyadenylation to be controlled by the sequences in the target gene. (C) Knock-in using a splice acceptor sequence, followed by Cre-mediated marker removal, the transgene is in-frame with the target locus. Fusion proteins can be avoided by placing IRES sequences 5′ to the transgene (*Figure 9*).

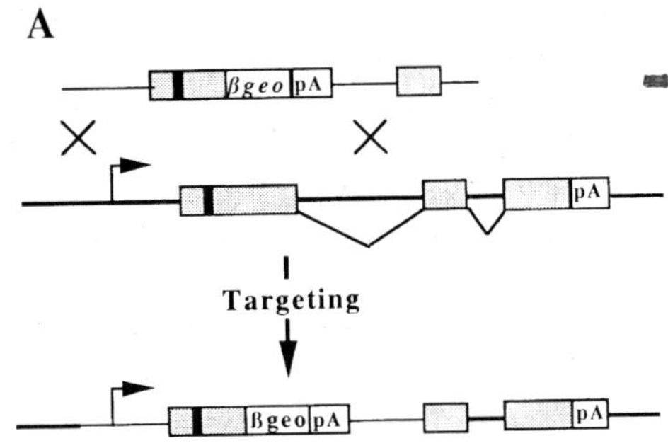
A
ßgeo
pA
pA
Targeting
ßgeo
pA
pA

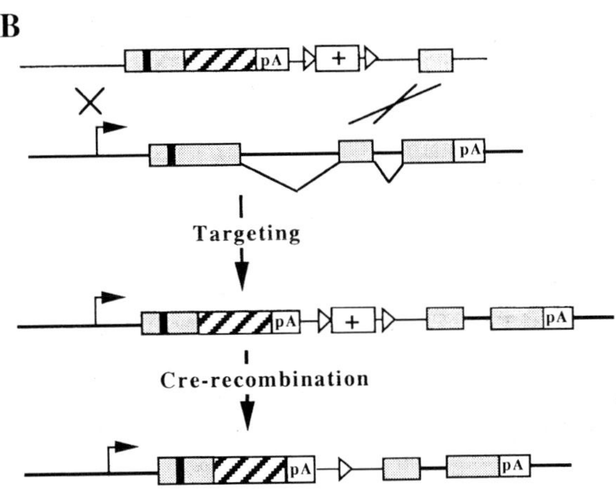
B
pA
+
pA
Targeting
pA
+
pA
Cre-recombination
pA
pA

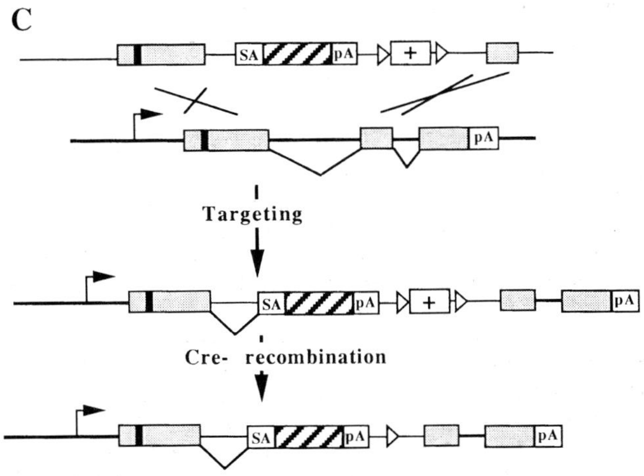
C
SA
pA
+
pA
Targeting
SA
pA
+
pA
Cre- recombination
SA
pA
pA

(Chapter 6). Using the splice acceptor sequence provides more versatility in the design of the vector, since any convenient restriction enzyme site in the intron can be used for cloning the gene of interest.

The three vector designs described above and depicted in *Figure 8* are based on the production of a fusion protein between the endogenous and knocked-in products. Although the fusion to a portion of the endogenous gene may be desired for proper subcellular localization, secretion fusion products may not always be functional, and the vector design is complicated by having to clone the transgene in-frame with the endogenous coding region. This may be circumvented by knocking-in the transgene into the 5′ untranslated region upstream of the endogenous translational start site (see *Figure 6*, Chapter 3). The considerations discussed earlier of the possibility of splicing around the inserted transgene if it is not placed in the first exon must be taken into account when designing all knock-in constructs.

A convenient way to circumvent the need for the generation of fusion products, however, is the use of internal ribosomal entry site (IRES) elements (for review, see ref. 49). The translation of mRNAs in eukaryotic cells begins with the association of ribosomes to the 5′-cap structure. Viruses of the picornavirus family, however, produce uncapped mRNAs whose translation is dependent on specific sequences in their 5′ untranslated regions. These IRES elements mediate cap-independent entry of the ribosome. The use of IRES elements to generate bicistronic mRNAs in mammalian cells was demonstrated by Pelletier and Sonenberg (50). In this study, mRNAs containing the coding regions for a selectable marker and a reporter gene, separated by an IRES element, were shown to express both proteins efficiently. The coding region upstream of the IRES is translated by cap-binding ribosomes, whereas the coding region downstream of the IRES is translated by IRES-binding ribosomes.

The most commonly used IRES element in mammalian cells is that of the encephalomyocarditis virus (51). IRES elements can be used in knock-in experiments to place the transgene under the control of an endogenous locus without disrupting its function (*Figure 9*). For this, the coding region of the transgene is cloned downstream from an IRES element and targeted to the 3′ untranslated region of the endogenous locus. IRES elements can also be used in targeting vectors to generate loss-of-function mutations in genes. Vectors can be designed in which promoter-less selectable markers are cloned downstream of IRES elements so that they will be translated when a bicistronic mRNA is produced from the targeted allele. This is analogous to the promoter trap vectors described in Section 4.2.2, but without the need for an in-frame fusion protein between the endogenous product and the selectable marker.

Similarly, IRES elements can be incorporated in knock-in targeting vectors, such as the ones depicted in *Figure 8*, to avoid the need to generate fusion proteins between the endogenous locus and the transgene. The use of IRES

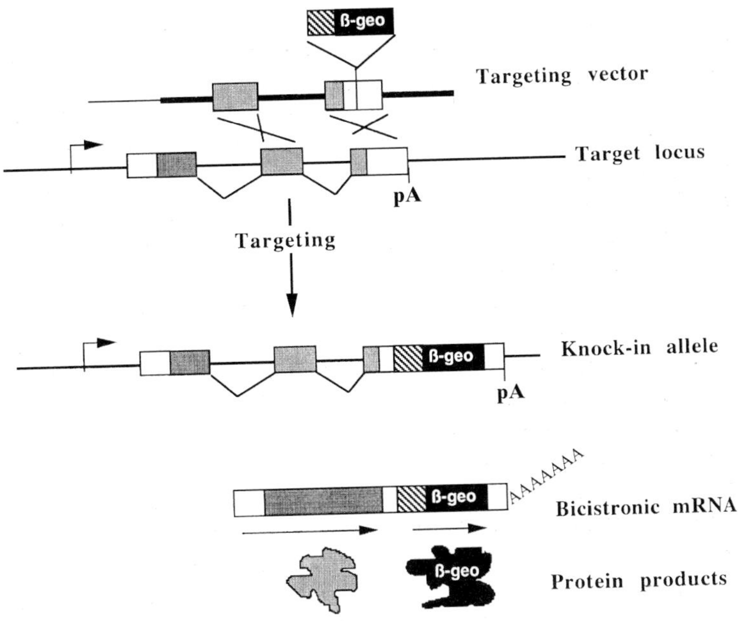

Figure 9. Generation of bicistronic mRNAs using an IRES element. The beta-geo fusion protein (48), is cloned downstream of an IRES element (striped box) and targeted to the 3′ untranslated region (white box) of the target gene (grey boxes). The chimeric mRNA produced drives the expression of both the endogenous and beta-geo proteins via cap- and IRES-mediated ribosome entry respectively. pA, polyadenylation signal.

elements greatly simplifies vector construction. Perhaps the most versatile combination that can be used in a knock-in vector is an IRES element downstream of a splice acceptor sequence and upstream of the transgene. This splice acceptor–IRES transgene combination can be cloned anywhere in an intron without regard for in-frame fusion after splicing. Although beyond the scope of this chapter, this combination is particularly useful for gene trap experiments using retroviruses (52) (see Chapter 6). Despite the obvious attractions of IRES elements, they do not always lead to a high level of translational activity.

8. Summary

The complexity and versatility of gene targeting vectors has increased with time. Vectors can be designed to generate virtually any genetic change in the genome, from single base pair substitutions to large scale chromosomal rearrangements. In most cases, several alternative methods can be used to generate the desired mutation. In this chapter we hope to have conveyed the impression that the most important aspect of any gene targeting experiment is vector design since this will ultimately determine the progress of downstream experimentation, not only in the generation of the recombinant allele, but also in the analysis of the resultant cells or animal. Clearly there are a number of methodologies that can be used singularly or in combination to manipulate the mammalian genome. The first, and perhaps most important step, is the design of the targeting vector. We hope this chapter offers the reader both a summary of the current methodologies as well as a guide for designing targeting vectors.

Acknowledgements

Work in the author's laboratory is supported by the National Institutes of Health and the Howard Hughes Medical Institute.

References

1. Lin, F.-L., Sperle, K., and Steernberg, N. (1985). *Proc. Natl. Acad. Sci. USA*, **82**, 1391.
2. Smithies, O., Gregg, R. G., Boggs, S. S., Koralewski, M. A., and Kucherlapati, R. S. (1985). *Nature*, **317**, 230.
3. Robberson, B. L., Cote, G. J., and Berget, S. M. (1990). *Mol. Cell. Biol.*, **10**, 84.
4. Zhang, H., Hasty, P., and Bradley, A. (1994). *Mol. Cell. Biol.*, **14**, 2404.
5. Joyner, A. L., Skarnes, W. C., and Rossant, J. (1989). *Nature*, **338**, 153.
6. Soriano, P., Montgomery, C., Geske, R., and Bradley, A. (1991). *Cell*, **64**, 693.
7. Hasty, P., Rivera-Perez, J., Chang, C., and Bradley, A. (1991). *Mol. Cell. Biol.*, **11**, 4509.
8. Ramirez-Solis, R., Rivera-Perez, J., Wallace, J. D., Wims, M., Zheng, H., and Bradley, A. (1992). *Anal. Biochem.*, **201**, 331.
9. Thomas, K. R. and Capecchi, M. R. (1987). *Cell*, **51**, 503.
10. Moens, C. B., Auerbach, A. B., Conlon, R. A., Joyner, A. J., and Rossant, J. (1992). *Genes Dev.*, **6**, 691.
11. Jasin, M. and Berg, P. (1988). *Genes Dev.*, **2**, 1353.
12. Hasty, P., Rivera-Perez, J., and Bradley, A. (1991). *Mol. Cell. Biol.*, **11**, 586.
13. te Riele, H., Robanus Maandag, E., and Berns, A. (1992). *Proc. Natl. Acad. Sci. USA*, **89**, 5128.
14. Thomas, K. R., Deng, C., and Capecchi, M. R. (1992). *Mol. Cell. Biol.*, **12**, 2919.
15. Modrich, P. and Lahue, R. (1996). *Annu. Rev. Biochem.*, **65**, 101.
16. Kolodner, R. (1996). *Genes Dev.*, **10**, 1433.
17. Rayssiguier, C., Thaler, D. S., and Radman, M. (1989). *Nature*, **342**, 396.
18. Selva, E. M., New, L., Crouse, G. F., and Lahue, R. S. (1995). *Genetics*, **139**, 1175.
19. de Wind, N., Dekker, M., Bern, A., Radman, M., and te Riele, H. (1995). *Cell*, **82**, 321.
20. Waldman, A. S. and Liskay, R. M. (1988). *Mol. Cell. Biol.*, **8**, 5350.
21. Schwartzberg, P. L. and Robertston, E. J. (1991). *Proc. Natl. Acad. Sci. USA*, **87**, 3210.

22. Jeannotte, L., Ruiz, J. C., and Robertston, E. J. (1991). *Mol. Cell. Biol.*, **11**, 578.
23. Donehower, L. A., Harvey, M., Slagle, B. L., McArthur, M. J., Montgomery, C. A., Butel, J. A., *et al.* (1992). *Nature*, **356**, 215.
24. Mansour, S. L., Thomas, K. R., and Capecchi, M. R. (1988). *Nature*, **336**, 348.
25. Reid, L. H., Gregg, R. G., Smithies, O., and Koller, B. H. (1990). *Proc. Natl. Acad. Sci. USA*, **87**, 4299.
26. Olson, E. N., Arnold, H. H., Rigby, P. W. J., and Wold, B. J. (1996). *Cell*, **84**, 1.
27. Kim, C., Epner, E., Forrester, W., and Groudine, M. (1992). *Genes Dev.*, **9**, 2203.
28. Fiering, S., Kim, C., Epner, E., and Groudine, M. (1993). *Proc. Natl. Acad. Sci. USA*, **90**, 8469.
29. Braun, T., Bober, E., Rudnicki, M. A., Jaenisch, R., and Arnold, H. H. (1994). *Development (Cambridge, UK)*, **120**, 3083.
30. Pham, C. T. N., MacIvor, D. M., Hug, B. A., Heusel, J. W., and Ley, T. J. (1996). *Proc. Natl. Acad. Sci. USA*, **93**, 13090.
31. Zimmer, A. and Gruss, P. (1989). *Nature*, **338**, 150.
32. Thomas, K. R. and Capecchi, M. R. (1986). *Cell*, **44**, 419.
33. Brinster, R. L., Braun, R. E., Lo, D., Avarbock, M. R., Oram, F., and Brinster, R. D. (1989). *Proc. Natl. Acad. Sci. USA*, **86**, 7087.
34. Lovell-Badge, R. H. (1987). In *Teratocarcinomas and embryonic stem cells: a practical approach* (ed. E. J. Robertson), pp. 153–81. IRL Press, Oxford.
35. Davis, A., Wims, M., and Bradley, A. (1992). *Mol. Cell. Biol.*, **12**, 2769.
36. Reid, L., Shesley, R., Kim, H.-S., and Smithies, O. (1991). *Mol. Cell. Biol.*, **11**, 2769.
37. Hasty, P., Ramirez-Solis, R., Krumlauf, R., and Bradley, A. (1991). *Nature*, **350**, 243.
38. Valancius, V. and Smithies, O. (1991). *Mol. Cell. Biol.*, **11**, 1402.
39. Holliday, R. (1964). *Genet. Res.*, **5**, 282.
40. Bollag, R. J. and Liskay, M. (1991). *Mol. Cell. Biol.*, **11**, 4839.
41. Bradley, A., Ramirez-Solis, R., Zheng, H., Hasty, P., and Davis, A. (1992). In *Post implantation development of the mouse*, pp. 256–76 (Ciba Foundation Symposium 165). Wiley, Chichester.
42. Hasty, P. and Bradley, A. (1993). In *Gene targeting: a practical approach* (ed. A. L. Joyner), pp. 1–31. IRL Press, Oxford.
43. Stacey, A., Schnieke, A., McWhir, J., Cooper, A., Colman, A., and Melton, D. W. (1994). *Mol. Cell. Biol.*, **14**, 1009.
44. Wu, H., Liu, X., and Jaenisch, R. (1994). *Proc. Natl. Acad. Sci. USA*, **91**, 2819.
45. Donoho, G., Jasin, M., and Berg, P. (1998). *Mol. Cell. Biol*, **18**, 4070.
46. Hanks, M., Wurst, W., Anson-Cartwright, L., Auerbach, A. B., and Joyner, A. L. (1995). *Science*, **269**, 679.
47. Gondo, Y., Nakamura, K., Nakao, K., Sasaoka, T., Ito, K., Kimura, M., *et al.* (1994). *Biochem. Biophys. Res. Commun.*, **202**, 830.
48. Friedrich, G. and Soriano, P. (1991). *Genes Dev.*, **5**, 1513.
49. Mountford, P. S. and Smith, A. G. (1995). *Trends Genet.*, **11**, 179.
50. Pelletier, J. and Sonenberg, N. (1988). *Nature*, **334**, 320.
51. Kim, D. G., Kang, H. M., Jang, S. K., and Shin, H. S. (1992). *Mol. Cell. Biol.*, **12**, 3636.
52. Zambrowicz, B., Friedrich, G. A., Buxton, E. C., Lilleberg, S. L., Person, C., and Sands, A. T. (1998). *Nature*, **392**, 608.

2

Site-specific recombination in cells and mice

SUSAN M. DYMECKI

1. Introduction

The use of site-specific recombinase systems has revolutionized our ability to genetically manipulate embryonic stem (ES) cells and mice. Recent advances using the Cre-*loxP* and Flp-*FRT* systems have now made it possible to generate '*clean*' germline mutations following a single gene targeting event, as well as to (in)activate genes in a *conditional* manner in the living mouse. Not only can target gene mutations be induced in a spatially and temporally restricted fashion, but lineage tracers can be activated in specific progenitor populations to chart cell fate directly in the wild-type or mutant mouse.

This chapter introduces site-specific recombination and details a variety of applications, many of which are extensions of the gene targeting vectors and manipulations presented by Hasty *et al.* in Chapter 1. Many of the mutagenesis techniques which exploit the Cre-*loxP* system have been compiled earlier in an excellent book by Torres and Kühn (1). In this chapter, I present the Flp-*FRT* system in addition to the Cre-*loxP* system, for individual or combined uses. Together, these surveys and protocols should provide a basis for a wide variety of studies on gene function *in vivo*. As novel recombinase-based applications continue to be developed, the possibilities for genome engineering appear without limit.

2. Site-specific recombinase systems: Cre-*loxP* and Flp-*FRT*

The simplest site-specific recombination systems are comprised of two elements: the recombinase enzyme and a small stretch of DNA specifically recognized by the particular recombinase. These two elements work together to either delete, insert, invert, or translocate associated DNA. Two such recombinase systems have been established in mice (2–5) providing the basic tools for *in vivo* genetic engineering: the Cre-*loxP* system from the bacteriophage P1 and the Flp-*FRT* system from the budding yeast *Saccharomyces cerevisiae*.

2.1 General properties

Both Cre and Flp are members of the λ integrase superfamily of site-specific recombinases (6) that cleave DNA at a distinct target sequence and then ligate it to the cleaved DNA of a second identical site to generate a contiguous strand. This recombination reaction is carried out with absolute fidelity, such that not a single nucleotide is gained or lost overall, and with no other requirements than the recombinase, the specific target DNA sequence, and some mono- or divalent cations (7). No cofactors or accessory proteins are required, making these recombinase systems quite adaptable for use in a wide variety of heterologous environments. The minimal target sites for Cre and Flp are each 34 bp (*Figure 1*), a size that is unlikely to occur at random in even the largest vertebrate genome, and yet small enough so as to be effectively neutral towards gene expression when positioned in chromosomal DNA for genetic manipulations. The orientation of these target sites relative to each other on a segment of DNA directs the type of modification catalyzed

---

Figure 1. Diagram of recombinase target sites. Target sites contain inverted 13 bp symmetry elements (horizontal arrows **a** and **b**) flanking an 8 bp A:T-rich non-palindromic core (shaded rectangle). One recombinase monomer binds each symmetry element, while the core sequence provides the site of strand cleavage, exchange, and ligation. The asymmetry of the core region, indicated by the shaded gradient, imparts directionality on the target site which, in turn, imparts directionality on the reaction: directly oriented sites lead to excision of intervening DNA, inverted sites causes inversion of intervening DNA (see *Figure 2* for details). By convention, nucleotides are numbered from the centre of the core. (A) The target site (18) recognized by Cre recombinase (*c*auses *re*combination) is called *loxP* (*lo*cus of cross-over (*x*) in *P*1) (12). For transgene and target vector construction, two features should be noted: first, there is only one complete open reading frame (ORF) through the *loxP* sequence (ORF +1) and secondly, an ATG (shown in bold) present in the core region is followed by an ORF in reading frame +2. (B) The *Fl*p *r*ecognition *t*arget sequence is designated FRT (19, 93, 94). The extra symmetry element **c**, although present in the 2μ plasmid, appears to be dispensable making the minimal *FRT* 34 bp (17). In contrast to the *loxP* site, the symmetry elements of the *FRT* site contain a single bp difference (large font). Solid dots show residues critical to Flp binding the symmetry element **a**; reciprocal sites on **b** are equally important (24). The length of the core region is critical for efficient recombination, while the actual sequence of the core can vary provided the cores of the two participating sites are identical. The pyrimidine-rich tracts extending out from the core region on opposite strands (shown in bold) give further asymmetry to the FRT. Features important to transgene and target vector construction include: an *Xba*I restriction site (underlined) and two ORFs through the *FRT* sequence. (C) Synaptic complex formed between recombinase molecules and two target sites. Although the two participating target sites are identical, one is distinguished here by the grey rectangle; the other, white. Recombinase monomers (ovals) bind to symmetry elements in cognate target site; two such target sites come together in a synaptic complex; cleavage, exchange, and ligation of one DNA strand forms a Holliday intermediate; cleavage, exchange, and ligation of the second strand resolves the Holliday intermediate into two recombinant molecules (7). In the final product, the symmetry arms of the two target sites have been exchanged and the core regions of the two recombinants are heteroduplex.

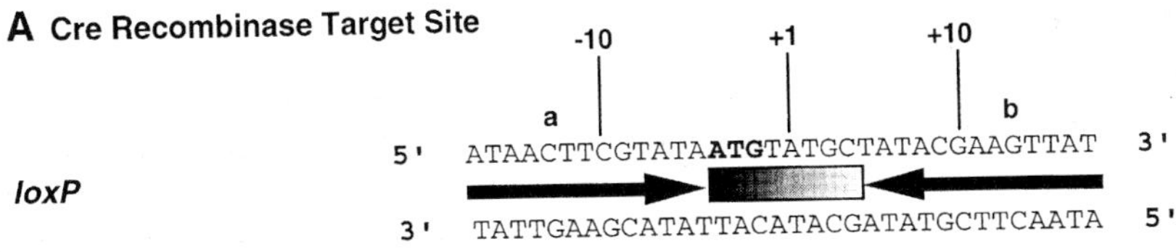

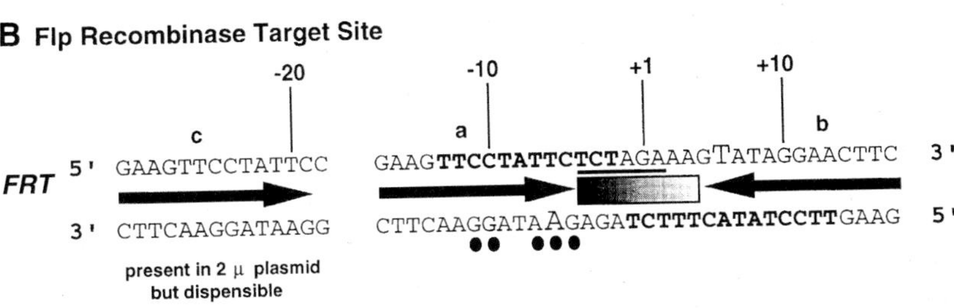

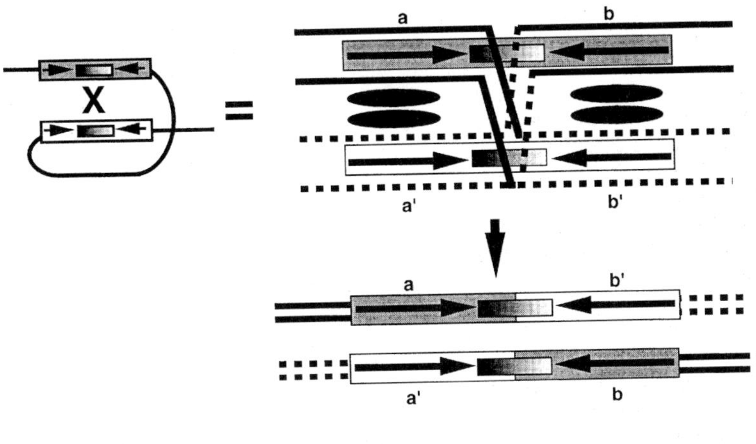

by the recombinase (illustrated in *Figure 2*): directly oriented sites lead to excision of intervening DNA, inverted sites cause inversion of intervening DNA, and target sites on separate linear molecules direct mutual exchange of regions distal to the sites (8). Remarkably, recombination is catalyzed regardless of DNA topology. Substrate DNA can be a supercoiled circle, a relaxed circle, or a linear chromosome (7). Moreover, efficient recombination can occur in terminally differentiated post-mitotic cells (9), as well as in dividing cells (10, 11). Together these properties (summarized in *Table 1*) make both Cre and Flp well suited for modifying or introducing DNA at specific sites in a variety of hosts, including mice and cultured mammalian cells.

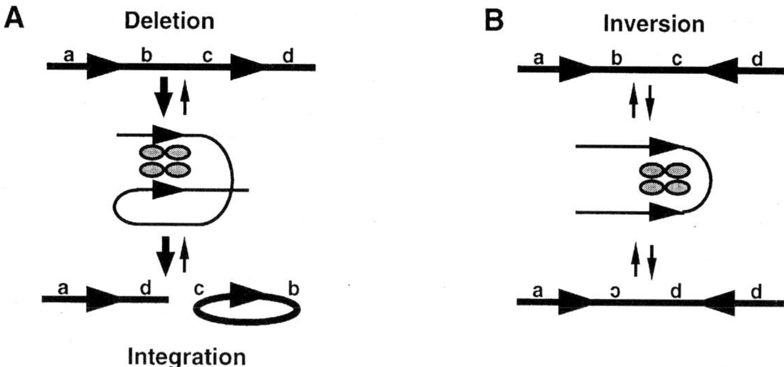

C Reciprocal Translocation Between Heterologous Chromosomes

e
f
e:f
f:e

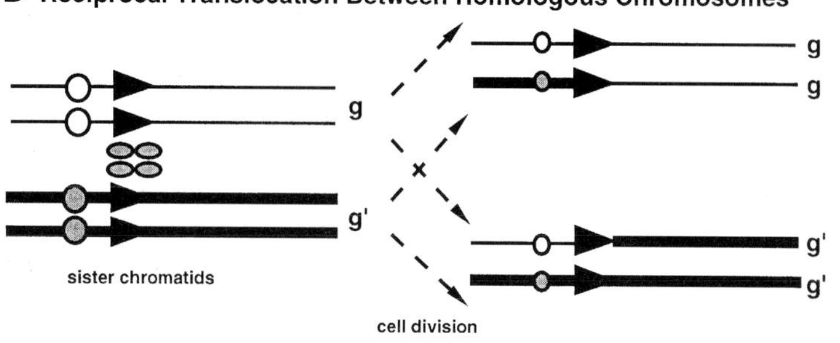

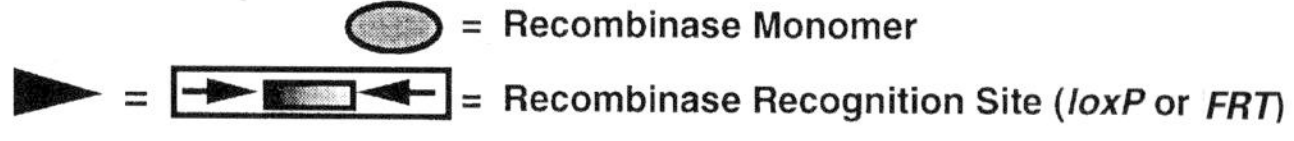

Cre recombinase is encoded by the bacteriophage P1 and catalyses recombination between sites called *loxP* (*lo*cus of cross-over (*x*) in *P*1) (*Figure 1A*) (12). In bacteria, Cre activity ensures stable maintenance of the P1 phage circular genome by resolving replicated dimeric chromosomes into monomers for segregation to each daughter during cell division. Dimer resolution occurs by the excision of monomers lying between directly repeated *loxP* sites (13).

In contrast to the prokaryotic origin of the Cre-*loxP* system, the Flp system is encoded by the 2μ plasmid from the eukaryote, *Saccharomyces cerevisiae* (14, 15). The natural function of Flp in *S. cerevisiae* is to increase the number of copies of the 2μ plasmid. This amplification is achieved by inverting the 2μ

Figure 2. Recombinase reactions catalysed by Cre or Flp. Black triangles indicate recombinase recognition target sites; black lines, chromosomal DNA, with orientation indicated by the letters **a**, **b**, **c**, and **d**; grey ovals, recombinase monomers; and arrows between recombinase substrates and products indicate the reversible nature of each reaction. Target sites align productively in one orientation, imparting directionality on the reaction. Flp and Cre will, therefore, (A) excise a circular molecule from between two directly repeated target sites separated by at least 82 bp (20), integrate a circular molecule into a linear molecule each possessing a target site, (B) invert the DNA between two inverted sites, and (C and D) exchange sequences distal to target sites present on two linear molecules (8). This latter type of intermolecular recombination can occur between heterologous chromosomes (indicated in C by the letters **e** and **f**), and between targets on homologous chromosomes after DNA replication (indicated in D by the letters g and g′) (60, 65, 95).

plasmid DNA lying between two inverted *FRT* repeats (*F*lp *r*ecognition *t*arget) (*Figure 1B*). As originally proposed by Futcher (16), partial replication of the 2μ plasmid yields a theta intermediate that contains two divergent replication forks and three *FRT* sites. Flp-mediated recombination inverts one replication fork with respect to the other such that the replication forks are now oriented in the same direction. As the forks 'chase' one another around the circle, plasmid multimers containing directly oriented *FRT* sites are generated. Flp then catalyses recombination across these direct *FRT* repeats to produce plasmid monomers at high copy.

As their disparate origin and recognition sequences would suggest, Cre and Flp show limited amino acid sequence homology. They do, however, share four absolutely conserved residues: arginine 191, histidine 305, arginine 308, and tyrosine 343 (the numbers refer to amino acid positions in Flp (17), with tyrosine 343 equivalent to tyrosine 324 in Cre). It is assumed that these conserved residues underlie a common mechanism of DNA cleavage and ligation.

2.2 Target sites and recombinase action

Although the nucleotide sequences are different, the overall structural elements comprising both the *loxP* (18) and *FRT* (19) sites are the same (*Figure 1*) and include inverted symmetry elements which act as binding sites

Table 1. Properties which make Cre and Flp well suited for genome engineering

Site specific:	→ ■■ ← 34 bp asymmetric recognition target sequence.
Conservative:	recombination occurs without any overall gain or loss of nucleotides.
Simple:	recombinase functions autonomously, no cofactors or ATP required.
Substrate topology:	supercoiled circular, relaxed circular, or linear DNA.
Heritable product:	resultant genetic alteration is stable and heritable.
Broad utility:	function in a wide range of cell types, including undifferentiated as well as post-mitotic cells. function over large distances (Mb).

for recombinase monomers, and an A:T-rich asymmetric core which is the site of strand cleavage, exchange, and ligation. The asymmetry of the core region is critical, imparting directionality on the target site which, in turn, imparts directionality on the recombination reaction. The end-result is that directly repeated sites specify excision of intervening DNA; target sites on one circular and one linear molecule direct integration of the circular DNA; inverted sites direct inversion of intervening DNA; and target sites on separate linear molecules instruct mutual exchange of regions distal to the target sites (*Figure 2*) (8). A detailed discussion on the mechanism of recombination is outside the domain of this chapter, but in outline: recombinase monomers bind to the symmetry elements of their cognate target sites; two such target sites (separated by at least 82 bp) (20) come together in a synaptic complex; cleavage, exchange, and ligation of one DNA strand forms a Holliday intermediate; cleavage, exchange, and ligation of the second strand resolves the Holliday intermediate into two recombinant molecules (*Figure 1C*) (7). In the final recovered product, the symmetry arms of the two target sites have been exchanged and the core regions of the two recombinants are heteroduplex. For an *intra*chromosomal deletion reaction (*Figure 2A*), this means a heteroduplex target site remains in the chromosome following DNA excision, making the reaction reversible. The *inter*molecular integration is, however, energetically less favoured. Moreover, in many *in vivo* situations the circular extrachromosomal product is lost, making the deletion reaction effectively irreversible.

2.3 Making recombination events irreversible

In theory, each site-specific recombination reaction is reversible. To bias recombination in a single direction, two approaches have been explored:

(a) Limiting recombinase activity to a brief pulse.
(b) Using two different mutant target sites which, following recombination, leave one of the two resultant heteroduplex sites inactive.

The first approach has proved most useful in *E. coli* (21), *Drosophila* (22), and the mosquito, *Aedes aegypti* (23), where heat shock promoters can rapidly up-regulate recombinase expression and where the parameters of recombinase persistence and decay have been accurately determined. This type of quick regulation is more difficult to achieve in mammalian cells, however potential methodologies include: inducible promoters, ligand-activated recombinase fusion proteins, introduction of non-selectable plasmids that express the recombinase and then are subsequently lost, and injecting purified recombinase protein.

The second strategy to shift the reaction equilibrium involves using mutant target sites. Although met with only moderate success in lower organisms (21, 23), the approach may, none the less, prove useful in certain mammalian applications and therefore deserves mention. Nucleotides in the *FRT* site which influence Flp binding and recombination efficiency (24) are indicated by black dots in *Figure 1B*. A mutation, such as in position –10 results in a very moderate decrease in recombination efficiency when present in symmetry element *a* or *b* (+10) alone, but nearly inactivates an FRT site if present in both *a* and *b* elements of a single target site. Therefore two FRT sites, one with a mutation in symmetry element *a* and the other in symmetry element *b*, should recombine with near wild-type efficiencies, but generate a heteroduplex product that carries both the *a* and *b* mutations; this double mutant site should be effectively inactive for further recombination (*Figure 3*). For example, integrations that resist subsequent excisional recombination could be catalysed (25) (*Figure 3*). Mutant *loxP* sites have been similarly exploited in plants (26).

3. Cre, Flp, FlpL, and Flpe: distinguishing properties suggest specific applications

The Cre-*loxP* and Flp-*FRT* site-specific recombination systems add great potential to *in vitro* and *in vivo* genomic manipulations (summarized in *Table 2*). Full realization of this potential, however, depends on the actual applied characteristics of each recombinase system in the different cell types of the mouse. Despite their mechanistic similarity, the applied properties of Cre and Flp are likely to differ given their disparate origin (bacteria versus yeast, respectively), as well as their different amino acid sequence and DNA recognition sites. Two very basic differences between bacteria and yeast are the temperature at which maximal growth occurs (~ 37 °C for bacteria; ~ 30 °C for yeast) and the nature of their chromatin (prokaryotic versus eukaryotic). Therefore, when employed in the mouse, Cre and Flp may be differentially affected by mouse temperature or by variations in chromatin structure surrounding recombinase target sites.

To help distinguish those *in vivo* applications best accomplished by Cre and those better served by Flp, Stewart and colleagues carefully determined the *in vitro* temperature–activity profiles for each recombinase (27). Cre recombinase was found to be active over a wider range of temperatures *in vitro* than Flp, with maximal performance at 42 °C. Stewart and colleagues (27) also documented a mutation (F70L) in a commercially available plasmid (Stratagene) which renders Flp even more sensitive to temperature, bringing *in vitro* recombinase activity down by one-third to one-tenth of wild-type; this variant will be referred to as FlpL. To improve the activity of Flp at higher temperatures and therefore its effectiveness for genetic manipulations in mammalian cell culture and in mice, Buchholz *et al.* (28) identified four amino acid changes (P2S, L33S, Y108N, S294P) which together improve Flp activity fourfold over that of wild-type (wt) Flp at 37 °C and tenfold at 40 °C. This enhanced Flp enzyme is called Flpe and has activity in ES cell cultures equivalent to that of Cre (28) (results approximated in *Table 3*).

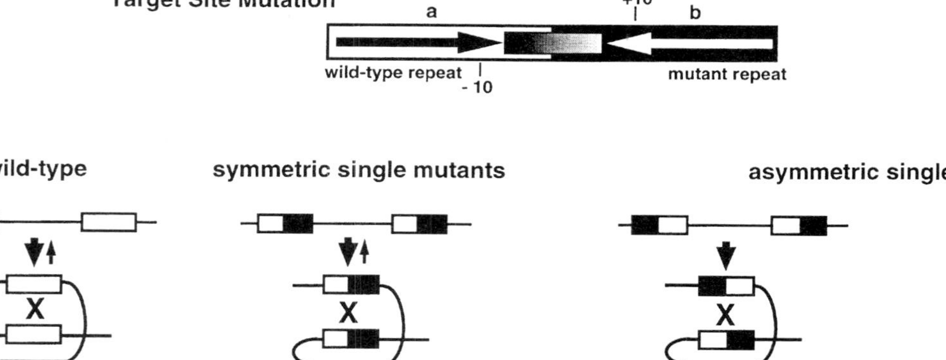

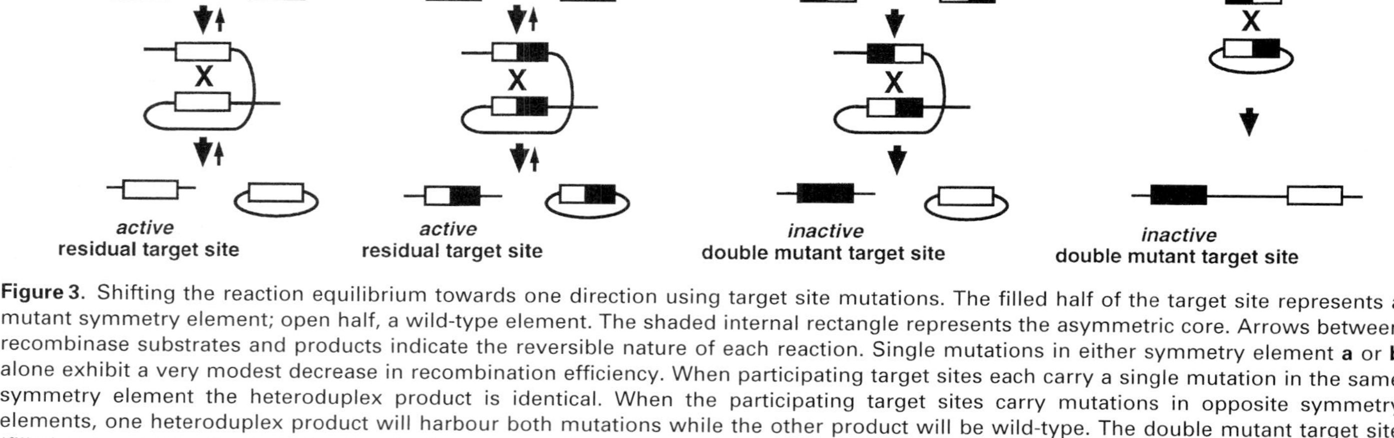

Figure 3. Shifting the reaction equilibrium towards one direction using target site mutations. The filled half of the target site represents a mutant symmetry element; open half, a wild-type element. The shaded internal rectangle represents the asymmetric core. Arrows between recombinase substrates and products indicate the reversible nature of each reaction. Single mutations in either symmetry element **a** or **b** alone exhibit a very modest decrease in recombination efficiency. When participating target sites each carry a single mutation in the same symmetry element the heteroduplex product is identical. When the participating target sites carry mutations in opposite symmetry elements, one heteroduplex product will harbour both mutations while the other product will be wild-type. The double mutant target site (filled rectangle) is effectively inactive for further recombination thereby shifting the equilibrium of the recombinase-mediated reaction to the product (17, 24, 25).

Table 2. Applications for recombinase reactions

Deletions
- Removal of selection marker genes
- Engineering subtle mutations
- Swapping sequences
- Conditional gene (in)activation
- Conditional gene repair
- Generation of deficiencies for gene mapping and mutant indentification

Integrations
- Isogenic gene transfer systems

Inversions
- Conditional gene expression
- Model human diseases associated with chromosomal inversions and chromosome loss

Translocations
- Model human diseases associated with chromosomal translocations
- Generation of mitotic clones (*Drosophila*)

Table 3. Temperature (°C) –activity (relative recombination) profiles for Cre and Flp[a]

Activity in ES cells at 37 °C	
Cre	++++++++++
Flp	+++
FlpL	+
Flpe	++++++++++

[a] See ref. 27 and 28.

As predicted from the temperature–activity profiles presented above, Cre appears to more effectively recombine a transgene array *in vivo* (2, 3, 29) than FlpL (5). However, when presented with an endogenous gene segment flanked by recombinase target sites, both FlpL (30) and Cre (31) can achieve near 100% recombination in a variety of tissues. Therefore despite the apparent *in vitro* temperature sensitivity of FlpL, it can function remarkably well *in vivo* provided sufficient levels of the recombinase are expressed (e.g. as when FlpL is expressed under the control of β-*actin* promoter and enhancer sequences) (30). Extrapolating from the cell culture data of Buchholz and colleagues described above, Flpe should show even greater recombinase activity *in vivo* than FlpL, most likely making Flpe and Cre of equal utility for genomic engineering in the mouse. Moreover, the three Flp variants (FlpL, Flp, Flpe), each with distinct dose–response properties at 37°C which together span over an order of magnitude, should provide the basis for a powerful arsenal of tools *in vivo*. It is also noteworthy that both Cre (1, 32, 33) and FlpL (5) function in the germline and can be used to generate transmissable modifications of *loxP*-flanked ('floxed') or *FRT*-flanked ('flrted') DNA sequences (e.g. selection markers and/or essential gene segments).

loxP and *FRT* sites positioned in a variety of genetic loci have served as effective recombination substrates in cell culture and the mouse. It is therefore likely that most chromosomal loci will be susceptible to recombinase function. Understanding how local chromatin structure, DNA methylation, and transcriptional activity affects site-specific recombination will require targeting many more genes in a systematic way, paying particular attention to refractory loci.

Based on these findings, Cre and Flpe are recommended for those applications that require the highest efficiency, such as tissue-specific or inducible gene knock-outs (1, 27, 28) , while Flp and FlpL are recommended for applications that rely on tight regulation, such as cell lineage studies (30) and gain-of-function experiments (34) that require complete absence of recombination before the predetermined activation point (see Section 6). Cre (1) and Flpe (28) have been found to excise a chromosomal substrate in ES cells much more efficiently than Flp or FlpL (35) and should therefore be used for those genomic manipulations that need to be performed in ES cells rather than the mouse.

4. Extending conventional gene replacement schemes using site-specific recombination

For conventional gene targeting, the most common experimental strategy has been to inactivate the function of a target gene by introducing a positive selection gene by homologous recombination. As described by Hasty *et al.* in Chapter 1, targeting vectors of the replacement type are most frequently used for this purpose and typically include a positive selection gene flanked by genomic DNA, with a negative selection gene on one end (illustrated in *Figure 4A*). The positive selection gene serves two functions:

(a) To select for the rare stably transfected ES cell clones ($\sim 1/10^4$ cells electroporated with the targeting vector).

(b) To inactivate the target gene by mutating coding exons and/or regulatory sequences.

The negative selection drug simply enriches for the desired homologous recombination event over random integrations by killing cells which have retained a wt copy of the negative selection gene. The end-product using such a replacement vector is the permanent presence of the positive selection gene in the targeted locus.

Although experimentally required for isolating targeted ES cell clones, selection genes that are retained can affect the target locus in unexpected ways. As described in Chapters 1 and 7, the regulation and splicing of the targeted gene can be disrupted (36, 37) or the expression of neighbouring genes can be altered by the strong promoter elements frequently present in selection cassettes (38, 39). Indeed, endogenous genes lying over 100 kb from the inserted selection marker can be affected (40). The *cis* acting complications associated with selection gene promoters become even more confounding when gene targeting is used to study the actual regulation of target gene expression (in contrast to simple gene inactivation) (41) or when used to engineer subtle mutations where it is imperative that all other aspects of target gene expression remain wild-type. Because of the potential complications associated with selection markers in recombinant loci, techniques to remove such markers and thereby generate 'clean' mutations have been developed which involve a second gene targeting step. Although successful, these 'hit-and-run' and 'tag-and-exchange' (42) (also referred to as 'double replacement') (43, 44) approaches (see Chapter 1) subject recombinant ES clones to an additional round of selection which is not only time-consuming

Figure 4. Generating *clean* deletions of target genes in the mouse germline. Following the convention established in Chapter 1, the thick line represents vector homology to the target locus; the thin line, bacterial plasmid. The line of intermediate thickness represents the target locus in the chromosome. The black squares represent exons; the open box containing a +, a positive selection marker such as *neo*^{r}; the open rectangle containing a –, a negative selection marker such as *HSVtk*; the black rectangle containing a –, a second negative selection marker gene such as the diphtheria toxin A-chain gene (49). The positive and negative selection markers each contain promoter and polyadenylation sequences. Cross-over points occur within regions of homology as indicated by an X placed between dashed lines. (A) Conventional gene replacement (see Chapter 1). The replacement vector is linearized outside the target homology prior to transfection. The positive selection marker interrupts the target homology, allowing for replacement of up to 20 kb of the target locus (52). The negative selection marker is placed adjacent to one homology arm. Following transfection with the linearized replacement vector, ES cells undergo positive (G418) and negative (gancyclovir) selection. (B) Using site-specific recombination to remove the positive selection marker. The replacement vector is essentially as in (A), except the positive selection marker is flanked by direct *loxP* or *FRT* repeats (grey stippled ovals). The floxed or flrted selection marker is removed (step 2) either *in vitro* or *in vivo*. The excised circular extrachromosomal product is typically lost, making the deleted product stable and heritable. (C) Using negative selection to enrich for site-directed marker deletion. A second negative selection gene is incorporated into the floxed or flrted cassette, enabling drug enrichment for excised ES cell clones. (D) Generating larger deletions along with marker removal. One recombinase target site is positioned away from the positive selection gene and into the homology region such that site-specific recombination (*in vitro* or *in vivo*) eliminates the enclosed genomic sequence along with the selection marker. The frequency of co-transferring the positive selection marker and the recombinase target site (*loxP* or *FRT*) by homologous recombination declines with increasing distance; the result is that in a fraction of the selected (G418-resistant) ES clones, the cross-over point (indicated by the X) will lie between the positive selection marker and the second recombinase target site such that the selection marker will have integrated without the target site. To minimize screening a large number of ES clones, the positive selection marker and the second recombinase target site are separated by no more than 10 kb.

See over leaf

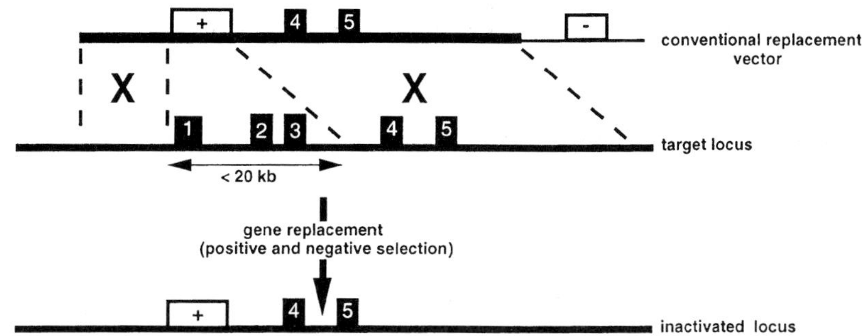

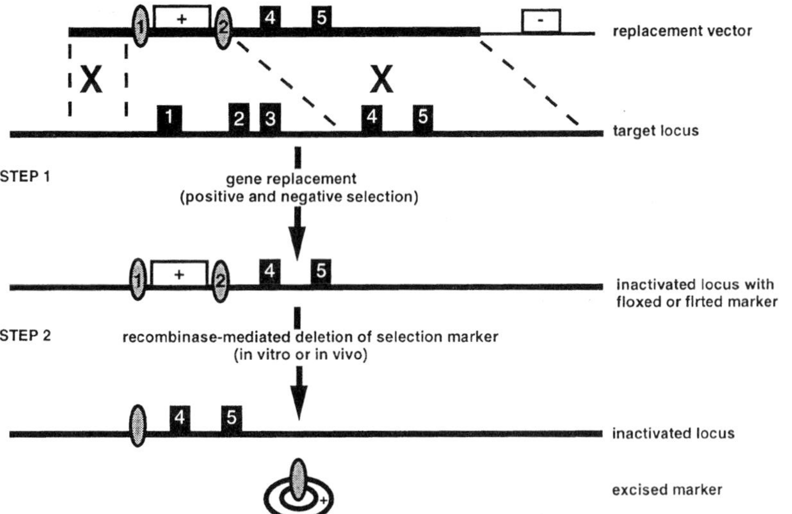

but can decrease ES cell pluripotency. Site-specific recombination performed after a single gene targeting step provides an alternative means for marker gene removal that, not only is easier, but can also be exploited to further modify the targeted locus. Recombinase-based experimental schemes for removing marker genes are presented in Section 4.1. Three additional techniques follow in Sections 4.2–4.4 which are centred around the simple removal of selection markers, but which go one step further to include the targeted introduction of larger gene deletions, subtle mutations, and heterologous sequences. The generation of large scale chromosomal rearrangements such as translocations, deletions, duplications, inversions, and chromosome gain or loss is discussed in Sections 4.5 and 4.6.

C Selection for Marker Gene Removal

replacement vector

target locus

STEP 1 gene replacement (positive and negative selection)

inactivated locus with floxed or flrted markers

STEP 2 recombinase-mediated deletion of selection marker (in vitro negative selection)

inactivated locus

excised marker

D Gene Deletion Along with Marker Removal

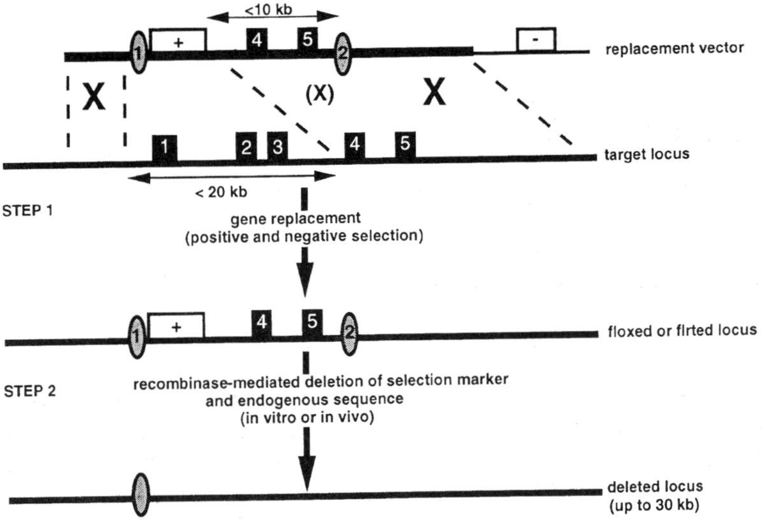

4.1 Removal of selection genes

One of the most straightforward ways to avoid selection gene interference and genetic ambiguity in gene targeting experiments is to use site-specific recombination to remove the selection gene after homologous recombination has been performed in ES cells and clones have been identified. The basic vector elements required for this second generation gene targeting strategy are essentially configured as for a conventional replacement vector; the major difference is that the positive selection gene (e.g. neomycin resistance gene, *neo*^{r}) is flanked by direct *loxP* or *FRT* repeats (see *Figure 4B*). Ideally, the targeting vector is designed so that excision of the selection gene results in a novel size DNA product that is readily detectable by PCR or Southern blot analysis. Additionally, consideration should be given to the precise localization of the floxed or flrted selection gene and whether the *loxP* or *FRT* site left in the targeted locus following marker excision might be used for genetic manipulations in the future (e.g. targeted integrations at the residual *loxP* or *FRT* site) (see Section 8).

For targeting vector construction, a floxed *neo*^{r} gene can be obtained from either plasmid pMMneoflox8 (1) or pL2neo (1), and a flrted *neo*^{r} gene can be obtained from pFRT$_2$neolacZ (45). Alternatively, individual recombination target sites can be inserted in a restriction site in the targeting vector as a double-stranded oligonucleotide or as a restriction fragment recovered from the following vectors: pGEMloxP (1), pGEM-30 (1), pBS246 (46), and pFRT$_2$ (45). Regardless of the source, it is important to confirm the orientation and functional integrity of the *loxP* or *FRT* sites in the final targeting vector, as subsequent experimental steps (e.g. ES cell transfection and culture, derivation of mice, and phenotypic analyses) represent a considerable investment of effort and reagents. Confirmation of functional *loxP* or *FRT* sites in the final gene targeting vector is accomplished by DNA sequencing as well as subjecting the targeting vector to a functional recombination assay. The latter has been facilitated by *E. coli* strains in which the *cre* or *FLP* genes have been integrated into the bacterial genome (47); the recombination competence of *loxP* or *FRT* sites engineered into any plasmid can be tested by transformation of the gene targeting vector into the appropriate bacterial strain (294-Cre or 294-Flp, respectively) followed by overnight growth at 37 °C. Isolation of plasmid DNA followed by simple restriction digestion and gel electrophoresis should show complete recombination of the targeting vector.

Having carefully designed and constructed the targeting vector, standard techniques incorporating both positive and negative selection are used to replace endogenous genomic sequence with the floxed or flrted *neo*^{r} cassette (see Chapter 3). After identifying properly targeted ES cell clones, the floxed or flrted *neo*^{r} is removed by recombinase-mediated deletion in ES cells or mice (*Figure 4B*, step 2). To delete floxed or flrted markers in culture, the recombinase is transiently expressed in the targeted ES cell clones. *Protocol 1* describes a technique to readily isolate ES cell subclones which have undergone recombinase-mediated removal of the *neo*^{r} gene. This technique should be performed with at least two independently targeted clones to ensure germline transmission of the mutation. This method was initially reported by Gu *et al.* (48) using the *loxP*-Cre system and later presented in more detail by Torres and Kühn (1).

Protocol 1. Recombinase-mediated deletion of a positive selection marker (neo^r) in ES cells[a]

Equipment and reagents

- Electroporation apparatus (Bio-Rad, Gene Pulser)
- Electroporation cuvettes (Bio-Rad, Cat. No. 165–2088)
- Tissue culture dishes: 9 cm, 96-well U- or V-bottom, and 48-well plates
- Homologous recombinant ES cell clones carrying a floxed or flrted neo^r gene
- ES cell medium (Chapter 3)
- PBS without Mg and Ca (Chapter 3)
- Recombinase expression plasmid, e.g. *pIC-Cre, pMC-Cre* (48, 1), or *pOG-Flpe, phACTB-Flpe* (28), respectively
- 0.05% trypsin/EDTA (e.g. Life Technologies, Cat. No. 25300–054)
- Plates containing neo^r mitomycin C-treated embryonic fibroblast feeder layers (Chapter 3)
- Geneticin (G418) (e.g. Life Technologies, Cat. No. 11811–049)

Method

1. Grow gene targeted ES cells in G418 for at least two passages to eliminate any cell clones that have randomly lost the neo^r gene. Use the same active G418 concentration (e.g. ~150–250 μg/ml) as employed in the original experiment to isolate the targeted clone.
2. Electroporate 0.5–1 × 10^7 gene targeted, G418-resistant, ES cells from step 1 with 5–30 μg of supercoiled recombinase vector (see Chapter 3).
3. Plate cells on dishes containing feeder cells at 5 × 10^6 cells/9 cm dish.
4. Change the medium the next day.
5. Two days after the electroporation (or as soon as colonies are established) wash cells twice with PBS and add 3 ml of trypsin/EDTA.
6. Incubate (37 °C, 5% CO_2) for 3–5 min or until cells begin to come off the plate.
7. Gently pipette the cells up and down to break cell clumps.
8. Add 7 ml ES cell medium and centrifuge (5 min, 270 *g*).
9. Aspirate supernatant, resuspend cells in ES cell medium, and plate at 10^3 cells/9 cm dish. The number of plates depends on the number of colonies you need to screen. One plate should yield approximately 100 colonies which should include 1–10 clones with the desired deletion. Plate out several dishes so that colonies are picked from different dishes. Freeze the remaining cells as a back-up (see Chapter 3) for later plating and colony isolation, if needed.
10. Approximately five days after re-plating at low density (or when colonies are established) replace the medium with PBS.
11. Pick colonies (see Chapter 3) into 50 μl of trypsin/EDTA in a 96-well V- or U-bottom plate and incubate for 3–5 min (37 °C, 5% CO_2).

Protocol 1. *Continued*

12. Gently pipette the cells up and down to disperse the colony.
13. Transfer half the cells to one well of a master and duplicate 48-well plate containing feeder cells and ES cell medium.
14. Add G418 to the duplicate plate to test for G418 sensitivity (*neo*^{r} deletion). The amount of G418 should be at least the same concentration of G418 used in the original experiment to isolate the targeted clone.
15. Incubate (37 °C, 5% CO_2) for three to five days.
16. Identify the desired G418-sensitive clones.
17. Expand sibling G418-sensitive clones from the master plate for the following manipulations: freezing, confirmation of G418 sensitivity, and genomic DNA isolation to verify the deletion event by Southern blot analysis (see Chapter 3 for details).

[a] Modified from the method given in ref. 1, with permission.

Rather than identify recombinase-mediated deletion events by sib-sensitivity to G418 as described in *Protocol 1*, an alternative approach incorporates both a positive and negative selection gene (e.g. the *HSVtk* gene) within the floxed or flrted cassette enabling enrichment for deleted subclones. The desired subclones are simply isolated by negative selection (e.g. in gancyclovir) and confirmed using molecular biology. However, given that deletion of a floxed *neo*^{r} gene is relatively efficient (reported to be 50% of the transient transfection efficiency) (1), it could be considered preferable to apply negative selection in the first gene targeting step (as diagrammed in *Figure 4A*) in order to enrich for the much more rare homologous recombinants. Alternatively, a different negative selection can be used at each step (e.g. diphtheria toxin for enrichment of the initial gene targeting event (49) and gancyclovir for selection of the recombinase-mediated *neo*^{r} deletion) (see *Figure 4C*). Vector size may however become impractical and multiple rounds of negative selection may prove deleterious to ES cell pluripotency.

One benefit of removing the positive selection marker in the targeted ES clone is that the second allele can be similarly mutated by electroporating excised cells with the initial targeting vector. The end-result is the generation of ES cells homozygous for the targeted mutation. Although homozygous mutant ES cells can be generated by other means, this recombinase-based method is reliable and circumvents the need for using increasing concentrations of the selection agent (Chapter 3) or for using two targeting vectors with different positive selection genes which could then necessitate the use of feeder cells resistant to both selection drugs.

Because both the Cre-*loxP* and Flp-*FRT* systems have been shown to function well in the mouse germline, floxed or flrted selection markers can also be removed in mice. In this scheme, the targeted gene harbouring the floxed or flrted marker is first transmitted through the germline via chimeras (Chapters 4 and 5) and the resulting mutant heterozygous mice (or germline transmitting chimeras) are then crossed to a transgenic strain of mice that expresses either Cre or Flp early in embryonic development or in the developing germline (1, 5, 32, 33). These recombinase mice are referred to as '*deleter*' strains. F_1 progeny carrying the targeted allele and the recombinase transgene will harbour alleles that have undergone site-specific deletion of the selection gene in a variety of tissues, including germ cells. F_2 progeny derived from germ cells which harbour a deleted target gene can then be identified by PCR or Southern blot analyses of DNA isolated from tail biopsies; the goal is to identify F_2 animals carrying a deleted target gene independent of the recombinase transgene. If the final aim is to generate a mouse strain lacking the positive selection marker, this type of *in vivo* approach is preferable as it reduces the total number of ES cell manipulations. A twist on this strategy is to use in the initial gene targeting experiment, ES cells harbouring a germline-specific recombinase transgene, such as *Prm1::cre* (33). The positive selection gene is then excised following germline transmission of the targeted allele from chimeric mice. It is worth noting that mice derived from ES cells harbouring an *HSVtk* gene transmit the targeted locus through the germline at lower frequencies than if *HSVtk* is not present. Consequently if the *in vivo* approach is to be used for marker deletion, *HSVtk* should not be included within the floxed or flrted cassette.

Although less straightforward, floxed and flrted sequences can be removed in zygotes (50, 51): supercoiled plasmid encoding the recombinase is injected into zygotes from ES cell-derived mice.

4.2 Generating larger deletions along with marker gene removal

Conventional single step gene targeting techniques can be used to delete and replace up to approximately 20 kb of genomic sequence (52). Small target loci can therefore be completely eliminated, ensuring gene inactivation. Many target loci however are significantly larger, requiring judicious deletion of 5′ sequences to produce null mutations by conventional techniques, while simultaneously avoiding generation of molecules with partial, complete, or dominant-negative function. The larger the 5′ deletion, the more likely to generate the desired null allele. By adding a site-specific recombination step to a conventional gene targeting experiment, the deletion can be extended by ~ 10 kb, allowing for gene deletions of up to ~ 30 kb. This method was first employed by Gu *et al.* (48) to delete J_H segments and the intron enhancer in the Ig heavy chain locus of targeted ES cells.

Here the targeting vector is essentially as described in Section 4.1 except that the second recombinase target site is placed away from the selection marker and into the homology region of the vector such that site-specific recombination will eliminate the enclosed endogenous sequence together with the marker gene (*Figure 4D*). In practice, the positive selection marker and the second recombinase target site should be separated by 10 kb or less, as the frequency with which both of these heterologous sequences are integrated declines as the distance between them increases. In a fraction of the selected (G418-resistant) ES cell clones, the cross-over point in the homology arm will lie between the positive selection gene and the second recombinase target site such that the selection marker will have integrated without the target site. Torres and Kühn (1) report a 70% frequency of co-transfer over a distance of 4 kb; 35% over 7 kb; 2% over 10 kb. Therefore to circumvent analysing a large number of homologous recombinant ES cell clones, the second *loxP* or *FRT* site should be positioned within 10 kb of the selection marker. PCR and Southern blot strategies to identify proper gene replacement events are essentially as for conventional gene targeting (see Chapters 1 and 3); the major differences are that the presence of all recombinase target sites must be confirmed by Southern blot analyses or PCR, and deleted alleles must be distinguished from non-deleted. This involves using probes capable of detecting incorporation of the second *loxP* or *FRT* site, and restriction enzyme digests that differentiate the deleted product of site-specific recombination from the original floxed or flrted gene.

As described in Section 4.1, the floxed or flrted marker and genomic sequences can be deleted following transient recombinase expression in recombinant ES cells or by breeding ES cell-derived mice to transgenic mice that express Cre or Flp in the germline.

4.3 Engineering subtle mutations

The combined gene targeting/site-specific deletion strategies described in Sections 4.1 and 4.2 can be modified to introduce a small non-selectable (e.g. point) mutation into the target locus or to swap endogenous for heterologous sequences (Section 4.4). In contrast to gene inactivation studies, here the experimental aim is to study the effect of the engineered point mutation or the inserted heterologous sequence. Consequently, extra measures need to be taken to ensure that the remainder of the targeted locus is wt with respect to gene regulation and function. These considerations are as follows:

(a) To minimize potential interference by the residual *loxP* or *FRT*, recombinase target sites should be positioned within introns or downstream of the polyadenylation (polyA) signal sequence of the gene; regions close to enhancer elements or splice sites should be avoided.

(b) Do not place the floxed or flrted selection marker directly next to the engineered mutation.

(c) To confirm that the residual *loxP* or *FRT* site is in fact neutral to gene regulation and function, consider generating a control mouse from ES cells that harbour only the residual *loxP* or *FRT* site without the targeted modification.

(d) Sequence key regions of the initial targeting vector to ensure that no unintended mutations have been introduced during construction.

The targeting vector for introducing a non-selectable mutation is essentially as described in Section 4.1, except that a small mutation is placed into one homology arm and the floxed or flrted positive selection marker is inserted into a large intron or downstream of the polyA signal sequence (*Figure 5*).

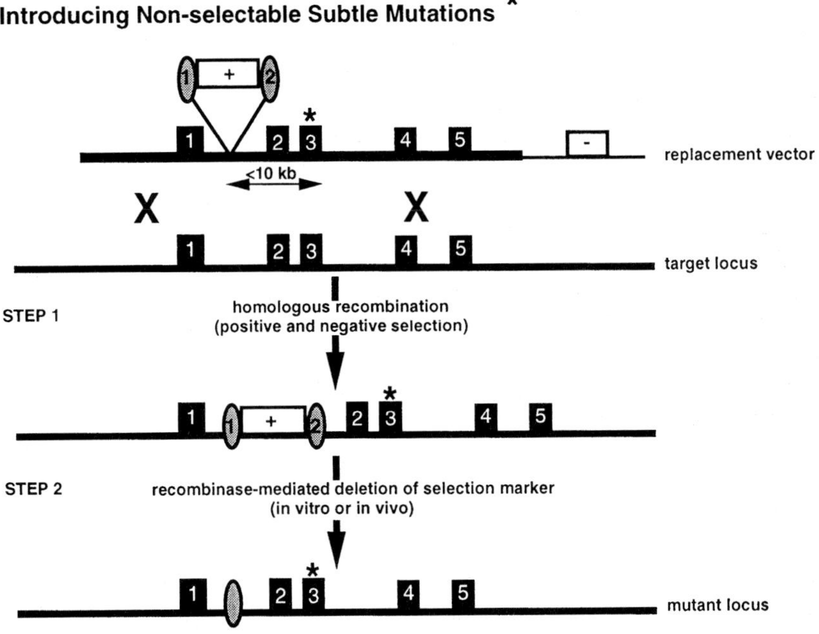

Figure 5. Using homologous recombination coupled with site-specific recombination to introduce non-selectable mutations. The lines, boxes, and ovals are as indicated in *Figure 4*. The small non-selectable mutation is designated by an asterisk; a novel restriction site is incorporated into the engineered mutation to facilitate distinguishing the mutant and wt loci. To minimize potential interference from the residual *loxP* or *FRT* site (left after step 2), the floxed or flrted selection marker is positioned within an intron or downstream of the polyA sequence. As the frequency of co-transfer by homologous recombination declines with increasing distance, the mutation and positive selection marker should be separated by no more than 10 kb. Recombinase-mediated deletion of the selection marker (step 2) is as described in *Figure 4*.

Table 4. Generating subtle mutations with gene targeting techniques

Advantages	Disadvantages
Hit-and-run[a]	
• No heterologous sequence remains in the target locus after the 'run' step.	• Intrachromosomal recombination event in step 2[a] is rare, requiring a second round of selection which could compromise ES cell pluripotency.
• Allows non-selectable mutation to be introduced anywhere in the cloned sequence.	
Site-specific recombination	
• Recombinase-mediated marker deletion in step 2[b] is a frequent event.[c]	• Single target site (34 bp) remains in locus after step 2[b] and may influence gene regulation or function.
• Recombinase-mediated marker deletion can be accomplished *in vivo*.[d]	

[a] Described by Hasty *et al.* in Chapter 1.
[b] See *Figure 5.*
[c] Deletion of a floxed *neo*^{r} gene occurs at a frequency at least 50% of the efficiency of the transient transfection using pIC-Cre or pGK-Cre plasmids (1).
[d] See *Figure 12.*

Because transfer of both the positive selection gene and the mutation declines with increasing distance, the mutation and selection marker should be separated by no more than 10 kb (described in Section 4.2). It is important to devise an easy means to distinguish the mutant and wt loci because some drug-resistant target cell clones will not contain the subtle mutation (due to both the point of cross-over in the homology arm and due to heteroduplex repair; see Chapter 1). This can be accomplished by incorporating a novel restriction site into the engineered mutation.

Prior to implementing Cre-*loxP* and Flp-*FRT* systems in mammalian cells, subtle mutations were most commonly generated using the hit-and-run approach described by Hasty *et al.* in Chapter 1; this technique relies on a single gene targeting event, followed by a selectable intrachromosomal recombination step to ultimately generate the modified locus. A comparison of the advantages and disadvantages between hit-and-run and recombinase-based methods is presented in *Table 4*. Choosing one particular approach will most likely hinge on finding a neutral position for the residual *loxP* or *FRT* site.

4.4 Swapping sequences

The recombinase-based strategy for deleting genomic sequence along with the selection marker can be modified so that the end-result is a swap of endogenous sequence with heterologous sequence. The targeting vector for this type of 'knock-in' strategy is as described in Section 4.2 except that heterologous sequence is directly linked to the selection marker to create a single large region of non-homology (see *Figure 6A*). Site-specific recombination eliminates the enclosed endogenous sequence together with the selection marker, effectively swapping endogenous sequence for heterologous sequence. As in generating point mutations, sequence swapping experiments often require that proper regulation of the target locus be otherwise maintained. Confirming *loxP* or *FRT* neutrality therefore continues to be an important consideration. This approach was first used to swap six exons coding for the mouse Cγ1 gene with homologous human DNA to generate a mouse strain producing humanized chimeric antibodies (53).

If the region to be swapped is small, this can be achieved by simple gene replacement using a floxed or flrted positive selection marker linked to the heterologous sequence, without the need to position one of the recombinase target sites into a homology region of the vector (*Figure 6B*) (see Chapter 1, Section 7 for details of knock-in technique). This knock-in strategy was first used to engineer a mouse strain in which *En-1* coding sequence was replaced with *En-2* sequences (54). The end-result was rescue of *En-1* mutant brain defects, demonstrating that the primary difference between En-1 and En-2 stems from their divergent expression patterns rather than different biochemical properties.

It is important to emphasize that each of the above site-specific recombinase-based methods have the valuable potential of being able to induce the desired deletion, point mutation, or sequence swap *in vivo* and in a tissue- or stage-specific manner (see Section 5).

4.5 Engineering large scale chromosomal rearrangements *in vitro*

Gene modifications encompassing up to ~ 30 kb of DNA sequence can be produced by combining a single gene targeting experiment with Cre- or Flp-mediated DNA deletion (see Sections 4.1–4.4). Such targeted, single gene mutations have been invaluable both for assaying gene function and for generating mouse models of many human diseases (55). Many human disorders, however, do not result from a mutation in an individual gene, but rather result from a large chromosomal rearrangement such as a translocation (55), duplication (56), inversion (57), deletion (58), or chromosome gain or loss (59). In order to model in the mouse human diseases associated with large chromosomal rearrangements, Ramirez-Solis, Bradley, and colleagues have combined two gene targeting steps with *loxP*-Cre recombination to generate large DNA rearrangements in ES cells in culture (60). This methodology results, not only in the modelling of a new class of human genetic disorders associated with chromosomal rearrangements, but also facilitates screening for recessive disease genes through the generation of mice with predetermined regions of haploidy.

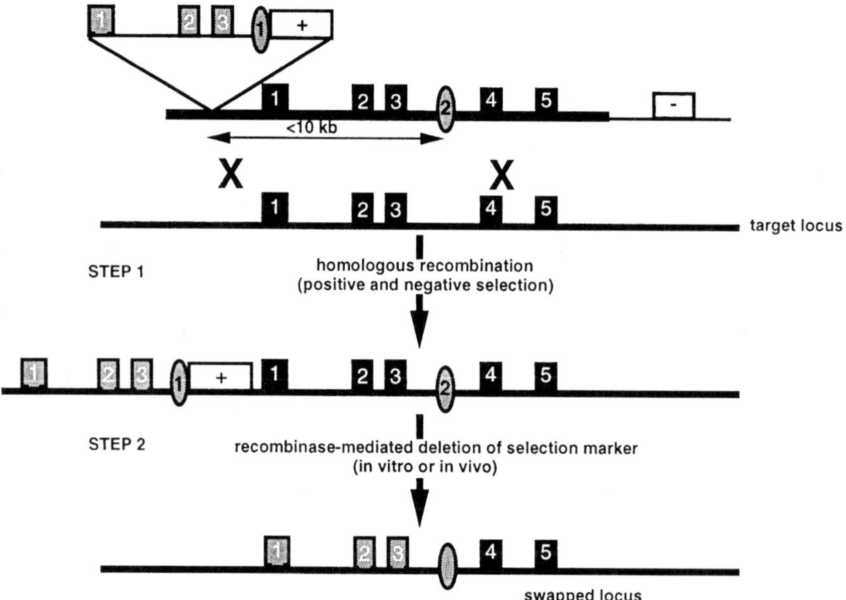

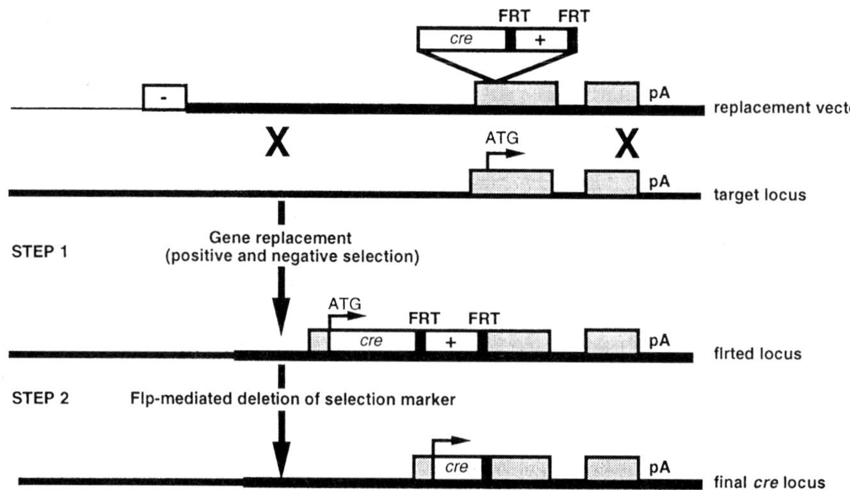

Note: The reciprocal set-up could also be used, knock-in *FLP* and flank the selection marker with *loxP* sites.

Figure 6. Using homologous recombination coupled with site-specific recombination to swap or knock-in gene sequences. (A) Sequence swap. The grey stippled squares represent heterologous sequence to be swapped with endogenous sequence, otherwise the lines, boxes, and ovals are as indicated in *Figure 4*. The heterologous sequence is directly linked to the positive selection gene to generate a single large region of non-homology. One recombinase target site (*loxP* or *FRT*) is positioned away from the selection marker and into the homology region such that site-specific recombination eliminates the enclosed endogenous sequence along with the marker, effectively swapping endogenous for heterologous sequence. To minimize potential interference from the residual *loxP* or *FRT* site (left after step 2), the floxed or flrted selection marker is positioned within an intron or downstream of the polyA sequence. Recombinase-mediated deletion of the selection marker and adjacent DNA (step 2) is as described in *Figure 4*. (B) Generating a tissue-specific Cre mouse using a knock-in approach. The lines and boxes are as indicated in *Figure 4* except the stippled rectangle represents the target gene controlling recombinase expression; the horizontal arrow and ATG, the translation start site for the target gene; the small black rectangles, *FRT* sites. The coding region of *cre* should replace the initiation ATG of the target gene, and the fusion transcript should use the splice donor/splice acceptor and polyA sequences of the target locus. In step 1, ES cells undergo positive and negative selection following transfection with the linearized replacement vector. Because the selection marker required in step 1 frequently affects host gene expression, it is flanked by direct *FRT* repeats for deletion by Flp in step 2. To generate Flp-expressing mice by this approach the selection marker should be flanked by direct *loxP* repeats and deleted in step 2 using Cre.

Because recombination between target sites separated by 90 kb or more are rare events (10^{-5}) in somatic cells (60), it is necessary to provide a strong selection for the rearranged product in cell culture. This can be achieved by reconstituting a positive selection gene following Cre-mediated DNA rearrangement. One such experimental strategy devised by Ramirez-Solis *et al.* (60) is depicted in *Figure 7*. In outline, a positive selection gene (e.g. the *Hprt* minigene) is divided into two non-functional fragments: a 5′ fragment and a 3′ fragment, each sharing a central intron which contains *loxP* sites in the same relative orientation. In this configuration, each individual fragment is non-functional. Following Cre-mediated recombination between the two *loxP* sites the minigene fragments become juxtaposed, reconstituting the positive selection gene. Such cells are now resistant to the selective drug (HAT for *Hprt*; see Chapter 1) and can be isolated. If this experiment is performed in ES cells, the isolated HAT-resistant cell clones can be used to generate chimeric mice (see Chapter 3).

The first step in this paradigm involves positioning the two non-functional halves of the positive selection gene at predetermined and distant sites in the genome, either on the same or different chromosomes. This is done by conventional gene targeting methods using either replacement or insertion vectors (see Chapters 1 and 3). The outcome of the Cre-mediated recombination between *loxP* sites will depend on the relative orientation and location of the two halves of the positive selection gene. If the two minigene fragments are located on different chromosomes but in the same orientation relative to

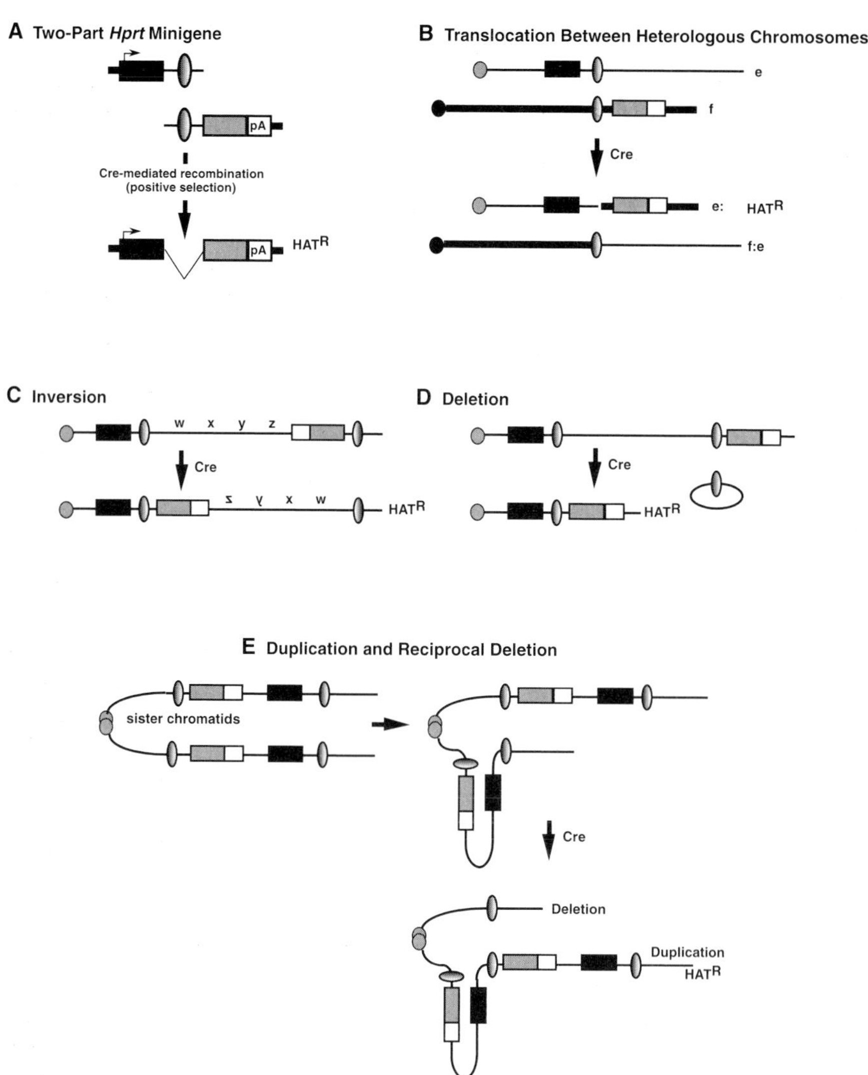

the centromeres, a chromosomal translocation will take place (*Figure 7B*) (60–62). If the two fragments are located on the same chromosome or in each homologue of an autosome then an inversion, deletion, or duplication will take place depending on the relative orientation of the two halves of the selection gene (*Figures 7C–E*) (60, 63). If the two portions of the selection gene and associated *loxP* sites are in opposite orientations, Cre-mediated recombination can also generate dicentric and acentric chromosomes which get lost over the course of cell division (60, 64).

From the set of translocation, duplication, deletion, and inversion events

Figure 7. *loxP*/Cre-mediated large scale chromosomal engineering. *loxP* sites are represented by stippled ovals with the orientation indicated by the shaded gradient; the 5′ and 3′ components of the *Hprt* minigene are depicted as dark and light shaded boxes, respectively. (A) *Hprt* minigene system used to select for Cre-mediated recombination. Recombination between *loxP* sites reconstitutes a functional *Hprt* minigene, rendering cells resistant to HAT (amniopterin, hypoxanthine, thymidine). (B) Strategy for generating translocations by targeting both halves of the *Hprt* minigene to different chromosomes (indicated by the letters **e** and **f**) but in the same orientation relative to each centromere. Cre recombination exchanges sequences distal to the *loxP* sites so as to generate a functional *Hprt* minigene and cell clones resistant to HAT. If the two chromosomes are homologues, one will have a deletion and one a duplication (see *Figure 8*). (C) Strategy for generating inversions by targeting both halves of the *Hprt* minigene to the same chromosome in opposite orientations. Chromosome orientation is indicated by the letters **w**, **x**, **y**, and **z**. (D) Strategy for generating deletions by targeting both halves of the *Hprt* minigene to the same chromosome in the same orientation. (E) Strategy for generating duplications by targeting both halves of the *Hprt* minigene to the same chromosome in the same orientation but wrong order. Cre-mediated recombination takes place after DNA replication. Sister chromatids are shown joined by their centromeres (circles). The duplication chromosome is selected in HAT and the deletion chromosome is segregated. (Figure adapted from the work of Hasty, Abuin, and Bradley, with permission.)

which can be catalysed by Cre (*Figures 7B–E*), the actual chromosomal configuration of each HAT-resistant cell clone must be determined. In brief, junction fragments generated by the different recombinase-mediated rearrangements should be analysed by Southern blot and/or PCR. Additionally, retention or loss of the selection markers used in the initial gene targeting of the two halves of the *Hprt* minigene will predict different rearrangements. For a detailed discussion on the analysis and confirmation of each chromosomal configuration illustrated in *Figure 7*, see refs 60 and 61.

4.6 TAMERE: engineering duplication and deletion chromosomes *in vivo*

In contrast to selecting for rare Cre-mediated chromosomal rearrangements *in vitro* (Section 4.5), it is also possible to engineer chromosomal deletions and concomitant duplications *in vivo* using a Cre-mediated chromosomal translocation protocol called *trans*-allelic targeted meiotic recombination (TAMERE) (65). As described in Section 4.5, *loxP*/Cre-mediated recombination between homologous chromosomes rarely occurs in somatic cells. To make such *trans*-allelic site-specific recombination is more efficient, Herault, Duboule, and colleagues (65) cleverly exploited the chromosome pairing which occurs during the zygotene stage of meiotic prophase, reasoning that the efficiency of Cre-mediated chromosomal translocations might greatly increase if the two substrate, *loxP*-containing, homologous chromosomes were closely aligned. To produce Cre in male spermatocytes during zygotene, Herault and colleagues used a transgene expressing Cre under the control of a

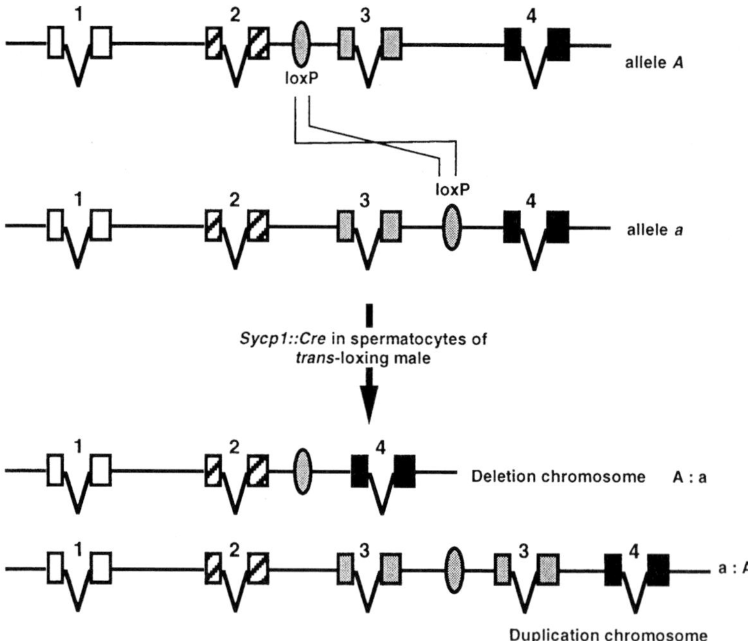

Figure 8. Chromosome engineering through Cre-mediated targeted meiotic recombination (TAMERE). Direct *loxP* repeats are represented by stippled ovals; boxes indicate exons; V-lines denote introns; hatched marks or shading group exons into genes labelled by number. Allele *A* has a *loxP* site between genes 2 and 3; allele *a*, between genes 3 and 4. Following chromosome pairing in meiotic prophase and Cre expression through the *Sycp1::cre* transgene, *trans*-allelic recombination occurs, generating deletion and duplication chromosomes (65).

minimal *Sycp1* (synaptonemal complex protein 1 gene) (66, 67) promoter. The TAMERE strategy is depicted in *Figure 8*, and was first used to generate reciprocal duplication and deletion chromosomes for 6 kb of the *Hoxd* locus which included the *Hoxd12* gene (65).

The first step in TAMERE involves generating mice carrying *loxP* sites positioned in the same orientation relative to the centromere but at distant sites on homologous chromosomes. This is done by conventional gene targeting methods (see Chapters 1 and 3). The second step involves generating male mice trans-heterozygous for the two *loxP*-containing alleles and hemizygous for the transgene *Sycp1::Cre*. These animals are referred to as '*trans*-loxing' males because *trans*-allelic recombination occurs during meiotic prophase to generate two novel alleles which segregate in haploid germ cells:

one allele containing a duplication of the DNA lying *in trans* between the *loxP* sites, and the other lacking this DNA as a result of reciprocal deletion. By outcrossing *trans*-loxing males with wt females and genotyping progeny, Herault *et al.* report a TAMERE frequency of approximately 12%, with individual males exhibiting different frequencies of *trans*-loxing, ranging from 5% in the lowest case up to 20% for the most efficient male (65). The maximum distance between *loxP* sites *in trans* that will still allow efficient Cre-mediated recombination is currently under investigation. TAMERE will likely prove to be a very powerful technique to, not only assay gene function, but to study phenotypes dependent on gene dose and to engineer 'rescue' chromosomes for haplo-insufficient loci which cannot be bred to homozygosity (65).

5. *Conditional* gene targeting

As presented in Chapter 1, conventional gene targeting generates a modified allele in *all* cells of the mouse from fertilization on. The modification typically involves permanently replacing a gene segment with a positive selection marker. As described in Section 4, a second generation of gene targeting approaches have been developed to remove selection markers via site-directed recombination in ES cells or the mouse germline. Like conventional gene targeting, the recombinase-modified locus is present in *all* cells of the mouse from the onset of development. A third generation of gene targeting strategies is presented in this section which encompasses not only recombinase-based marker removal, but more importantly uses site-specific recombination to modify genes in discrete cells of the living mouse. This methodology has been termed '*conditional*' gene targeting and more specifically refers to a gene modification in the mouse that is restricted to either certain cell types (tissue-specific), to a particular stage within development (temporal-specific), or both (1, 4).

5.1 *Conditional* gene knock-out versus *conditional* gene repair

The regional and temporal specificity provided by *conditional* gene targeting expands analyses of gene function in three very powerful ways:

(a) A widely expressed gene can be tested for function in a particular cell lineage without being influenced by gene loss in adjacent tissues, as the rest of the embryo is genetically wild-type.

(b) By *inducing* the gene modification at a particular stage in development, the organism does not have an opportunity to adapt to the genetic alteration as the wt gene product was previously present. Compensatory responses which can obscure interpretation of conventional germline mutations are eliminated, providing more precise relationships between genotype and phenotype.

(c) *Conditional* gene targeting can also be used to investigate gene function at late embryonic stages or in the adult if null mutations lead to a severe or lethal phenotype earlier during embryonic development.

In these ways the Cre-*loxP* and Flp-*FRT* systems have and will continue to revolutionize studies of gene function in the mouse. Practical considerations and experimental design are discussed below.

The first example of conditional gene targeting in the mouse involved inactivating the DNA polymerase β gene (*pol*β). *pol*β, encoding a ubiquitously expressed DNA repair enzyme, is required for embryonic development. To circumvent this lethality and allow tests of *pol*β function in T cell receptor gene rearrangements, Rajewsky and colleagues used the Cre-*loxP* system to delete the promoter and first exon of *pol*β specifically in T lymphocytes (4). This landmark work illustrates three key points:

(a) The binary design of a conditional knock-out experiment.

(b) The viability of the conditional mutant allowing gene function to be analysed.

(c) How an informative outcome relies on the highest efficiency of recombination in the designated lineage.

For a *conditional* gene targeting experiment, two separate mouse strains are typically generated and intercrossed (see *Figure 9*). One mouse expresses the recombinase (Cre in the above example) in selected lineages or tissues (e.g. T cells); the other mouse line carries a gene segment (e.g. promoter and first exon of *pol*β) flanked by recognition target sites (*loxP* sites). In offspring, cells expressing the recombinase delete the target gene segment, while the target gene remains functional in cells of all other tissues where Cre is not expressed. This binary design lends great versatility to the system, as different Cre transgenics can be crossed to the same floxed locus to study gene function in a variety of cell types, or conversely, the function of different floxed genes can be studied in one cell type by crossing to the same Cre transgenic.

As illustrated in the above example, a real breakthrough achieved by *conditional* gene targeting is that mutant animals can be viable, allowing the result of mutating the target gene to be studied in the cell type of interest. In general, viability depends on the lineage under study and, in particular, on adequate restriction of recombinase activity (and therefore mutagenesis) to the designated lineage. Unintended early or ubiquitous recombinase expression can lead to gene deletion in many tissues, approximating conventional (rather than the desired *conditional*) mutagenesis. Techniques to precisely determine the activity profile for Cre and Flp transgenics are described in Section 5.4.

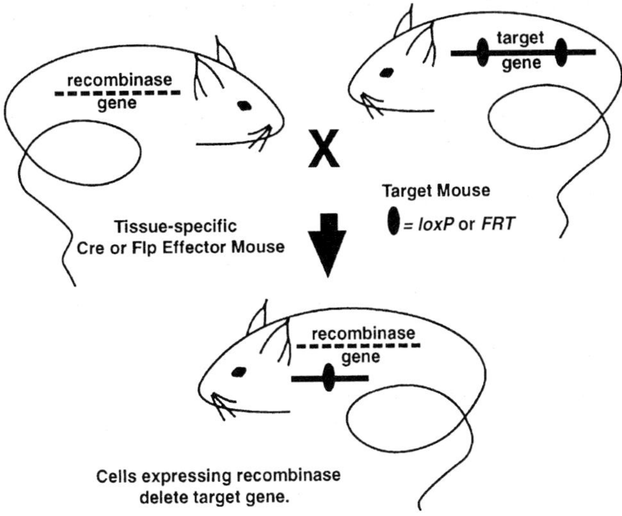

Figure 9. Binary scheme for recombinase-based gene modifications. One mouse expresses the recombinase in selected lineages or tissues ('effector' mouse); the other mouse carries a gene segment flanked by recognition target sites ('target' mouse). In offspring, cells expressing the recombinase delete the gene segment.

Two critical factors determining the success of a conditional knock-out experiment are:

(a) That the promoter driving recombinase expression is sufficiently active in all cells of the desired type.
(b) That recombination of the target locus occurs efficiently in the designated cell lineage.

If excision is mosaic so that the mutation is induced in only a fraction of the desired cells (as a result of either or both of these factors), it is possible that wt cells will have a selective (survival or growth) advantage over mutant cells, effectively obscuring the loss-of-function phenotype. Effective lineage-specific gene deletion is even more critical if the target gene codes for a secreted molecule, where expression of the gene product by just a few wt cells could be sufficient to rescue the mutant defect. Cell type-specific knock-outs are therefore best suited to study genes which act *cell-autonomously*. The half-life of the mRNA and/or protein encoded by the target locus can also limit the effectiveness of this approach; if the gene product is very stable, its reduction will lag behind the induced genetic alteration, potentially masking the mutant defect.

If a gene is likely to act non-autonomously or encode for a very stable gene product, and if it appears difficult to achieve sufficient recombinase activity in the designated lineage, an alternative approach to *conditional gene inactivation* involves *conditional repair* of a previously mutated locus (*Figure 10*).

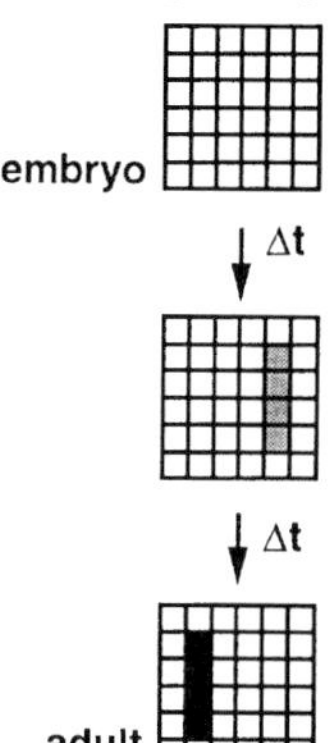

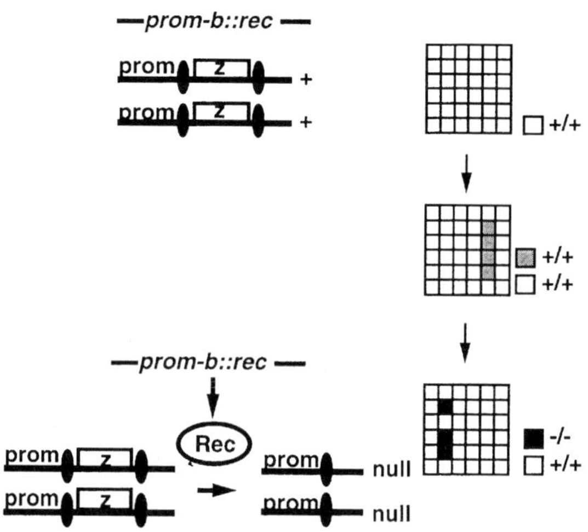

Figure 10. *Conditional* gene targeting: knock-out versus repair strategies for studying gene '*z*' function. Each grid represents a mouse, cartooned developing from embryo to adult; Δt, an increment of time during which development progresses. (A) Hypothetical gene *z* is expressed, first in embryonic lineage *a* (grey squares), and later in adult lineage *b* (black squares). (B) Targeted disruption of gene *z* in the germline. Gene *z* function is required in lineage *a* for viability; subsequent death of the embryo precludes studying the role of gene *z* later in adult lineage *b*. (C, D) Two recombinase-based strategies for bypassing embryonic lethality associated with the germline mutation of gene *z* thereby allowing study of gene *z* in lineage *b*. Relevant genotypes are diagrammed to the left of each grid. Open rectangle with 'z' indicates an essential region of gene *z*; black line, chromosomal DNA; black oval, recombinase recognition target site (either as direct *loxP* or *FRT* repeats); prom, gene *z* promoter sequence; *prom::rec,* transgene or knock-in locus (see Section 5.4 and *Figure 6B*) where regulatory elements from gene *a* or *b* control recombinase expression in lineage *a* or lineage *b*, respectively; and Rec, recombinase

B Germline Knock-Out of Gene *Z*

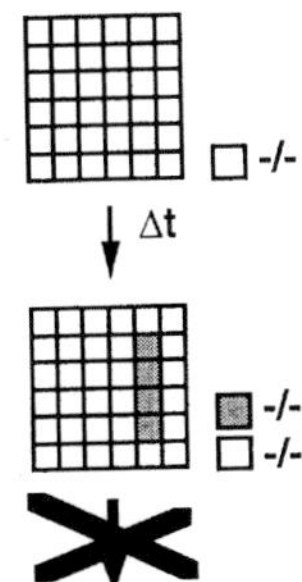

Gene *z* required in lineage *a* for viability.

Lethality precludes study of gene *z* function in lineage *b*.

D Conditional Gene *Z* Repair in Lineage *a*

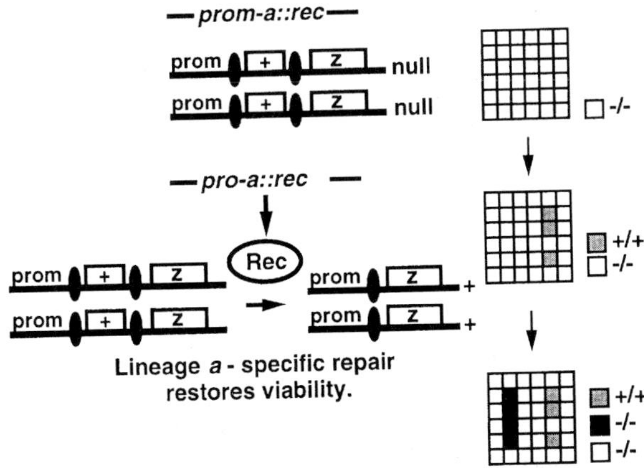

Lineage *b*: 100% mutant (■).

protein (Cre or Flp). (C) Restricting gene disruption to lineage *b*. Two recombinase target sites flank an essential region of gene *z* without affecting wt (+) gene *z* function. By expressing recombinase in lineage *b* in mice homozygous for the flanked gene *z*, site-specific excisional recombination inactivates gene *z* only in cells of lineage *b* in the adult. (D) Restricting gene repair to lineage *a*. Here gene *z* is inactivated in the germline by the insertion of a positive selection marker (open rectangle containing +). The positive selection marker, containing a promoter and tandomly repeated polyA sequences and flanked by two directly repeated recombinase target sites, is positioned between the promoter and translation start to disrupt gene *z* expression. By expressing recombinase in lineage *a*, excisional recombination removes the disrupting marker thereby repairing gene *z* to wt just in *a*-type cells; the rest of the embryo remains homozygous null. Sufficient repair occurs in cells of lineage *a* so that mice progress into adulthood; because repair is restricted to lineage *a*, lineage *b* remains 100% nullizygous.

Table 5. *Conditional* gene knock-out versus *conditional* gene repair

	Advantages	Disadvantages
Knock-out		
	• Mouse is wt until lineage B-specific[a] mutation.	• Often have mosaic excision in lineage B:[a] wild-type lineage B cells could gain selective advantage and mask phenotype.
	• Phenotype is not confounded by compensatory mechanisms.	• May be inappropriate for genes acting non-autonomously.
	• Does not require identification of primary lineage A.[a]	• May be uninformative if encoded mRNA and/or protein is very stable.
Repair		
	• 100% of lineage B is mutant.	• Requires prior identification of lineage A.[a]
	• Suitable for any gene, whether cell-autonomous or non-autonomous.	• Requires sufficient rescue of lineage A such that viability is restored.
	• Suitable even if encoded mRNA or protein is very stable.	• Except for repaired lineage A, entire anima is mutant—therefore compensatory mechanisms could mask phenotype.
		• Residual *loxP* or *FRT* site must not affect gene regulation or function, as goal is reconstitution to wt.
		• Inappropriate to study a late role in the tissue that produces early lethality (e.g. descendants of lineage A).

[a]See *Figure 10.*

Rather than mutating a wild-type target gene to a null allele in a lineage-specific fashion, the *repair* approach reconstitutes a null allele to wild-type in a lineage-specific fashion. Gene reconstitution in the appropriate lineage (i.e. the first lineage where gene function is required) should rescue viability so that the effects of the null mutation can be studied later in development in other tissues (see *Figure 10D*). The advantage is that all cells in the later tissue of interest are mutant, independent of recombination efficiency, making this approach appropriate for studying any gene, regardless of its mode of action (autonomous or non-autonomous). The main disadvantage of this approach is that all cells in the organism carry the mutation until repair occurs in the designated lineage. The wt gene product is therefore largely absent, making cellular adaptation or compensation a potential problem. The advantages and disadvantages of each approach are listed in *Table 5.*

5.2 Engineering target genes for lineage-specific mutagenesis

Conditional gene inactivation using the Cre-*loxP* or Flp-*FRT* system involves four steps:

(a) Placing recombinase target sites into the ES cell genome using a gene replacement-type strategy.
(b) Removing the positive selection marker gene by recombinase-mediated excision.
(c) Transmitting the floxed or flrted target gene through the mouse germline.
(d) Inactivating the floxed or flrted locus in a conditional manner in mice by expressing recombinase from a transgene.

The essential components of a targeting vector for conditional gene inactivation are listed below and illustrated in *Figure 11*. Requirements a–d are common to generic targeting vectors, while e–g relate to marker gene removal and *conditional* gene inactivation.

(a) Two arms of isogenic DNA homologous to the genomic target locus.
(b) A positive selection gene *inserted* into a genomic sequence lying between the homology arms.
(c) A negative selection gene positioned at the end of a homology arm.
(d) A unique restriction site to linearize the vector outside the homologous sequences.
(e) Direct repeats of recombinase target sites flanking the entire gene if small, or essential 5′ gene segments if large.
(f) Recombinase target sites flanking the positive selection marker.
(g) To minimize potential interference with wt gene function, recombinase target sites and floxed or flrted positive selection markers should be positioned within introns.

For vector construction, a floxed *neo*^{r} gene can be obtained from either plasmid pMMneoflox8 (1) or pL2neo (1), and a flrted *neo*^{r} gene can be obtained form pFRT$_2$neolacZ (45). Individual target sites can be inserted in a restriction site in the targeting vector as a double-stranded oligonucleotide or can be retrieved from pGEMloxP (1), pGEM-30 (1), pBS246 (46), and pFRT$_2$ (45) as restriction fragments. As previously mentioned (Section 4), it is important to confirm the orientation and functional integrity of the *loxP* and/or *FRT* sites in the final targeting vector. This is accomplished by DNA sequencing as well as subjecting the targeting vector to recombination in *E. coli* (47).

A strategy developed by Rajewsky and colleagues (4) for generating a *loxP*-flanked target gene is diagrammed in *Figure 11A* and *B*. Standard techniques incorporating both positive and negative selection (*Figure 11A*) are used to enrich for targeted ES cell clones in step 1. ES cell clones carrying the desired homologous recombination event are identified by PCR and Southern blot analyses; care should be taken to confirm not only that homologous recombination events occurred in both homology arms and that no gene

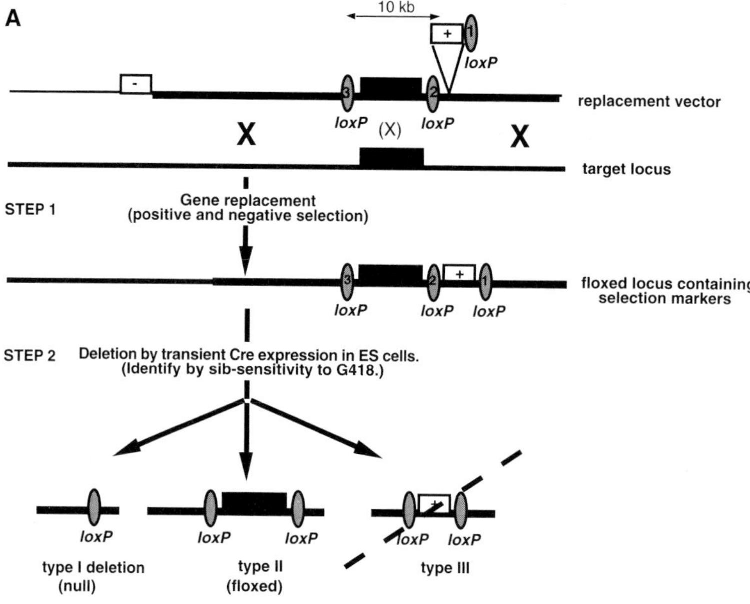
Cre-loxP Strategies for Generating a floxed Gene and a Deleted Gene in ES Cell Culture
A
10 kb
+
-
loxP
replacement vector
X
(X)
X
target locus
STEP 1
Gene replacement
(positive and negative selection)
floxed locus containing
selection markers
STEP 2
Deletion by transient Cre expression in ES cells.
(Identify by sib-sensitivity to G418.)
type I deletion
(null)
type II
(floxed)
type III

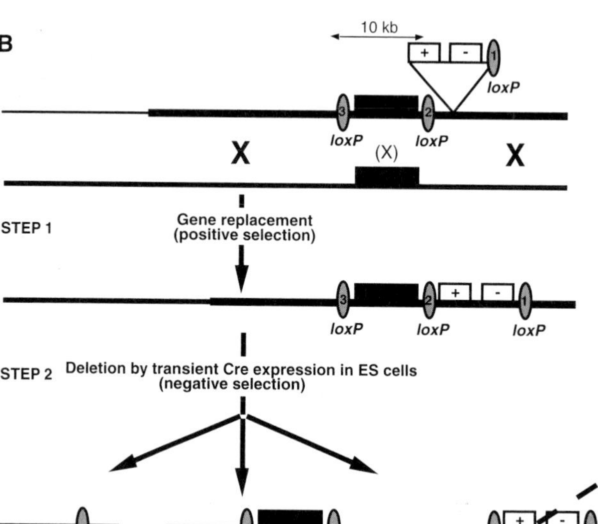
B
10 kb
+
-
loxP
X
(X)
X
STEP 1
Gene replacement
(positive selection)
STEP 2
Deletion by transient Cre expression in ES cells
(negative selection)
type I deletion
(null)
type II
(floxed)
type III
(select against)

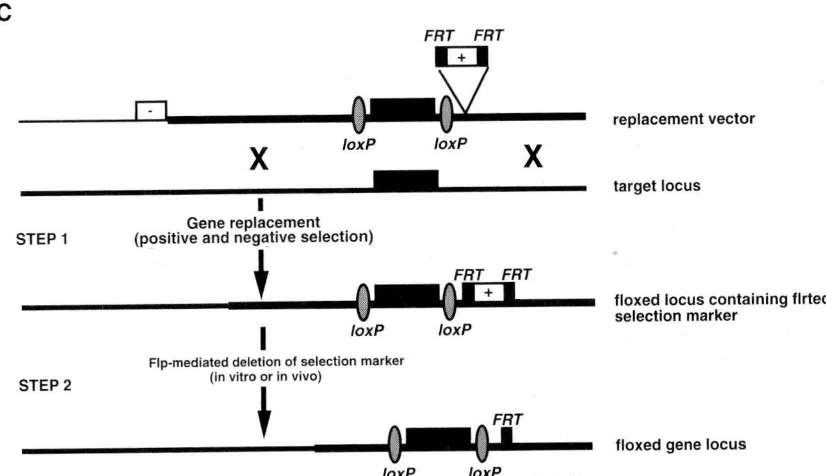

Note: The reciprocal set-up could also be employed, flanking the gene with *FRT* sites and the selection marker with *loxP* sites.

Figure 11. Strategies for generating a *loxP*-flanked gene for conditional gene targeting. As in *Figure 4*, the thick line represents vector homology to the target locus; the thin line, bacterial plasmid. The line of intermediate thickness represents the target locus in the chromosome. The black rectangle represents an essential gene segment; the open box containing a +, a positive selection marker such as *neo*^{r}; the open rectangle containing a –, a negative selection marker such as *HSVtk*; grey stippled ovals, *loxP* sites. The positive and negative selection marker each contain promoter and polyA sequences. The Xs represent cross-over points. (A and B) Two-step strategies using the Cre-*loxP* recombination system. Three different types of ES cell subclones are generated: one harbouring the desired floxed gene (type II deletion), one a null allele (type I deletion), and the other, a floxed selection marker (type III deletion). (A) Using positive and negative selection in step 1 to provide a strong selection for the targeted homologous recombination product. Following transfection with the linearized replacement vector, ES cells undergo positive (G418) and negative (gancyclovir) selection in step 1. Approximately 50% of gene-targeted ES clones identified in step 1 will lack the third *loxP* site as a result of cross-over events (demarcated by X) specifically within the *loxP*-flanked genomic sequence (1). ES cell clones that retain all three *loxP* sites are identified by Southern blotting and then transiently transfected with a Cre-encoding plasmid (step 2). Type I and type II subclones are distinguished from type III by sib-sensitivity to G418 (see *Protocol 1*); type I subclones are distinguished from type II by PCR and Southern analyses. (B) Using positive selection in step 1 to enrich for the targeted homologous recombination product and negative selection in step 2 to isolate type I and II Cre-mediated deletions. Whether using approach (A) or (B), if type I deletions predominate or difficulties arise in maintaining the pluripotency of ES subclones following two rounds of transfection, a dual recombinase approach can be employed. (C) Two-step strategy using both the Cre-*loxP* and Flp-*FRT* recombination systems. Small black rectangles represent direct *FRT* repeats. Here the flrted positive selection marker is removed by Flp in step 2. The reciprocal target site arrangement can be used for Flp-mediated mutagenesis, with direct *FRT* repeats flanking the gene segment and direct *loxP* repeats flanking the selection marker.

duplication events occurred, but that all *loxP* sites are properly positioned. Because co-integration of heterologous sequences declines with distance (1), the third *loxP* site should be placed within 10 kb of the positive selection marker. In step 2, transient expression of Cre is used in ES cells to delete the positive selection marker to generate the desired floxed allele, also referred to as a type II deletion. As all three *loxP* sites are directly oriented in the targeted allele, two additional recombination products can be produced, both of which represent null alleles: a type I deletion in which both the selection marker and gene segments are deleted, and a type III deletion in which just the gene segments are deleted while the selection marker is retained. The type I product reflects the final genetic modification to be achieved in a *conditional* manner *in vivo* and therefore serves as a useful control. The type III deletion should also represent a null allele, but is confounded by the presence of the selection marker, and is therefore of less practical use. ES cell subclones isolated in step 2 are screened for G418 sensitivity to identify type I and type II deletions (see *Protocol 1*); subclones are further characterized using Southern blot analyses to distinguish each allele. Type I and type II deletions can also be isolated by negative selection (rather than by sib-sensitivity to G418) provided that a negative selection gene has been inserted adjacent to the positive selection marker (see *Figure 11B*). As described in Section 4, it is preferable to apply negative selection in step 1 in order to enrich for the much more rare homologous recombination events. Therefore the strategy presented in *Figure 11A* is preferred over that of *11B*. Alternatively, two different negative selection genes could be employed: one inserted into the floxed *neo*^{r} cassette and the other positioned at the end of one homology arm (see Section 4.1). Using either approach, the Rajewsky laboratory has reported isolating type I deletions at equal frequency to type II deletions (1). There have been, however, loci where type I deletions were found 10–100 times more frequently than type II deletions (1). Therefore, in practice, it could be impossible to isolate an ES subclone harbouring the desired type II floxed allele. An effective solution lies in exploiting the Flp-*FRT* system along with Cre-*loxP*.

To ensure isolation of an ES cell subclone carrying a type II floxed target locus, the positive selection marker can be flanked with direct *FRT* repeats instead of *loxP* sites (the target gene nonetheless remains flanked by direct *loxP* sites) (see *Figure 11C*). The positive selection gene can then either be deleted directly in the targeted ES cell clones by transient expression of Flpe (as described in Section 4.1), or the floxed locus containing the flrted selection marker is first transmitted through the germline and the resulting mice are then crossed to a transgenic that expresses Flp early in embryonic development such as the *hACTB::FLP deleter* strain (5) (step 1 in *Figure 12A*). F_1 progeny will harbour the desired type II floxed gene locus (minus the flrted selection marker) in a variety of tissues including germ cells; F_2 progeny carrying the type II allele in all tissues are identified and used in subsequent conditional mutagenesis experiments (e.g. used in step 2 of *Figure 12A*). An advantage of the latter *in vivo* approach is that the ES cells only have to be manipulated once.

5.3 Engineering target genes for lineage-specific repair

This repair methodology relies on generating a null targeted allele that can be restored to wt in a *conditional* manner. One strategy is to insert a floxed or flrted STOP cassette between the promoter and translation start sequence of the target locus so as to completely disrupt target gene expression (*Figure 13*). A STOP cassette developed by Lakso *et al.* (2) includes spacer DNA, the small intron and polyA signal from SV40, a gratuitous ATG translation start, and a 5′ splice donor signal to abolish correct expression from any residual transcription of the desired downstream gene. Positive selection markers, with their own transcriptional termination/polyA signals have also served as effective STOP sequences in transgenes (5, 11, 45) and should function similarly in endogenous loci. The essential components of a targeting vector for conditional gene repair are as listed in Section 5.2; the major difference is that only two recombinase target sites are incorporated into the gene targeting vector, with subsequent excision mediated in the animal by crossing to a recombinase-expressing transgenic (*Figure 12B*). Because the goal is gene reconstitution, it is important that the *loxP* or *FRT* site retained after STOP deletion is in fact neutral to gene regulation and function. If the gene of interest is expressed in ES cells, thought should be given to generating control ES cell subclones that have undergone marker deletion so as to confirm restoration of wild-type gene expression.

Because *conditional repair* is suitable to study most genes, it will likely become a frequently exploited methodology; to date, however, it has not yet been widely tested *in vivo*. As described in *Table 6*, this method requires *a priori* knowledge of the first lineage in which the target locus is required for viability, as well as genomic regulatory sequences to direct recombinase expression, and therefore recombinase-mediated repair, to that lineage. These requirements may initially be limiting; however, given the power of this approach it should be seriously evaluated and considered as recombinase-based methodologies mature.

An effective variation of this repair approach has been to use recombinase-mediated DNA excision to repair a hypomorphic genotype to wt (68, 69). When the *neo*^{r} is inserted into an intron it frequently down-regulates target gene expression through aberrant mRNA splicing to cryptic splice acceptor sites in *neo*^{r} (36, 37). If *neo*^{r} is flanked by *loxP* or *FRT* sites, it can be removed to restore wild-type gene function either in ES cell culture (69), the mouse germline (68), or in a lineage-specific fashion. It is important to note than an inverted *neo*^{r} sequence encodes a cryptic splice acceptor site which is stronger than when *neo*^{r} is inserted in the same orientation as the target gene; thus, for

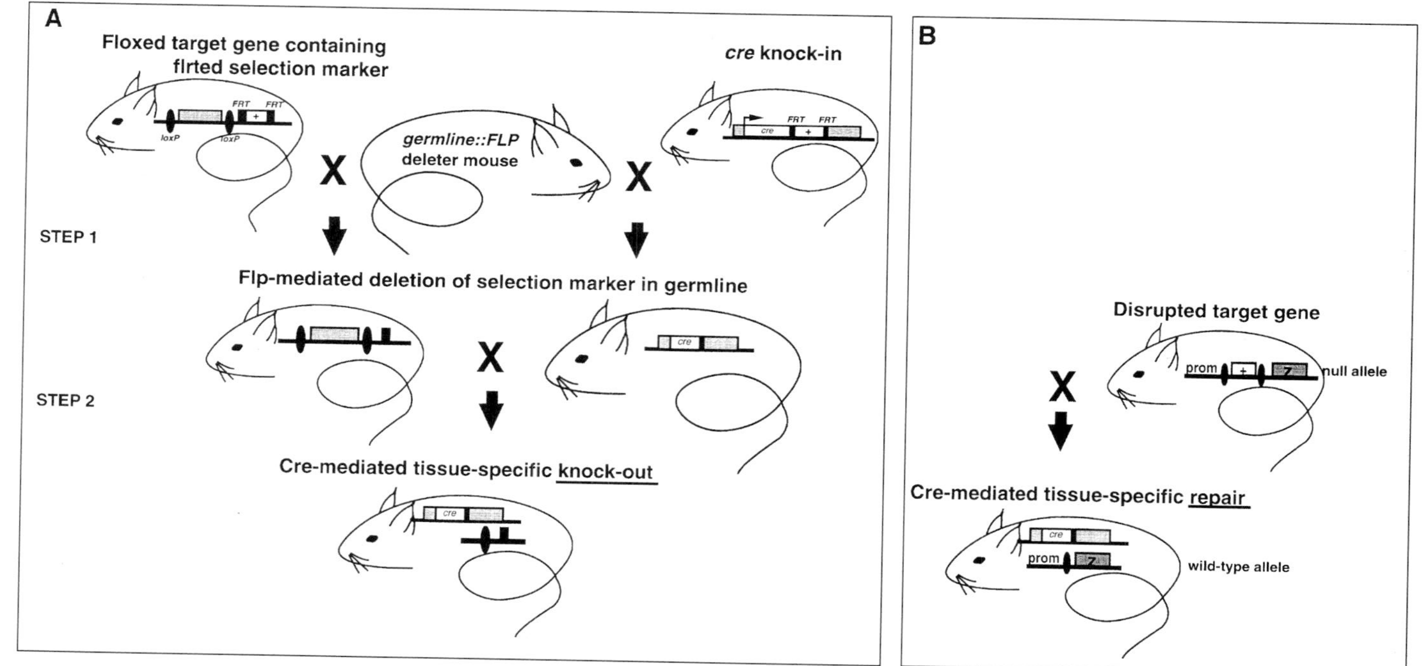

Figure 12. Overview of mating schemes for conditional gene knock-out or repair. (A) Gene knock-out scheme. The *FRT*-flanked selection marker in the floxed target locus (upper mouse on left) or the *cre* knock-in locus (upper mouse on right) is removed in step 1 by crossing to a *deleter* mouse expressing Flp in the germline (upper mouse in the middle). Although not shown, offspring having deleted the *neo*^{r} gene are bred against an *isogenic* strain in order to isolate the desired floxed or *cre* loci (middle mice) away from the *FLP* transgene. Conditional gene targeting is achieved in offspring (bottom mouse) from the cross in step 2. (B) Gene repair scheme. The disrupting selection marker is maintained in the target locus until step 2. Conditional repair to wt is achieved in offspring (bottom mouse) from the cross in step 2. In both (A) and (B), only cells expressing the recombinase are either mutated or repaired, respectively. The reciprocal arrangement of Cre-*loxP* and Flp-*FRT* systems can also be employed.

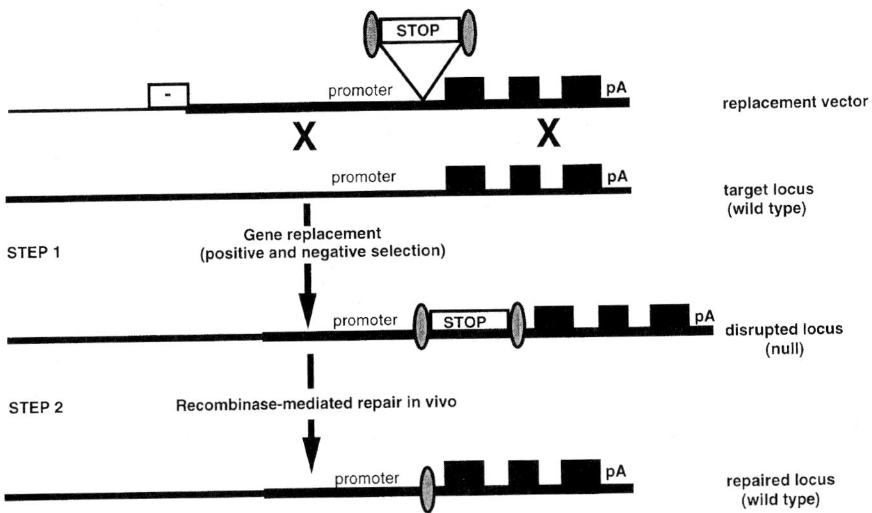

Figure 13. Generating a null target allele that can be repaired to wt by site-specific recombination. The lines and boxes are as indicated in *Figure 4*; grey stippled ovals represent direct *loxP* or FRT repeats. A floxed or flrted STOP cassette is inserted between the promoter and translation start sequence of the target locus so as to completely disrupt target gene expression. Care must also be taken to ensure that the residual *loxP* or FRT site (after step 2) is neutral with respect to target gene function.

genes requiring higher levels of interference to produce informative phenotypes an inverted neo^r sequence should be used (69).

5.4 Generating recombinase mice

Critical to all *conditional* gene targeting experiments (mutagenesis or repair) is the ability to restrict recombinase activity to a specific tissue or stage during development, and to express sufficient recombinase to achieve the highest efficiency of recombination. Recombinase transgenics can be generated in one of two ways:

(a) Zygote injection of a transgene that can express the recombinase.
(b) Targeted insertion of the recombinase gene into an endogenous coding region (a knock-in approach).

The following are guidelines for the construction of a recombinase transgene for pronuclear injection.

(a) Couple the appropriate promoter/enhancer sequences to the *cre* or *FLP* coding sequence.

Table 6. Generation of recombinase mice

	Advantages	Disadvantages
Knock-in		
	• Proper control of recombinase expression.	• More laborious to generate mice.
	• Minimizes mosaicism in recombinase expression.	• Could necessitate maintaining mice as heterozygotes.
	• Does not require prior isolation of defined promoter/enhancer sequences.	• Many genes are expressed in numerous tissues.
Transgenics		
	• Straightforward to generate mice by zygote injection.	• Often mosaic expression of recombinase.
	• Frequently can maintain as homozygotes	• Requires prior isolation of promoter/enhancer sequences.
		• Requires screening many lines to obtain correct expression pattern and level.

(b) Bracket the translation start codon with Kozak consensus nucleotides for efficient translation of the recombinase.

(c) Depending on the experimental goal, higher levels of recombinase activity may be achieved by appending the Cre or Flp with a nuclear localization sequence.

(d) Incorporate splice donor/splice acceptor sites and a polyA signal sequence to maximize transgene expression.

(e) Consider flanking the transgene with insulator sequences (70, 71) to minimize the effects of integration site on transgene expression.

(f) Unique restriction sites to isolate the transgene (including the insulator sequences if present) away from the bacterial plasmid backbone.

Endogenous or engineered promoter/enhancer combinations can be used to drive recombinase expression depending on the experimental purpose, the greatest limitation is often simply their availability. For transgene construction, the Cre-encoding sequence can be isolated from plasmid pGKcre^{NLS}bpA (1) or pBS118 (46); wt Flp-encoding sequence can be isolated from pOG44Flp (11) or pFlp (45), FlpL-encoding sequence from pOG44 (Stratagene) or pFlpL (45), and Flpe sequence from pOG::Flpe (28), phACTB::Flpe (28), or pFlpe (Dymecki, unpublished data). Intron/exon regions recovered from the SV40 genome, the β-*globin* locus (72), and the human growth hormone gene (73) have been used successfully to increase transgene expression. Insulator sequences from the chicken β-globin upstream region (70, 71) can be placed

5′ and 3′ to the transgene to increase the frequency of position-independent expression and therefore will likely decrease the total number of founder lines which need to be produced and screened for the desired expression pattern.

If possible, the functional integrity of each recombinase transgene should be assayed in cell culture prior to pronuclear injection. This requires a cell line capable of activating expression of the recombinase transgene being tested. If such a cell line is available, transgene function is assayed by transient co-transfection with a recombinase substrate plasmid that expresses an easily detectable reporter, e.g. β-galactosidase (βGal) or green fluorescent protein (GFP), following site-specific recombination (see *Protocol 2*). This type of recombinase substrate is called an *in situ* '*indicator*', as cells having undergone a site-specific recombination event are either marked blue by X-Gal stain (Chapter 6) if the activated reporter gene encodes βGal, or fluoresce green if the reporter gene encodes GFP (Chapter 5). An *indicator* transgene, as diagrammed in *Figure 14*, contains the following elements:

(a) A reporter gene which has been functionally silenced by insertion of a *loxP*- or *FRT*-STOP cassette (see Section 5.3).

(b) Promoter and enhancer sequences capable of driving transgene expression in a wide range of cell types.

(c) Kozak consensus nucleotides surrounding the initiator ATG for efficient translation of the reporter.

(d) Splice donor/splice acceptor sites (optional) and polyA sequences to maximize reporter expression following recombinase-mediated activation.

In situ lacZ indicator plasmids to assay Cre activity are typically comprised of the chicken β-*actin* promoter driving a *loxP*-disrupted *lacZ* gene and include pCAG-CAT-Z (50), pcAct-XstopXlacZ (74, 75), and pfloxLacZ (76); an alternative indicator contains the SV40 promoter driving a *loxP*-disrupted *lacZ* gene and is designated pSVlacZT (1). Substrates for Flp include pNEOβGAL (SV40 early promoter driving an *FRT*-disrupted *lacZ* gene) (11), pFRTZ (human β-*actin* promoter/enhancer sequences driving an *FRT*-disrupted *lacZ* gene) (5, 45), and pHMG::FRTZ (mouse HMG CoA reductase promoter/enhancer sequences driving an *FRT*-disrupted *lacZ* gene) (Dymecki, unpublished data). In each case, a plasmid mimicking the predicted recombination product (see *Figure 14*) should be transfected in parallel as a positive control. This not only provides an estimate of the transfection efficiency, but also confirms the activity of the promoter driving reporter expression. For Flp transgenes these include pFRTβGAL (11), pFRTZ-active (5, 45), and pHMG::FRTZ-active (Dymecki unpublished data), respectively.

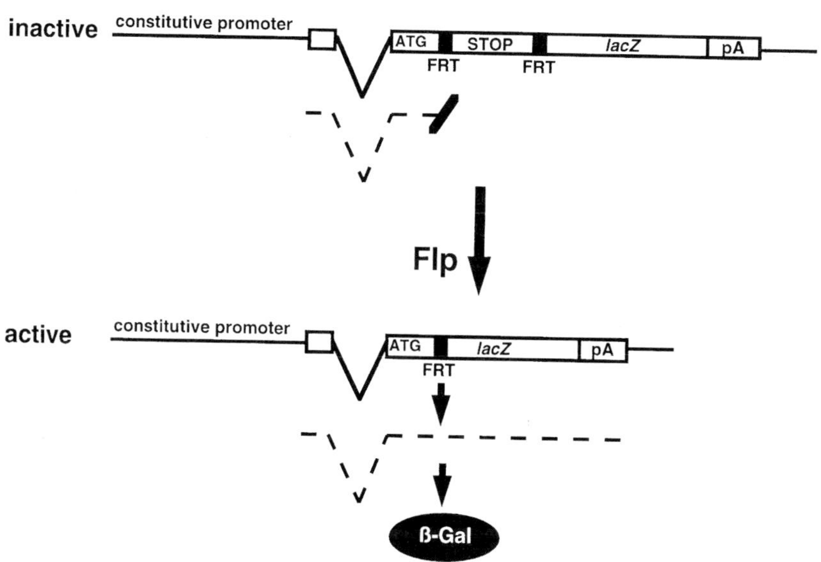

Figure 14. *Indicator* pathway. Diagrammed on top is an *FRT*-disrupted (inactive) *lacZ* fusion transgene designed to optimize expression of *lacZ* and detection of βGal following Flp-mediated excisional recombination (11, 45). The *FRT*-cassette contains transcription and translation STOP sequences and is positioned to disrupt the *lacZ* ORF. The translation start is indicated by ATG. Because there are no ATG codons to start translation of functional βGal downstream from the cassette, βGal activity is strictly dependent on Flp-mediated excisional recombination. The dashed line represents the mRNA encoded by each transgene. X-Gal detection of βGal activity '*indicates*' recombinase positive cells. Reporter genes other than *lacZ* (e.g. the genes encoding green fluorescent protein or alkaline phosphatase) can be similarly employed. Reciprocal Cre *indicator* transgenes are described in Section 5.4.

The most significant advantage in generating recombinase mice by zygote injection lies in the ease of introducing the *cre* or *FLP* transgene into the mouse germline. There are, none the less, drawbacks to this approach that need to be considered (see *Table 6*). First, recombinase expression is frequently mosaic, with only a subset of the expected cell population expressing the transgene. When used in a conditional gene targeting experiment, recombinase catalysed gene deletions will be similarly, or even more, mosaic. Depending on the nature of the experiment, this may or may not be tolerated. A second drawback of using transgenics is that they require previous isolation and characterization of the promoter and enhancer sequences from the gene of interest; these may not be readily available.

Protocol 2. Testing functional integrity of recombinase transgenes by transient co-transfection[a] with an indicator transgene

Equipment and reagents

- Tissue culture dishes
- Round-bottom polystyrene test-tubes, e.g. 12 × 75 mm (Falcon, Cat. No. 2003)
- Recombinase transgene (tg)
- Indicator tg[c] (e.g. pFRTZ, for testing Flp-encoding tg[d]; pSVlacZT, for testing Cre-encoding tg[e])
- Control 1: tg mimicking predicted recombination product (e.g. pFRTZ-product[d])
- Control 2: tg containing the same promoter sequence as driving recombinase expression but lacking the recombinase-encoding sequence[f]
- Control 3: tg containing the same promoter sequence as driving indicator expression but lacking the indicator-encoding sequence[f]
- Cell line capable of activating expression of both the recombinase tg being tested and the indicator tg
- DNA: supercoiled plasmid resuspended in dH_2O[b]
- Standard growth medium (with and without antibiotics)
- Dulbecco's modified Eagle's media (DMEM) (Life Technologies, Cat. No. 11960) or appropriate media
- 10% fetal bovine serum (FBS) (e.g. Life Technologies, Cat. No. 26140)
- 2 mM L-glutamine (e.g. Life Technologies, Cat. No. 25030)
- Antibiotics, e.g. penicillin–streptomycin (Life Technologies, Cat. No. 15140)
- Reduced serum medium without antibiotics (RSM) (e.g. Opti-Mem I; Life Technologies, Cat. No. 31985–021/-013)
- PBS without Mg and Ca
- Lipofection reagent (e.g. LipofectAmine; Life Technologies, Cat. No. 18324–012/-020)

Method

1. Using standard growth media with antibiotics, aliquot 2–4 × 10^5 cells into each of five 3.5 cm dishes (or 6-well plate).
2. Incubate (37 °C, 5% CO_2) until 50–75% confluent (~ 18–24 h; optimal cell density will vary with cell type).
3. 2 h before lipofection, wash cells with PBS and replace with standard growth media *without* antibiotics.
4. Prepare the following solutions in 12 × 75 mm sterile polystyrene tubes:
   (a) Solution A. For each transfection, dilute 0.5–2 μg DNA (total) to 100 μl using RSM, keeping the total amount of DNA and copies of promoter sequence constant between each sample.
   (b) DNA samples:
   - Dish 1 (negative control, assesses endogenous βGal activity): no DNA.
   - Dish 2 (control for recombinase tg): 1 μg recombinase tg + 1 μg control 3.
   - Dish 3 (experimental test of function): 1 μg recombinase tg + 1 μg indicator tg.
   - Dish 4 (control for indicator tg): 1 μg control 2 + 1 μg indicator tg.
   - Dish 5 (positive control, transfection efficiency): 1 μg control 2 + 1 μg control 1.

Protocol 2. *Continued*

(c) Solution B. For each transfection, dilute 10–50 μg (actual amount will vary depending on the cell line) LipofectAmine to 100 μl using RSM.

5. Combine solution A with solution B, mix gently, and incubate at room temperature for 45 min.
6. Wash cells with 2 ml of RSM.
7. For each transfection, add 0.8 ml of RSM to the tube containing the DNA–liposome complexes. Mix gently.
8. Aspirate RSM from cells and overlay with DNA–liposome–RSM solution (0.8 ml).
9. Incubate (37 °C, 5% CO_2) for 16 h (optimal time varies with cell type).
10. Replace the RSM with 3 ml standard growth medium.
11. Incubate (37 °C, 5% CO_2) for 24–48 h.
12. Assay cells for βGal activity by X-Gal stain (see Chapter 6).
13. Let stain develop for 24 h (37 °C, humidified chamber) in order to detect even weak expressing cells.
14. Count the number of blue/non-stained cells and normalize to the transfection efficiency.

[a] Although lipofection is described, the optimal means of transfection will vary according to the cell type.
[b] The EDTA present in TE can decrease the efficiency of lipofection.
[c] It may be helpful to generate a stable transformant harbouring an integrated copy of the indicator tg so that many recombinase tgs can be assessed by single construct-transfection rather than by co-transfection.
[d] Described in ref. 5.
[e] Described in ref. 1.
[f] Plasmids which control for the total amount of DNA and promoter copies present in each transfection.

In contrast to a transgenic approach, the knock-in strategy using gene targeting permits the use of any endogenous sequence to drive recombinase expression; 5′ genomic clones for gene targeting are the only required reagents. Most importantly, by inserting the recombinase gene into an endogenous transcription unit, expression should be optimally regulated with minimal mosaicism. This general strategy was first described by Rickert *et al.* (77) where the *cre* gene was introduced into the CD19 locus to direct B lymphocyte-specific mutagenesis in mice. For a detailed discussion of gene knock-in vectors and methodologies see Chapter 1. Because selection markers, with their own transcriptional control elements, can interfere with expression of the target locus (and therefore the inserted recombinase transgene) removal is critical. Because even low levels of the recombinase may result in marker excision in ES cells and thereby preclude positive selection, it may be prudent to flank the marker with *FRT* sites for a *cre* knock-in, and *loxP* sites for a *FLP* knock-in (see *Figure 6B*). Alternatively, ES cells can be used for the initial gene targeting experiment which carry a germline-specific recombinase transgene, such as *Prm1::cre* (33); the marker is then removed upon germline transmission of the knock-in allele from chimeric mice.

Whether generated by making transgenics or by gene replacement techniques, it is absolutely critical that the recombinase activity profile be adequately determined for each recombinase mouse strain. Analyses should include:

(a) Detecting recombinase mRNA by *in situ* hybridization or RT-PCR.

(b) Visualizing recombinase protein by immunohistochemistry or Western blot analyses.

(c) Assaying recombinase activity by crossing to an *indicator* mouse strain.

In situ detection of recombinase mRNA can readily be achieved by standard methods (78) using antisense probes directed against the *cre* or *FLP* sequence. Antibodies that specifically recognize either Cre (Berkeley Antibody, Cat. No. MMS106P) or Flp (27) protein have been generated, each proficient for Western blot analysis. The anti-Cre antibodies can also be used for flow cytometry (1) and immunohistochemistry on sections (74).

5.5 Generating indicator mice

Once the recombinase expression profile is determined, a comparison should be made with the actual activity profile, as detecting recombination events is frequently the more sensitive assay for recombinase expression due to clonal expansion of early recombination events. The activity profile also provides the most relevant prediction of where gene mutagenesis or repair will occur in the final conditional gene targeting experiment. To determine the activity profile, recombinase mice are crossed to a mouse carrying an *indicator* transgene such as the *loxP*- or *FRT*-disrupted *lacZ* transgenes described above and illustrated in *Figure 14*. Offspring carrying both the recombinase and indicator transgenes are then analysed for gene deletion by histochemical detection of βGal in tissue sections and whole embryos. The most informative *indicator* strains are obviously those that express the activated reporter in a wide range of cell types. Because reporter expression is dependent on transgene copy number and integration site, as well as the promoter/enhancer used, a large number of founders should to be generated in order to identify an optimal *indicator* line. βGal *indicator* mouse lines for Cre activity that contain the chicken β-*actin* promoter driving a *loxP*-disrupted *lacZ* transgene include *CAG-CAT-Z* (50), *cAct-XstopXlacZ* (74, 75), and *floxLacZ* (76). The first two strains are restricted *indicators*, as histochemical detection of βGal is limited to certain brain regions, heart, and muscle in both embryos and adults;

the latter strain exhibits somewhat broader expression. Current Flp indicator mouse lines include various *HMG::FRTZ* transgenics capable of marking neural derivatives (Dymecki, unpublished data). Additional indicator strains have recently been developed that involve targeting *loxP*-disrupted reporters to various ubiquitously expressed endogenous mouse genes (79, 80).

Histochemical detection of recombination (e.g. by X-Gal stain) will present merely a subset of events due to restricted *lacZ* expression in all but the last set of *indicator* mouse lines. Therefore recombinase activity can also be monitored directly at the DNA level. Recombined target (trans)genes also can be detected by Southern blot and/or PCR analysis of genomic DNA isolated from various tissues. Although this method of detecting recombined target genes does not provide single cell resolution, it does eliminate the variability in *lacZ* transgene expression associated with promoter activity, copy number, and transgene integration site.

5.6 Breeding schemes for *conditional* gene targeting

Examples of breeding schemes for conditional gene targeting are presented in *Figure 15*. After two generations, mice harbouring the recombinase transgene and homozygous for the floxed gene represent ~ 12.5% of the offspring. This percentage assumes that the *cre* recombinase and floxed loci are not linked. The percentage of informative F_3 offspring can be increased to ~ 20% by intercrossing (*cre*/+, floxed/+) F_2 animals; in this situation it is important that any F_3 progeny homozygous for the recombinase locus (e.g. *cre/cre*, +/+) are wild-type so that the observed lineage-specific (*cre/cre*, Δ/Δ) phenotype stems solely from deletion of the floxed locus and is not a result of a *cre* transgene insertional mutation (insertional mutations occur in ~ 1/20 zygote injections).

As shown in *Figure 15B*, a null allele can be incorporated into the breeding scheme. By placing the floxed allele over the null allele in F_3 animals, less demand is place on the recombinase — only one target gene needs to undergo excisional recombination to generate a homozygous null cell. Again, by intercrossing F_2 (*cre*/+, floxed/+) with F_2 (*cre*/+, null/+) animals, the percentage of informative offspring can be increased to ~ 20%. Bear in mind that many other breeding strategies can be developed, depending on the viability and fertility of each genotype.

6. Switching-on transgenes in the living mouse

The binary approach central to *conditional* gene targeting (*Figure 9*) can also be exploited to activate any heterologous transgene in a spatially- and temporally-restricted fashion *in vivo*. One mouse strain expresses the recombinase in selected lineages or tissues; the other mouse carries a transgene functionally silenced by insertion of a *loxP*- or *FRT*-flanked STOP cassette in front of it (see Section 5.3). In offspring carrying both transgenes, cells expressing the recombinase will activate expression of the transgene by excising

A **Breeding Scheme to Generate Conditional Mutation**

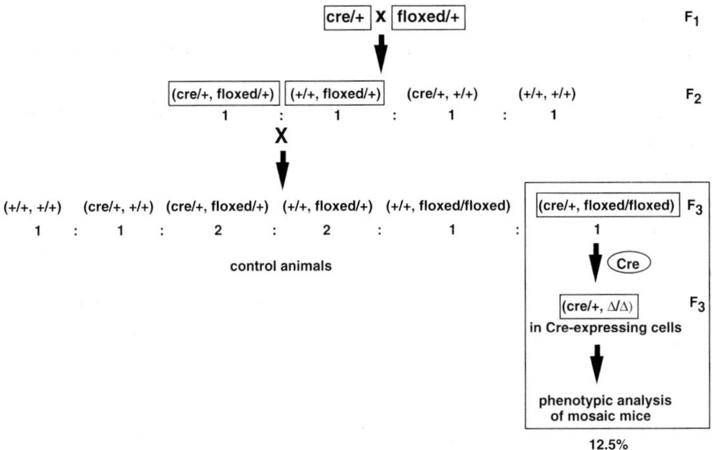

B **Alternative Breeding Scheme Incorporating a Null Allele.**

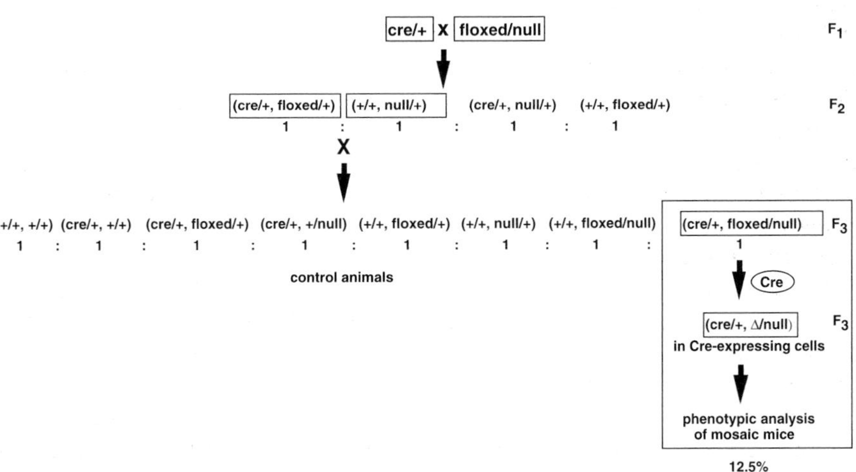

Figure 15. Examples of breeding schemes for conditional gene targeting. The symbol Δ indicates the deleted (null) allele generated by recombination of the floxed locus; +, represents the wt locus. (A) Heterozygous breeding scheme. After two generations, mice harbouring the recombinase locus and homozygous for the floxed gene represent ~ 12.5% of the offspring (highlighted by large box). This percentage assumes that the *cre* recombinase and floxed loci are not linked. The percentage of informative F_3 offspring can be increased to ~ 20% by intercrossing (*cre*/+, floxed/+) F_2 animals. (B) Incorporation of a null allele into the breeding scheme. By placing the floxed allele over the null allele in F_3 animals, less demand is place on the recombinase — only one target needs to undergo excisional recombination to generate a nullizgyous cell.

the disrupting STOP cassette. The transgene should remain silent in lineages where the recombinase has never been expressed. Activation of the *indicator* transgene described in Section 5 is one example of this type of approach.

Recombinase-regulated transgene expression is especially valuable if transgenic mice made by conventional methods (zygote injection) are sufficiently impaired so as to preclude establishing a line for study. Experiments best served by this approach are therefore oncogenes, cell cycle regulators, many developmental genes, or toxins for lineage-specific ablations (81). This methodology was first used to activate expression of the oncogenic SV40 large T antigen to induce tumours specifically in the eye lens (2). This general paradigm can also be exploited to fate-map specific progenitor populations directly in the mouse, this application is described below in Section 7.

7. Fate-mapping by site-directed recombination

Genetic mechanisms controlling cell fate specification during embryonic development have largely been defined through studies in *Drosophila*, *C. elegans*, and now the zebrafish *Danio rerio*. Understanding how cell identity is specified in the more complex mouse embryo has been a greater challenge, requiring the integration of data from multiple vertebrate experimental systems. Due to the relative ease of injecting lineage tracers and heterologous tissues in the chick, avian systems have been used to map the origin of various cell types within embryonic tissues. Gene expression profiles in both chick and mouse have then been compared to these fate-maps, and hypotheses of gene function proposed which can then be tested by targeted mutagenesis in mice. Although fruitful, this paradigm is limited in two ways. First, fate-mapping and mutagenesis experiments are performed in different organisms, precluding direct investigations on how mutations influence cell fate. Secondly, the spatial and temporal expression patterns of many developmentally important genes are surprisingly complex, making it difficult to establish relationships between early gene expression and later cell identity. For example, expressing and non-expressing cells are often intermingled within territories, precise patterns may vary between rostral and caudal regions, and expression is frequently transient.

To circumvent these limitations, site-specific recombination strategies are being developed for charting cell fate directly in the mouse. The strengths of these new lineage tagging methodologies are that:

(a) Progenitor cells and fully differentiated descendant cells, although temporally distinct, can be linked regardless of intervening cell migrations and morphogenetic movements.

(b) Progenitor cells can be defined by transient expression of a given gene, thereby relating early gene expression to later cell identity.

(c) Relevant mouse mutants can be assayed for the effect of the mutation on marked lineages by crossing into recombinase-tagged mouse strains (discussed in Section 7.2).

7.1 Tagging cell lineages in the mouse

The basic elements comprising a recombinase-based fate-mapping system are very similar to the *indicator* system for determining the activity profile of Cre and Flp transgenics (Section 5, *Figure 14*). The major difference is that for fate-mapping, the target transgenic not only needs to 'indicate' a recombination event in a given cell, but to '*remember*' or '*provide a permanent record*' of that recombination event in a heritable manner far after the time of recombinase expression. 'Remembering' is achieved by transmitting a permanently recombined and activated reporter to all descendant cells—daughters, granddaughters, etc. The operative feature here is *permanent reporter expression* in a given cell lineage. Success therefore hinges on using a promoter with *constitutive* and, ideally, *ubiquitous* activity to drive the *indicator* transgene. As described in Section 5.5, these constitutively active regulatory sequences are placed upstream of the reporter (e.g. *lacZ*, the GFP gene, or the alkaline phosphatase gene), but separated by a STOP segment of DNA flanked by direct recombinase recognition sites. The reporter is not expressed unless the STOP DNA is excised by the recombinase. Following such an excision event, the reporter is switched-on and cell marking becomes heritable and dependent on constitutive reporter expression only. Candidate promoters to drive the *indicator* transgene which have been evaluated include chicken β-*actin*, human β-*actin*, *HMG-coA reductase*, *RNA polymerase II*, various synthetic promoters, as well as genomic sequences associated with various ubiquitously expressed mouse genes (the latter can be used to create *indicator* mice using knock-in strategies).

The important features of this scheme are illustrated in *Figure 16* and are as follows:

(a) A stable and heritable genetic change is produced in a progenitor population in response to transient recombinase expression.

(b) This progenitor-specific genetic alteration provides a non-invasive way to mark progenitors and their daughter cells.

(c) Recombinase function itself has no phenotypic effect on development.

In practice, reporter modification and activation becomes restricted *in vivo* to a specific progenitor population by crossing *indicator* mice to mice expressing the recombinase under control of a progenitor-specific enhancer element. Double transgenic progeny are then analysed for reporter expression starting from the time the recombinase is transiently expressed in the designated progenitor cells. This is done by detection of βGal (GFP or alkaline phosphatase, depending on the *indicator* transgene) in histological sections or whole-mounts from a series of staged embryos. Both Cre and Flp

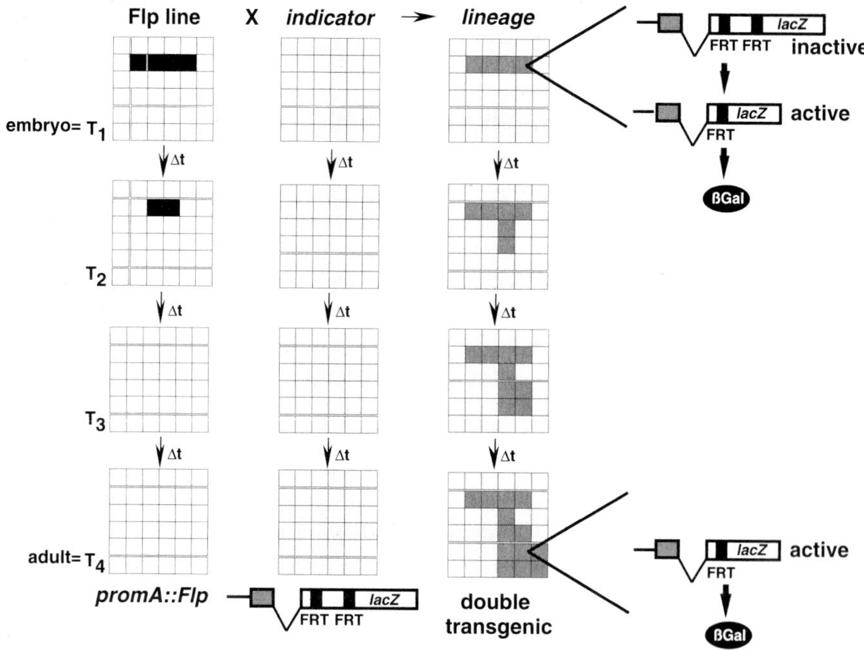

Figure 16. Binary scheme for mapping cell fate. Each grid represents a mouse, and is cartooned developing from an early embryo (T1) to adult (T4); Δt, an increment of time during which development progresses. The left column illustrates a mouse line expressing *FLP* in the transient and progenitor-specific pattern of gene *A* (small black squares); the middle column (light grey), a mouse line harbouring the *indicator* transgene diagrammed at the bottom and in *Figure 14*. The *indicator* transgene not only indicates a Flp-mediated recombination event by the gain of reporter activity, but 'remembers' that event by transmitting a permanently activated reporter to all progeny cells. Permanent activation requires using a promoter (grey hatched rectangle) with constitutive and ubiquitous activity to drive expression of the *indicator* transgene. Fate-mapping involves crossing the progenitor-specific Flp strain to the *indicator* strain. Double transgenic offspring (illustrated in the right column) undergo reporter activation in a progenitor cell-specific manner, progeny cells inherit the activated reporter (dark grey squares). In this way, gene *A* activity in progenitor cells is linked to a pattern of fates these cells and their descendants adopt in the adult.

are suitable for fate-mapping applications, as they both are capable of producing restricted and heritable gene deletions in response to transient progenitor-specific recombinase expression in the mouse embryo (30, 75).

To date, the development of optimal *indicator* mouse lines is still in progress, as identifying promoter and enhancer sequences that can constitutively drive reporter expression as cell differentiation and development proceeds is a sizeable task. Nevertheless, a *c*β-*STOP-lacZ* indicator mouse line has been successfully used by Zinyk and colleagues (75) to fate-map *in situ* the mouse midbrain–hindbrain constriction. Lower resolution fate-maps can be generated by following *indicator* transgene deletions at the DNA level by Southern hybridization or PCR. This type of approach has been employed using Flp recombinase to fate-map cells which arise from the dorsal neural tube and which express the *Wnt1* gene (30). The map resolution obtained by direct DNA assay is, of course, limited to the resolution of tissue dissection.

In these types of fate-mapping experiments, the progenitor population to be tracked is defined by the pattern of recombinase expression. Consequently, enhancers with limited and transient expression are most useful, enabling a spatially and temporally defined group of progenitor cells in the embryo to be traced into adult tissues. This is in contrast to recombinase knock-ins which can have the problem of continuous and changing expression patterns (reflecting the dynamic expression of the wt locus), making fate-mapping of a particular group of early cells difficult.

7.2 Combining recombinase-mediated fate-mapping with mouse mutations to yield high resolution studies of gene function

To date, the majority of vertebrate lineage data has come from the chick and frog, and although this information has been extremely valuable in extrapolating predictions for mouse embryogenesis, it is none the less of limited use given the lack of genetics in these organisms. Recombinase-based methods for mapping cell fate directly in the mouse, a vertebrate with rich genetic resources, presents an advantage. Relevant mutants can be screened for their effect on cell fate simply by crossing the recombinase-tagged mouse strains into the many available mouse mutants (targeted, spontaneous, gene trapped, or chemically induced).

8. Targeted integrations: isogenic cell and mouse lines for structure–function studies

Zygote injection and ES cell transfection, two of the most commonly employed techniques for making transgenics, can yield a variable number of copies of the test transgene integrated at a random site in the chromosomal DNA. The consequent phenotype observed, is therefore, not only the result of the injected transgene, but also the result of transgene copy number, local chromatin structure, DNA methylation, and transcriptional effects of neighbouring loci on the transgene. To extricate the transgene-associated phenotype from position effects, a large number of transgenic lines are typically generated and analysed for any given construct—the phenotype common to all lines is attributed to the transgene. This requirement makes transgenic promoter analyses and structure–function studies laborious and expensive. Site-specific recombination offers a solution, as a series of modified transgenes can be targeted to the same chromosome position and in the same orientation, virtually eliminating confounding position effects.

By exploiting the intermolecular integration reaction catalysed by Cre or Flp (*Figure 2A*), a single chromosomal *loxP* or *FRT* element can serve as a landing site for the integration of exogenous DNA. Such experiments have been successfully accomplished in mammalian cells using either the Cre (82) or Flp (11) systems, providing a potential means to generate isogenic cell lines and mouse strains. In practice, a cell line or zygote with a *loxP* or *FRT* site must be made. Then a *loxP*- or *FRT*-containing construct is introduced together with a vector for transient Cre or Flp expression, respectively. The recombination recognition site on the incoming vector recombines with the chromosomal recombinase recognition site, resulting in the integration of the vector DNA (see *Figure 17*). The efficiency of site-specific integration appears to be little better than homologous recombination. Such low efficiency is a disadvantage for two reasons. First, it means that to generate isogenic mouse strains, targeted integrations must be performed in ES cells where selection and screening can be employed. Secondly, the retained positive selection gene may not necessarily prove neutral as described in Section 4.1. Consequently, to realize the full potential of targeted integrations, the positive selection marker should be removed following identification of recombinant cell clones. As described in Section 4, this requires the use of a second site-specific recombination system. For example, if Cre recombinase is to mediate transgene integration, Flp recombinase should be used to remove the selection cassette.

The low efficiency of targeted integration likely stems from the reversibility of the reaction, with the intramolecular *excision* reaction occurring at a much higher frequency than integration. As described earlier, two strategies can be employed to bias the reaction towards integration:

(a) Limit recombinase activity to a brief pulse to avoid subsequent excision (methods for inducible recombinase expression are discussed in Section 9).

(b) Use two asymmetrically mutant recombination target sites so that following recombination, one of the two sites is inactive (*Figure 3*) (25).

9. Inducible recombination: gaining temporal and spatial control over genome modifications

As described in Section 5, cell type-specific, constitutive recombinase expression can provide *spatial* control over recombinase-mediated gene modifications. Three approaches have recently been developed which add *temporal* control. The first method involves an *inducible promoter* to express the recombinase, the second approach exploits *post-translational* activation of

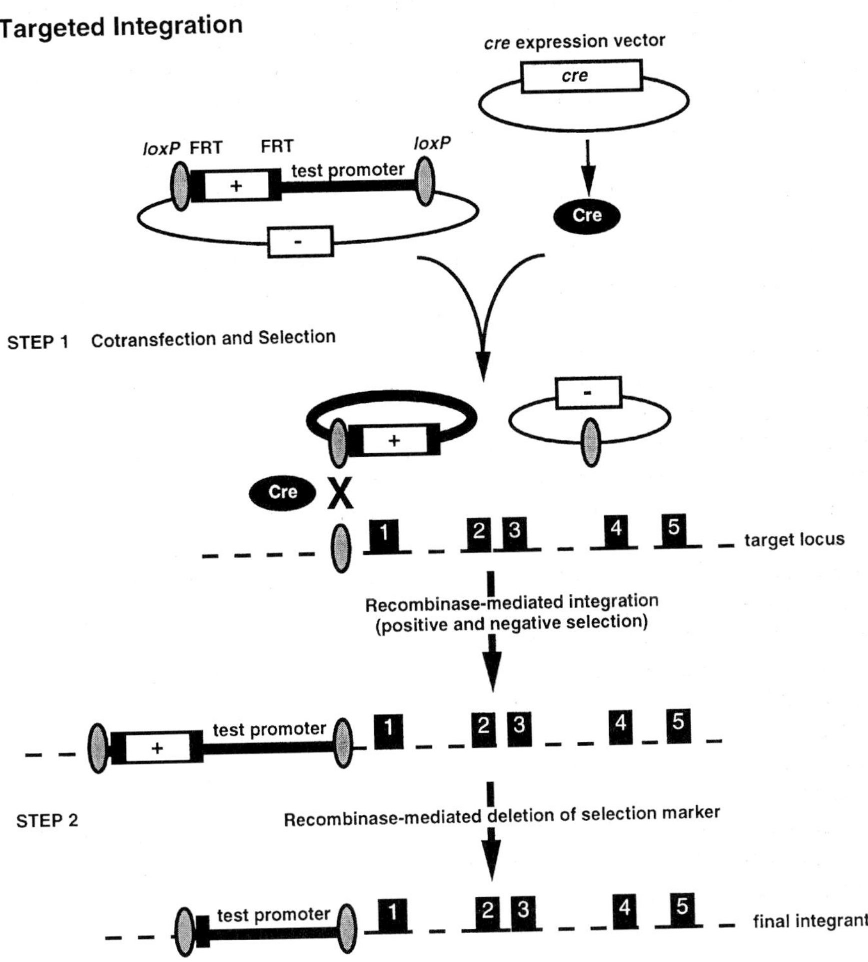

Figure 17. Using site-specific integration to target modified DNA sequences to the same chromosome position. In this example, the intermolecular integration reaction catalysed by Cre is used to insert modified promoter sequences upstream of a target gene. The 'test' plasmid contains a floxed cassette comprised of the modified promoter sequence directly linked to a flrted positive selection marker. ES cells are co-transfected with the test plasmid and a *cre* expression vector. Cre-mediated excisional recombination reduces the test plasmid into two circular DNAs (one carrying the test promoter sequence and positive selection marker, and one harbouring the plasmid backbone and negative selection marker). Cre-mediated recombination inserts the circular DNA into the target locus at the *loxP* site. The positive and negative selection enriches for integration of the desired test sequences, versus DNA carrying the plasmid backbone. As the positive selection marker required in step 1 frequently affects host gene expression, it is removed by Flp-mediated deletion in step 2.

a constitutively transcribed 'fusion recombinase', and the third technique uses a recombinase-expressing adenovirus vector. These are outlined in turn below.

In the first report of *temporal* gene targeting, Kühn *et al.* (31) describe using the interferon-α/β inducible promoter of the *Mx1* gene to drive *cre* expression. In double transgenic animals, Cre-mediated deletion of a floxed target gene was induced following systemic administration of either interferon-α/β or polyinosinic-polycytidylic acid to activate *cre* expression. Efficient gene deletion was achieved in a wide range of tissues, but not all tissues, within two days of treatment, even in organs composed mainly of resting cells, such as the liver. This *Mx1::cre* transgenic should, therefore, prove useful in experimental situations where it is critical that the wild-type gene product under study be present throughout development until the time of induction, but where spatial restriction of the genetic alteration is not required.

To gain both *spatial* and *temporal* control over recombinase transcription, various transactivator systems are being tested in mice. One of the first to be characterized has been the tetracycline (tet) responsive system (83); although met with only modest success, it serves as a good example. The DNA-binding and tetracycline-binding domains of the tetracycline repressor (tetR) are fused to the acidic domain of the herpes simplex viral protein 16 (VP16) to create a chimeric tetR::VP16 protein. This chimeric transactivator is used to regulate expression of a *cre* transgene. The promoter driving *cre* expression is engineered to contain tet-operator (tetO) DNA sequences to which the tetR::VP16 fusion protein binds in the absence of tet and activates transcription. Therefore in the presence of tet, *cre* expression should be repressed (as the tetR::VP16 transactivator does not bind tetO DNA); in the absence of tet *cre* should be expressed and recombine a floxed target gene. In practice, *cre* expression has not been fully repressed by tet, so that unintended recombination occurs. Other transactivators such as the mutant (or reverse) tetR may provide tighter temporal induction. Although this type of methodology is still maturing, it is worth noting that it requires the generation of mouse strains harbouring three functional transgenes (the *transactivator* gene, the *tetO::recombinase* transgene, and the floxed or flrted target).

An alternative and seemingly simpler means to gain *spatial* and *temporal* control over genome modifications involves using a recombinase–steroid receptor fusion protein that can be activated at will by systemic administration of hormone. This approach first involved demonstrating that recombinase–steroid receptor fusion proteins were inactive in the absence of cognate steroid hormone, but responded rapidly to hormone administration. Having established such a 'switch' for regulating recombinase activity in tissue culture cells and in transgenic mice (84–89), the second advance involved restricting expression of the fusion recombinase to a given cell type in mice through tissue-specific promoter and enhancer elements. Using the immunoglobulin heavy chain enhancer linked to the SV40 promoter, Stewart and colleagues (88) constitutively expressed a Cre-fusion specifically in B lymphocytes. The fusion transgene encoded a mutant oestrogen receptor ligand binding domain (EBD) linked to the C-terminus of Cre; the mutant EBD was insensitive to the endogenous hormone β-oestradiol, but responsive to the synthetic oestrogen antagonist 4-OH tamoxifen. In double transgenic animals, deletion of the floxed target occurred specifically in B cells and only after intraperitoneal injection of 4-OH tamoxifen. Therefore two regulatory modes have been combined in a single recombinase transgene to achieve quite precise mutagenesis in the mouse.

Whether employing the *transcriptional* or *post-translational* approach, the following points should be considered when designing an inducible gene targeting experiment:

(a) Maximum recombination can require repeated administration of the inducer over three to five days (89), therefore the kinetics of induction are reasonable but not rapid.

(b) Efficiency of gene deletion depends on the dose of the inducer.

(c) Recombination in the absence of inducer is typically low ($< 1\%$).

(d) Toxicity associated with the current battery of inducers is low but should be carefully monitored.

(e) Current reagents have limitations for *in utero* induction (see ref. 89) due to toxicity to embryos at high doses, but can be used if inefficient excision is acceptable at early embryonic stages.

The third approach to regulate gene modification involves using an adenoviral vector expressing Cre or Flp recombinase (90). Spatio-temporally controlled recombination is achieved after injection of virus into specific tissues. With the guiding aid of high frequency ultrasound, recombinase-expressing virus can even be injected into the amniotic fluid, brain ventricles, or specific regions of the developing mouse embryo after embryonic day 8.5 (91, 92).

10. Tools: Cre-*loxP* and Flp-FRT vectors and mice

Published vectors central to recombinase-based genetic manipulations in the mouse are summarized in *Table 7* (Cre-*loxP*) and *Table 8* (Flp-*FRT*); more detailed descriptions can be obtained from the associated references. Information on many of the Cre- or *loxP*-containing plasmids can also be found on the K. Rajewsky web page: `www.genetik.uni-koeln.de/gene_targeting/`. A summary of Cre expressing mouse strains can be obtained from the database established by A. Nagy: `www.mshri.on.ca/develop/nagy/cre.htm`. Similar Flp transgenic databases will soon follow (Dymecki).

Table 7. Cre-*loxP* constructs for genome engineering in mammalian cells

Construct	Promoter	Splice/pA	Coding sequence	Recombinase target sites	Selectable markers	Remarks	Reference
pBS185	CMV	/MT-1 pA	*cre*			Life Technologies[a]	46
pBS118	RSV	/SV40 pA	*cre*			Life Technologies[a]	46
pIC-Cre	MC1	/SV40 pA	*cre*			Rajewsky web page[b]	48
pMC-Cre	MC1	/SV40 pA	*cre*			Rajewsky web page[b]	48
pGK-CreNLSbpA	PGK	/bGH pA	*cre*			Rajewsky web page[b]	1
pCAGGS-Cre	Chicken β-actin		*cre*				50
pCMVCre-ER^T	CMV	rabbit β-globin /SV40 pA	*cre–oestrogen receptor* (G521R) fusion			Hormone responsive	85
pGK-CreED4	PGK	/bGH pA	*cre–oestrogen receptor* (G521R) fusion			Hormone responsive	87
pANCreMer	Human β-actin	/SV40 pA	*cre–oestrogen receptor* (G521R) fusion			Hormone responsive	96
pCre-PR	SV40	/SV40 pA	*cre–progesterone receptor* (hPR891) fusion			Hormone responsive	87
pBS246				Direct *loxPs* flanking MCS		Life Technologies[a]	46
pBS302				Direct *loxPs* flanking STOP cassette		Life Technologies[a]	46
pSF1		/SV40 pA		*loxP*	*neo*^{r}	*lox-neo* fusion shuttle vectors distributed by Life Technologies	46
pBS226	CMV			*loxP*			
pGEMloxP				*loxP*		Rajewsky web page[b]	1
pGEM-30				*loxP*		Rajewsky web page[b]	1
pMMneoflox8				Direct *loxPs* flanking	*neo*^{r}	Rajewsky web page[b]	1
pL2neo				Direct *loxPs* flanking	*neo*^{r}	Rajewsky web page[b]	1

pSVlacZT	SV40		*loxP*-disrupted *lacZ*	Direct *loxPs* flanking	*neo*^{r}	*Indicator* transgene	1
pCAG-CAT-Z	Chicken β-actin		*loxP*-disrupted *lacZ*	Direct *loxPs* flanking *CAT*		*Indicator* transgene	50
pcAct-XstopXlacZ	Chicken β-actin	3′UTR	*loxP*-disrupted *lacZ*	Direct *loxPs* flanking *STOP*		*Indicator* transgene	74

[a] Distributed by Life Technologies.
[b] Rajewsky web page: www.genetik.uni-koelin.de/gene_targetiing.
bGH pA: bovine growth hormone gene polyadenylation sequence, CMV: major immediate early promoter of human cytomegalovirus, *neo*^{r}: neomycin phosphotransferase, *HSV-tk*: herpes simplex virus thymidine kinase gene, MC1: synthetic mutant polyoma enhanced, *HSV-tk* promoter, MCS: multicloning site, NLS: SV40 large T antigen-derived nuclear localization signal, PGK: phosphoglycerate kinase I promoter, RSV: long terminal repeat of Rous sarcoma virus, SV40 pA: SV40 polyadenylation sequence.

Table 8. *FLP-FRT* constructs for genome engineering in mammalian cells

Construct	Promoter	Splice/pA	Coding sequence	Recombinase target sites	Selectable markers	Remarks	Reference
pOG44	CMV	Synthetic intron /SV40 pA	*wt FLP*			Stratagene[a]	11
pOG44*FLPL*	CMV	Synthetic intron /SV40 pA	*mutant FLP* [F70L]			Stratagene[a]	
pOG44*FLPE*	CMV	Synthetic intron /SV40 pA	*enhanced FLP*			[P2S, L33S, Y108N, S294P]	28
p*FLP*		Synthetic intron /SV40 pA	*wt FLP*			Flanked by MCS	45
p*FLPL*		Synthetic intron /SV40 pA	*mutant FLP* [F70L]			Flanked by MCS	45
p*FLPE*		Synthetic intron /SV40 pA	*enhanced FLP*			Flanked by MCS [P2S, L33S, Y108N, S294P]	*b*
p*hACT*β*::FLP*	Human β-actin	Synthetic intron /SV40 pA	*wt FLP*				5
p*hACT*β*::FLPL*	Human β-actin	Synthetic intron /SV40 pA	*mutant FLP* [F70L]				5
p*hACT*β*::FLPE*	Human β-actin	Synthetic intron /SV40 pA	*enhanced FLP*			[P2S, L33S, Y108N, S294P]	28 *b*
p*FLPL-EBD*^{G400V}	CMV	Synthetic intron /SV40 pA	*Flp–oestrogen receptor* (G400V) fusion				84
p*FRT*$_2$				Direct *FRT*s		Flanking MCS	45
p*FRT*$_2$*lacZ*		/SV40 pA	*FRT*-disrupted (NLS) *lacZ*	Direct *FRT*s		Flanking MCS	45
p*FRT*$_2$*neo.lacZ*		/SV40 pA	*FRT*-disrupted (NLS) *lacZ*	Direct *FRT*s flanking	*neo*^{r}		45

p*FRT*$_2$*lacZ*		/SV40 pA	FRT-disrupted (NLS) *lacZ*	Direct *FRTs* flanking MCS			45
p*FRT.lacZ*		/SV40 pA	(NLS) *lacZ*	Single *FRT* in-frame to *lacZ*			45
pFRTZ	Human β-actin	Human β-actin /SV40 pA	FRT-disrupted (NLS) *lacZ*	Direct *FRTs* flanking	*neo*^{r}		5
pFRTZ-product	Human β-actin	Human β-actin /SV40 pA	(NLS) *lacZ*	Single *FRT* in-frame to *lacZ*			5
pHMG::FRTZ	HMG CoA reductase	HMG CoA /SV40 pA	(NLS) *lacZ*	Direct *FRTs* flanking	*neo*^{r}		*b*
pHMG::FRTZ -product	HMG CoA reductase	HMG CoA /SV40 pA	(NLS) *lacZ*	Single *FRT* in-frame to *lacZ*			*b*
pNEOβGAL	SV40 early	SV40 intron /SV40 pA	FRT-disrupted *lacZ*	Direct *FRTs* flanking	*neo*^{r}	Strategene[a]	11
pFRTβGAL	SV40 early	SV40 intron /SV40 pA	(NLS) *lacZ*	Single *FRT* in-frame to *lacZ*		Strategene[a]	11
pOG45				*FRT*	*neo*^{r}	Strategene[a]	11

[a] Distributed by Stratagene.
[b] Dymecki, unpublished data.
bGH pA: bovine growth hormone gene polyadenylation sequence, CMV: major immediate early promoter of human cytomegalovirus, *neo*^{r}: neomycin phosphotransferase, *HSV-tk*: herpes simplex virus thymidine kinase gene, *HSV-tk* promoter, MCS: multicloning site, NLS: SV40 large T antigen-derived nuclear localization signal, PGK: phosphoglycerate kinase I promoter, RSV: long terminal repeat of Rous sarcoma virus, SV40 pA: SV40 polyadenylation sequence.

Acknowledgements

I thank all colleagues who contributed unpublished data, especially Francis Stewart; Raul Torres and Ralf Kühn for their protocol on Cre-mediated deletions in ES cells and for putting together such a useful manual on conditional gene targeting; Paul Hasty, Alejandro Abuin, and Allan Bradley for their input on Section 4.5; and Alexandra Joyner for many useful comments on this chapter. I would also like to acknowledge support from the Helen Hay Whitney Foundation, the John Merck Fund, and the NIH.

References

1. Torres, R. M. and Kuhn, R. (1997). In *Laboratory protocols for conditional gene targeting* (ed. R. M. Torres and R. Kuhn), p. 167. Oxford University Press, Oxford.
2. Lakso, M., Sauer, B., Mosinger, B., Lee, E. J., Manning, R. W., Yu, S.-H., *et al.* (1992). *Proc. Natl. Acad. Sci. USA*, **89**, 6232.
3. Orban, P. C., Chui, D., and Marth, J. D. (1992). *Proc. Natl. Acad. Sci. USA*, **89**, 6861.
4. Gu, H., Marth, J. D., Orban, P. C., Mossmannand, H., and Rajewsky, K. (1994). *Science*, **265**, 103.
5. Dymecki, S. (1996). *Proc. Natl. Acad. Sci. USA*, **93**, 6191.
6. Argos, P., Landy, A., Abremski, K., Egan, J. B., Haggard-Ljungquist, E., Hoess, R. H., *et al.* (1986). *EMBO J.*, **5**, 433.
7. Stark, W. M., Boocock, M. R., and Sherratt, D. J. (1992). *Trends Genet.*, **8**, 432.
8. Kilby, N. J., Snaith, M. R., and Murray, J. A. H. (1993). *Trends Genet.*, **9**, 413.
9. Agah, R., Frenkel, P. A., French, B. A., Michael, L. H., Overbeek, P. A., and Schneider, M. D. (1997). *J. Clin. Invest.*, **100**, 169.
10. Sauer, B. and Henderson, N. (1988). *Proc. Natl. Acad. Sci. USA*, **85**, 5166.
11. O'Gorman, S., Fox, D. T., and Wahl, G. M. (1991). *Science*, **251**, 1351.
12. Sternberg, N. and Hamilton, D. (1981). *J. Mol. Biol.*, **150**, 467.
13. Austin, S., Ziese, M., and Sternberg, N. (1981). *Cell*, **25**, 729.
14. Gerbaud, C., Fournier, P., Blanc, H., Aigle, M., Heslot, H., and Guerineau, M. (1979). *Gene*, **5**, 233.
15. Broach, J. R. and Hicks, J. B. (1980). *Cell*, **21**, 501.
16. Futcher, A. B. (1986). *J. Theor. Biol.*, **119**, 197.
17. Sadowski, P. D. (1995). *Prog. Nucleic Acid Res. Mol. Biol.*, **51**, 53.
18. Hoess, R. H., Ziese, M., and Sternberg, N. (1982). *Proc. Natl. Acad. Sci. USA*, **79**, 3398.
19. Jayaram, M. (1985). *Proc. Natl. Acad. Sci. USA*, **82**, 5875.
20. Hoess, R., Wierzbicki, A., and Abremski, K. (1985). *Gene*, **40**, 325.
21. Huang, L.-C., Wood, E. A., and Cox, M. M. (1991). *Nucleic Acids Res.*, **19**, 443.
22. Golic, K. G. and Lindquist, S. (1989). *Cell*, **59**, 499.
23. Morris, A. C., Schaub, T. L., and James, A. A. (1991). *Nucleic Acids Res.*, **19**, 5895.
24. Senecoff, J. F., Rossmeissl, P. J., and Cox, M. M. (1988). *J. Mol. Biol.*, **201**, 405.
25. Araki, K., Araki, M., and Yamamura, K. (1997). *Nucleic Acids Res.*, **25**, 868.
26. Albert, H., Dale, E. C., Lee, E., and Ow, D. W. (1995). *Plant J.*, **7**, 649.

27. Buchholz, F., Ringrose, L., Angrand, P.-O., Rossi, F., and Stewart, A. F. (1996). *Nucleic Acids Res.*, **24**, 4256.
28. Buchholz, F., Angrand, P.-O., and Stewart, A. F. (1998). *Nature Biotechnol.*, **16**, 657.
29. Lakso, M., Pichel, J. G., Gorman, J. R., Sauer, B., Okamoto, Y., Lee, E., *et al.* (1996). *Proc. Natl. Acad. Sci. USA*, **93**, 5860.
30. Dymecki, S. M. and Tomasiewicz, H. (1998). *Dev. Biol.*, **201**, 57.
31. Kühn, R., Schwenk, F., Aguet, M., and Rajewsky, K. (1995). *Science*, **269**, 1427.
32. Lewandoski, M., Wassarman, K. M., and Martin, G. R. (1997). *Curr. Biol.*, **7**, 148.
33. O'Gorman, S., Dagenais, N. A., Qian, M., and Marchuk, Y. (1997). *Proc. Natl. Acad. Sci. USA*, **94**, 14602.
34. Vooijs, M., van der Valk, M., te Riele, H., and Berns, A. (1998). *Oncogene*, **17**, 1.
35. Jung, S., Rajewsky, K., and Radbruch, A. (1993). *Science*, **259**, 984.
36. Jacks, T., Shih, T. S., Schmitt, E. M., Bronson, R. T., Bernards, A., and Weinberg, R. A. (1994). *Nature Genet.*, **7**, 353.
37. Carmeliet, P., Ferreira, V., Breier, G., Pollefeyt, S., Kieckens, L., Gertsenstein, M., *et al.* (1996). *Nature*, **380**, 435.
38. Lerner, A., D'Adamio, L., Diener, A. C., Clayton, L. K., and Reinherz, E. L. (1993). *J. Immunol.*, **151**, 3152.
39. Ohno, H., Goto, S., Taki, S., Shirasawa, T., Nakano, H., Miyatake, S., *et al.* (1994). *EMBO J.*, **13**, 1157.
40. Pham, C. T., MacIvor, D. M., Hug, B. A., Heusel, J. W., and Ley, T. J. (1996). *Proc. Natl. Acad. Sci. USA*, **93**, 13090.
41. Fiering, S., Kim, C. G., Epner, E. M., and Groudine, M. (1993). *Proc. Natl. Acad. Sci. USA*, **90**, 8469.
42. Askew, G. R., Doetschman, T., and Lingrel, J. B. (1993). *Mol. Cell. Biol.*, **14**, 4115.
43. Stacey, A., Schnieke, A., McWhir, J., Cooper, J., Colman, A., and Melton, D. W. (1994). *Mol. Cell. Biol.*, **14**, 1009.
44. Wu, H., Liu, X., and Jaenisch, R. (1994). *Proc. Natl. Acad. Sci. USA*, **91**, 2819.
45. Dymecki, S. M. (1996). *Gene*, **171**, 197.
46. Sauer, B. (1993). In *Manipulation of transgenes by site-specific recombination: use of Cre recombinase* (ed. P. M. Wassarman and M. L. DePamphilis), Vol. 225, p. 890. Academic Press, San Diego.
47. Buchholz, F., Angrand, P. O., and Stewart, A. F. (1996). *Nucleic Acids Res.*, **24**, 3118.
48. Gu, H., Zou, Y. R., and Rajewsky, K. (1993). *Cell*, **73**, 1155.
49. Yagi, T., Ikawa, Y., Yoshida, K., Shigentani, Y., Takeda, N., Mabubuchi, I., *et al.* (1990). *Proc. Natl. Acad. Sci. USA*, **87**, 9918.
50. Araki, K., Araki, M., Miyazaki, J.-I., and Vassalli, P. (1995). *Proc. Natl. Acad. Sci. USA*, **92**, 160.
51. Ludwig, D. L., Stringer, J. R., Wight, D. C., Doetschman, H. C., and Duffy, J. J. (1996). *Transgenic Res.*, **5**, 385.
52. Zhang, H., Hasty, P., and Bradley, A. (1994). *Mol. Cell. Biol.*, **14**, 2404.
53. Zou, Y.-R., Muller, W., Gu, H., and Rajewsky, K. (1994). *Curr. Biol.*, **4**, 1099.
54. Hanks, M., Wurst, W., Anson-Cartwright, L., Auerbach, A., and Joyner, A. L. (1995). *Science*, **269**, 679.
55. Smithies, O. (1993). *Trends Genet.*, **9**, 112.
56. Roa, B. B. and Lupski, J. R. (1994). *Adv. Hum. Genet.*, **22**, 117.

57. Denny, C. T., Hollis, G. F., Hecht, F., Morgan, R., Link, M. P., Smith, S. D., *et al.* (1986). *Science*, **234**, 197.
58. Popescu, N. C. and Zimonjic, D. B. (1997). *Cancer Genet. Cytogenet.*, **93**, 10.
59. Hassold, T., Abruzzo, M., Adkins, K., Griffin, D., Merrill, M., Millie, E., *et al.* (1996). *Environ. Mol. Mutagen.*, **28**, 167.
60. Ramirez-Solis, R., Liu, P., and Bradley, A. (1995). *Nature*, **378**, 720.
61. Smith, A. J. E. (1995). *Nature Genet.*, **9**, 376.
62. Van Deursen, J., Fornerod, M., Van Rees, B., and Grosveld, G. (1995). *Proc. Natl. Acad. Sci. USA*, **92**, 7376.
63. Li, Z. W., Stark, G., Gotz, J., Rulicke, T., Gschwind, M., Huber, G., *et al.* (1996). *Proc. Natl. Acad. Sci. USA*, **93**, 6158.
64. Lewandoski, M. and Martin, G. R. (1997). *Nature Genet.*, **17**, 223.
65. Herault, Y., Rassoulzadegan, M., Cuzin, F., and Duboule, D. (1998). *Nature Genet.*, **20**, 381.
66. Meuwissen, R. L., Offenberg, H. H., Dietrich, A. J., Riesewijk, A., van Iersel, M., and Heyting, C. (1992). *EMBO J.*, **11**, 5091.
67. Vidal, F., Sage, J., Cuzin, F., and Rassoulzadegan, M. (1998). *Mol. Reprod. Dev.*, **51**, 274.
68. Meyers, E. N., Lewandoski, M., and Martin, G. R. (1998). *Nature Genet.*, **18**, 136.
69. Nagy, A., Moens, C., Ivanyi, E., Pawling, J., Gertsenstein, M., Hadjantonakis, A. K., *et al.* (1998). *Curr. Biol.*, **8**, 661.
70. Chung, J. H., Whiteley, M., and Felsenfeld, G. (1993). *Cell*, **74**, 505.
71. Pikaart, M. J., Recillas-Targa, F., and Felsenfeld, G. (1998). *Genes Dev.*, **12**, 2852.
72. Brinster, R. L., Allen, J. M., Behringer, R. R., Gelinas, R. E., and Palmiter, R. D. (1988). *Proc. Natl. Acad. Sci. USA*, **85**, 836.
73. Chaffin, K. E., Beals, C. R., Wilkie, T. M., Forbush, K. A., Simon, M. I., and Perlmutter, R. M. (1990). *EMBO J.*, **9**, 3821.
74. Tsien, J. Z., Chen, D. F., Gerber, D., Tom, C., Mercer, E. H., Anderson, D. J., *et al.* (1996). *Cell*, **87**, 1317.
75. Zinyk, D. L., Mercer, E. H., Harris, E., Anderson, D. J., and Joyner, A. L. (1998). *Curr. Biol.*, **8**, 665.
76. Akagi, K., Sandig, V., Vooijs, M., Van der Valk, M., Giovannini, M., Strauss, M., *et al.* (1997). *Nucleic Acids Res.*, **25**, 1766.
77. Rickert, R. C., Roes, J., and Rajewsky, K. (1997). *Nucleic Acids Res.*, **25**, 1317.
78. Wilkinson, D. G. (1992). In *In situ hybridization: a practical approach* (ed. D. G. Wilkinson), p. 75. Oxford University Press, Oxford.
79. Soriano, P. (1999). *Nature Genet.*, **21**, 70.
80. Michael, S. K., Brennan, J., and Robertson, E. J. (1999). *Mech. Dev.*, **85**, 35.
81. Grieshammer, U., Lewandoski, M., Prevette, D., Oppenheim, R. W., and Martin, G. R. (1998). *Dev. Biol.*, **197**, 234.
82. Fukushige, S. and Sauer, B. (1992). *Proc. Natl. Acad. Sci. USA*, **89**, 7905.
83. St-Onge, L., Furth, P., and Gruss, P. (1996). *Nucleic Acids Res.*, **24**, 3875.
84. Logie, C. and Stewart, F. (1995). *Proc. Natl. Acad. Sci. USA*, **92**, 5940.
85. Metzger, D., Clifford, J., Chiba, H., and Chambon, P. (1995). *Proc. Natl. Acad. Sci. USA*, **92**, 6991.
86. Feil, R., Brocard, J., Mascrez, B., LeMeur, M., Metzger, D., and Chambon, P. (1996). *Proc. Natl. Acad. Sci. USA*, **93**, 10887.

87. Kellendonk, C., Tronche, F., Monaghan, A.-P., Angrand, P.-O., Stewart, F., and Schutz, G. (1996). *Nucleic Acids Res.*, **24**, 1404.
88. Schwenk, F., Kuhn, R., Angrand, P.-O., Rajewsky, K., and Stewart, A. F. (1998). *Nucleic Acids Res.*, **15**, 1427.
89. Danielian, P. S., Mucino, D., Rowitch, D. H., Michael, S. K., and McMahon, P. (1998). *Curr. Biol.*, **8**, 1326.
90. Wang, Y., Krushel, L. A., and Edelman, G. M. (1996). *Proc. Natl. Acad. Sci. USA*, **93**, 3932.
91. Olsson, M., Campbell, K., and Turnbull, D. H. (1997). *Neuron*, **19**, 761.
92. Liu, A., Joyner, A. L., and Turnbull, D. (1998). *Mech. Dev.*, **75**, 107.
93. Senecoff, J. F., Bruckner, R. C., and Cox, M. M. (1985). *Proc. Natl. Acad. Sci. USA*, **82**, 7270.
94. McLeod, M., Craft, S., and Broach, J. R. (1986). *Mol. Cell. Biol.*, **6**, 3357.
95. Chou, T. and Perrimon, N. (1992). *Genetics*, **131**, 643.
96. Zhang, Y., Riesterer, C., Ayrall, A. M., Sablitzky, F., Littlewood, T. D., and Reth, M. (1996). *Nucleic Acids Res.*, **24**, 543.

3

Production of targeted embryonic stem cell clones

MICHAEL P. MATISE, WOJTEK AUERBACH and ALEXANDRA L. JOYNER

1. Introduction

The discovery that cloned DNA introduced into tissue culture cells can undergo homologous recombination at specific chromosomal loci has revolutionized our ability to study gene function in cell culture and *in vivo*. In theory, this technique, termed gene targeting, allows one to generate any type of mutation in any cloned gene. The kinds of mutations that can be created include null mutations, point mutations, deletions of specific functional domains, exchanges of functional domains from related genes, and gain-of-function mutations in which exogenous cDNA sequences are inserted adjacent to endogenous regulatory sequences. In principle, such specific genetic alterations can be made in any cell line growing in culture. However, not all cell types can be maintained in culture under the conditions necessary for transfection and selection. Over ten years ago, pluripotent embryonic stem (ES) cells derived from the inner cell mass (ICM) of mouse blastocyst stage embryos were isolated and conditions defined for their propogation and maintenance in culture (1, 2). ES cells resemble ICM cells in many respects, including their ability to contribute to all embryonic tissues in chimeric mice. Using stringent culture conditions, the embryonic developmental potential of ES cells can be maintained following genetic manipulations and after many passages *in vitro*. Furthermore, permanent mouse lines carrying genetic alterations introduced into ES cells can be obtained by transmitting the mutation through the germline by generating ES cell chimeras (described in Chapters 4 and 5). Thus, applying gene targeting technology to ES cells in culture affords researchers the opportunity to modify endogenous genes and study their function *in vivo*. In initial studies, one of the main challenges of gene targeting was to distinguish the rare homologous recombination events from more commonly occurring random integrations (discussed in Chapter 1). However, advances in cell culture and in selection schemes, in vector construction using isogenic DNA, and in the application of rapid screening procedures have made it possible to identify homologous recombination events efficiently.

Since there are numerous publications available that describe basic tissue culture techniques in this chapter we will only describe techniques specific for ES cells. We present methods for introducing DNA into ES cells for homologous recombination, selection and screening procedures for identifying and recovering targeted cell clones with an improved method utilizing 96-well tissue culture plates, as well as a simple method for establishing new ES cell lines. A video illustrating some of these techniques is available from Cold Spring Harbor Press (3). In addition, to provide a working overview of the steps involved in generating mouse chimeras using gene targeting in ES cells, we have summarized key steps from this and other chapters in this volume in flowchart form in *Figure 1*.

2. Propagation and maintenance of ES cells

A number of different ES cell lines are available (see Chapter 4, *Table 1*) and it is advisable to use the recommended growth media and suggested culture conditions for each particular cell line. Since we have used primarily the W4 (W. Auerbach and A. L. Joyner, unpublished) and R1 (4) (see Chapter 5) cell lines, we will describe methods for these lines.

2.1 General conditions

A well equipped tissue culture facility is essential and should include the following:

- laminar flow cabinet
- humidified incubator (5% CO_2/95% air, maintained at 37 °C)
- inverted microscope with a range of phase-contrast objectives (× 4 to × 25)
- binocular dissecting microscope with a transmitted light source
- liquid nitrogen storage tanks
- table-top centrifuge (capable of at least 270 *g*)
- water-bath
- freezers at –70 °C and –20 °C, and refrigerator at 4 °C

All tissue culture procedures described must be carried out under sterile conditions using sterile plasticware and detergent-free glassware. Water quality is very important; we recommend a Millipore Q filtration system (Millipore) for water purification or ultrapure water that can be purchased from commercial suppliers (e.g. Gibco).

After preparation, ES cell medium should be stored at 4 °C. Before adding to cultured cells, warm to 37 °C **briefly**. Medium should be supplemented with additional 2 μM glutamine after 10–12 days of storage as glutamine is unstable.

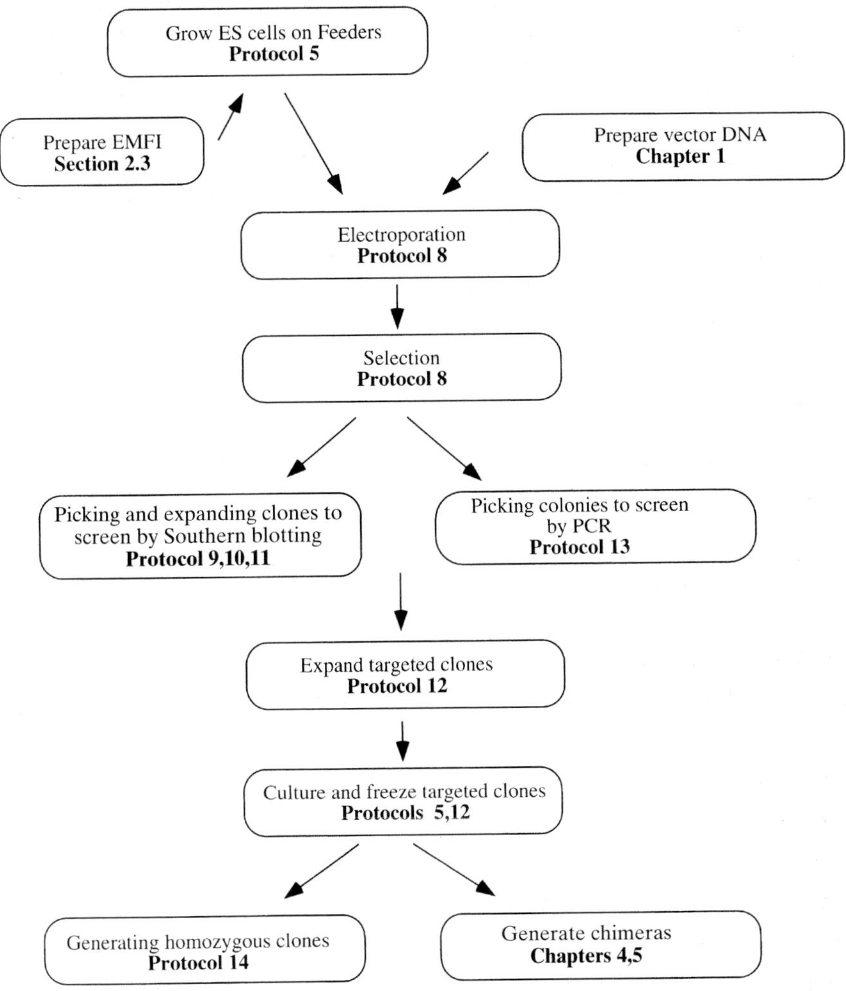

Figure 1. Schematic overview of steps involved in generating targeted ES cell clones.

2.2 ES cell culture media and solutions

W4 or R1 ES cell growth media should contain the following:

(a) Dulbecco's minimal essential medium (DMEM) with high glucose (Gibco, Cat. No. 11960–044).

(b) 0.1 mM non-essential amino acids (100 × stock) (Gibco, Cat. No. 11140–019), aliquoted and stored at 4 °C.

(c) 1 mM sodium pyruvate (100 × stock) (Gibco, Cat. No. 11360–013), aliquoted and stored at 4 °C.

(d) 10^{-4} M β-mercaptoethanol (Sigma, Cat. No. M-6250), aliquoted and stored at –20 °C.
(e) 2 mM L-glutamine (100 × stock) (Gibco, Cat. No. 25030–016), aliquoted and stored at –20 °C.
(f) 15% fetal bovine serum (FBS) (heat inactivated, 56 °C, 30 min.).
(g) Penicillin and streptomycin (final concentration 50 μg/ml each) (Gibco, Cat. No. 600–564AG), aliquoted and stored at –20 °C. (NB: not essential; see discussion Chapter 4.)
(h) 1000 U/ml LIF (Gibco, Cat. No. 13275–019) (see below).

ES cells can be maintained in an undifferentiated state by culturing them on feeder cell layers or on gelatinized plates (*Protocol 5*) with the addition of leukaemia inhibitory factor (LIF) (5–8). ES cells maintained without feeder layers in the presence of recombinant LIF are able to remain pluripotent, since in chimeras some cell lines can contribute to all somatic tissues, as well as to the germline (4, 9). ES cell lines can also be established on gelatin, although the few lines that have been established under such conditions have a high degree of chromosomal abnormalities (9). This suggests that ES cell lines established in LIF can not maintain their pluripotency when maintained for long periods without feeders. We therefore do not recommend extended growth of ES cells in LIF alone without feeders.

Recombinant LIF can either be purchased from commercial suppliers (e.g. Gibco), or obtained from supernatants of COS cells, or purified from bacteria transfected with recombinant LIF constructs (10). If such supernatants are used, each batch should be carefully tested for its ability to maintain ES cells in an undifferentiated state as in *Protocol 6*. Commercial suppliers of recombinant LIF usually recommend a concentration of 1000 U/ml. However, each ES cell line has particular requirements that should be used. We use 1000 U/ml LIF (Gibco) for culture of W4 and R1 ES cells on both gelatin or feeders.

The quality of FBS is very important for the maintenance of ES cells. We recommend that different batches be tested from different suppliers (see *Protocol 6*) for their ability to support growth of pluripotent ES cells. Some suppliers (e.g. Gemini, Gibco) sell pre-tested serum. Suitable sera batches should then be ordered in large quantities and stored at –80 °C for up to two years or –20 °C for 6–12 months.

We use PBS without Ca^{2+} and Mg^{2+} for washing embryos prior to preparation of primary embryonic fibroblast cells and for rinsing ES cell layers before trypsinization.

For routine passaging of ES cells we disaggregate them by using 0.05% trypsin (final concentration) dissolved in Tris–saline/EDTA (TE).

2.3 Production of fibroblast feeder layers

For long-term culture and maintenance, pluripotent ES cells should be grown on monolayers of mitotically inactivated fibroblast cells (1, 11). Primary embryonic fibroblast (EMFI) cells (11) or the STO fibroblast cell line (1, 2, 12) are the most commonly used feeder layers. Most ES cell lines have been established on EMFI or STO cells and it is recommended to maintain each ES cell line on the feeder type on which it was originally established. Many ES cell lines can be cultured on gelatinized tissue culture plates (see *Protocol 4*, step 1) in the presence of LIF (5, 6) for one passage before electroporation of targeting vector DNA or during selection following electroporation (*Protocol 8*).

The W4 ES cell line was established on EMFI feeders and so we only use EMFI cells for feeder layers (11). The major disadvantage of EMFI cells is their short life span (15–20 cell divisions), which necessitates that new stocks of frozen cells be made on a regular basis. Since each batch of EMFI cells that are prepared will have a different quality, they should be tested prior to usage for doubling time and the number of cell divisions before senescence. We have found no obvious difference between EMFI cells made from CD1 or Swiss Webster outbred, C57BL/6J inbred, or transgenic mice expressing the bacterial *neomycin phosphotransferase* gene (*neo*).

The following 2 protocols are for laboratories with a high volume of ES cell work that require fresh feeder plates regularly (each week). An alternative for small volume labs is described in *Protocol 3*.

Protocol 1. Preparation of EMFI cell stocks

Equipment and reagents

- 14–16 days post-coitus (dpc) pregnant mice
- Sterile dissecting instruments (washed in 70% ethanol)
- Mixture of 3 mm and 5 mm diameter glass beads (Conning), 5 ml of each, autoclaved in an Erlenmeyer flask
- Small (1–2 inch) stir bars (autoclaved)
- 50 ml Falcon tubes (sterile)
- 150 mm tissue culture (TC) dishes
- 100 mm Petri dishes (or TC dishes)
- Freezing vials (e.g. Nunc, Cat. No. 366 656)
- PBS without Ca^{2+} and Mg^{2+} (Speciality Media, Cat. No. BSS-1006B)
- 0.05% trypsin in saline/EDTA (Gibco, Cat. No. 25300–054).
- DMEM + 10% FBS
- DNase I at 100 μg/ml (10 mg/ml frozen stock in PBS) (Sigma, Cat. No. D4527)
- Trypan blue (Flow Labs, Cat. No. 16–910–49)
- 1 × freezing medium: 25% FBS, 10% DMSO in DMEM

Method

1. Sacrifice the pregnant mouse when embryos are about 15 dpc. Moisten the belly with 70% ethanol and dissect out the uterus (see Chapter 4).
2. Transfer the uterus to a 100 mm Petri dish containing PBS, and dissect the embryos away from the uterus and all the membranes (see Chapter 6). Transfer the embryos into a new dish containing PBS.
3. Remove heads and all internal organs (liver, heart, kidney, lung, and intestine).

Protocol 1. *Continued*

4. Wash eight to ten carcasses in a 50 ml Falcon tube in 50 ml PBS at least twice, to remove as much blood as possible.
5. Mince the carcasses in 3 ml trypsin/EDTA into cubes of about 1 mm in diameter with watchmaker's scissors in a container prepared by cutting ~ 1 inch from the bottom off a 50 ml Falcon tube with a heated scalpel.
6. Add trypsin/EDTA to a final volume of 20 ml and transfer into a 50 ml Falcon tube.
7. Add 200 μl DNase I (10 mg/ml stock) per 20 ml. Put sterile stirring bar and 5 ml of glass beads inside each tube. Place tubes in rack on a stirring plate and incubate at 37 °C.
8. Add an additional 10 ml trypsin/EDTA per tube and stir for another 30 min. Repeat again.
9. Decant cell suspension into a new 50 ml tube containing 3 ml of FBS (to stop trypsin activity).
10. Rinse glass beads twice with 3 ml of DMEM + 10% FBS, and add to same 50 ml Falcon tube. Pellet. Centrifuge cells at 270 *g* for 5 min. Resuspend the pellet in 10 ml DMEM + 10% FBS. If the cell pellet is viscous, add another 200 μl DNase I (10 mg/ml stock) and incubate for a further 30 min at 37 °C.
11. Combine cells from two tubes and count viable nucleated cells using trypan blue:
    (a) Remove 100 μl into an Eppendorf tube.
    (b) Add ~ one drop of trypan blue.
    (c) Pipette sample of stained cells (may need to dilute further) onto a haemocytometer and count stained nuclei.

    You should get about 5×10^7 to 10^8 viable cells from ten embryos.
12. Plate 5×10^6 **nucleated** cells per 150 mm tissue culture dish in 25 ml DMEM + 10% FBS, and culture overnight at 37 °C (5% CO_2 in air).
13. Change the medium after 24 h to remove cell debris.
14. After two to three days of culture the EMFI cells should form a confluent monolayer. Trypsinize each plate and re-plate onto five further plates (150 mm).
15. When the plates are confluent (usually after two or three days) freeze all the cells from each plate in one freezing vial in 1 ml of 1 × freezing medium and store at –70 °C for one day. Transfer the vials to liquid nitrogen.

Protocol 2. Preparation of EMFI feeder layers

Equipment and reagents

- Frozen vials of primary embryo fibroblasts (*Protocol 1*)
- Tissue culture dishes (Nunc)
- PBS without Ca^{2+} and Mg^{2+}
- 0.05% trypsin in saline/EDTA
- DMEM + 10% FBS
- Mitomycin C (e.g. Sigma, Cat. No. M-0503): stock 1 mg/ml in PBS stored in dark at 4°C and used within two weeks (mitomycin C is toxic; wear gloves and use caution when handling)

Method

1. Thaw a frozen vial of EMFI cells quickly at 37 °C.
2. Add cells to10 ml DMEM + 10% FBS and centrifuge (270 *g*, 5 min).
3. Decant supernatant, resuspend the cell pellet gently in 10 ml DMEM + 10% FBS, and split onto five 150 mm plates each containing a total of 25 ml DMEM + 10% FBS. Mix well.
4. Incubate cells at 37 °C, 5% CO_2.
5. When the cells form a confluent monolayer (approx. three days) each plate should either be:
   (a) Trypsinized, split onto five additional 150 mm dishes, and grown until they form a confluent monolayer (approx. three days), or
   (b) Directly treated with mitomycin C to inhibit cell growth and division.[a]
6. Remove the medium from the confluent plates and add 10 ml DMEM + 10% FBS containing 100 μl mitomycin C (1 mg/ml stock). Swirl plates to ensure an even distribution of medium.
7. Incubate cells at 37 °C, 5% CO_2 for 2–2.5 h.
8. Wash the monolayer of cells twice with 10 ml PBS per dish.
9. Add 5 ml trypsin/EDTA to each plate.
10. Incubate 37 °C, 5% CO_2 until the cells come off the plate (5–10 min).
11. Add 10 ml DMEM + 10% FBS to each plate and break any cell aggregates by gently pipetting.
12. Centrifuge cells (270 *g*, 5 min) and resuspend the pellet in DMEM + 10% FBS.
13. Count the cells and dilute to a concentration of 2×10^5 cells/ml.
14. Plate the cells immediately onto tissue culture dishes containing DMEM + 10% FBS. See *Table 1* for the appropriate cell densities and volumes of medium for different plate sizes.
15. Allow feeders to attach at least 2 h, but preferably overnight, before adding ES cells.

Protocol 2. *Continued*

16. Change the medium to ES cell medium (see Section 2.2) immediately before adding ES cells. Mitomycin C treated EMFI feeders can be used for up to seven days with medium changes every three to four days.

[a] An alternative to steps 6–12 is to trypsinize the cells, rinse them, resuspend them in media, and treat them with 6000–10 000 rads of gamma irradiation.

Table 1. Volume of EMFI cells plated on different size tissue culture dishes

Plate diameter (cm)	Area (cm²)	Relative size	ml medium per plate[a]	Number of feeder plates from confluent 150 mm feeder plate
14	150	1	25	1
9	63	1/2.4	10	3–4
6	20	1/7.5	5	6–8
3.5	9.6	1/15.6	2	18–20
1.5	1.8	1/80	0.5/well	80 wells
96-well plate	0.45	1/350	200 μl/well	3–4 plates

[a] EMFI cells at 2×10^5 cells/ml or SNL STO cells at $2.5–3.5 \times 10^5$ cells/ml.

Protocol 3. Preparation of a stock of mitomycin C treated EMFI cells (for small volume laboratories)

Equipment and reagents

- See *Protocol 2*
- 1 × freezing medium: 25% FBS, 10% DMSO in DMEM
- Freezing vials (e.g. Nunc, Cat. No. 366 656)
- 0.1% gelatin (optional)

Method

1. Thaw a frozen vial of EMFI cells quickly at 37 °C.
2. Add cells to10 ml DMEM + 10% FBS and centrifuge (270 *g*, 5 min).
3. Decant supernatant, resuspend the cell pellet gently in 10 ml DMEM + 10% FBS, and split onto five 150 mm plates each containing a total of 25 ml DMEM + 10% FBS. Mix well.
4. Incubate cells at 37 °C, 5% CO_2.
5. When the cells form a confluent monolayer (approx. three days) each plate should be trypsinized, split onto five additional 150 mm dishes, and grown until they form a confluent monolayer (approx. three days).
6. Remove the medium from the confluent plates and add 10 ml DMEM

+ 10% FBS containing 100 μl mitomycin C (1 mg/ml stock). Swirl plates to ensure an even distribution of medium.

7. Incubate cells at 37 °C, 5% CO_2 for 2–2.5 h.
8. Wash the monolayer of cells twice with 10 ml PBS per dish.
9. Add 5 ml trypsin/EDTA to each plate.
10. Incubate 37 °C, 5% CO_2 until the cells come off the plate (5–10 min).
11. Add 10 ml DMEM + 10% FBS to each plate and break any cell aggregates by gently pipetting.
12. Centrifuge cells (270 *g*, 5 min) and resuspend the pellet in freezing medium. Freeze all the cells from each plate in one freezing vial in 1 ml of 1× freezing medium and store at –70 °C for one day. Transfer the vials to liquid nitrogen.
13. To make feeder plates thaw a frozen vial of mitomycin C treated EMFI cells quickly at 37 °C.
14. Add cells to10 ml DMEM + 10% FBS and centrifuge (270 *g*, 5 min).
15. Decant supernatant and resuspend the cell pellet gently in 30 ml DMEM + 10% FBS.
16. Seed cells directly into tissue culture plates. Depending of the size of the plates required put 10 ml/100 mm plate, 5 ml/60 mm plate, or 1.5 ml/35 mm plate.
17. Allow feeders to attach preferably overnight, before adding ES cells or at least 2 h if using gelatinized plates (*Protocol 4*, step 1).
18. Change the medium to ES cell medium (Section 2.2) prior to adding ES cells.

The STO cell line is a thioguanine/oubain-resistant subline of SIM mouse fibroblasts (13) that was established by Dr A. Bernstein (Mount Sinai Hospital, Toronto). This cell line was first found to support growth of pluripotent embryonal carcinoma (EC) cell lines (12, 14) and later to support growth of ES cell lines (1, 2). The STO cell line can be obtained from the American Type Culture Collection, but it is best to obtain a frozen aliquot from a laboratory using STO cells as a source of feeders for ES cells. This is because with many passages, the STO cell line changes and loses its ability to support optimal ES cell growth. For this reason, when you obtain a frozen vial of STO cells, it should be thawed, immediately expanded, and 20–40 vials containing 1 ml of cells frozen away (2×10^6 cells/ml in 1 × freezing medium). A new vial should then be thawed onto one 100 mm dish every four to six weeks and expanded. Rl cells were established on the SNL STO subline made by A. Bradley (15), although they can also be grown on EMFI feeders.

Protocol 4. Preparation of SNL STO feeder layers

Equipment and reagents

- SNL STO cells
- Tissue culture dishes
- PBS without Ca^{2+} and Mg^{2+}
- 0.05% trypsin in saline/EDTA
- DMEM + 10% FBS
- 0.1% gelatin (BDH, Cat. No. 440454B), autoclaved in H_2O, and stored at 4°C

Method

1. Prepare gelatinized plates (*in a hood*) as follows:
   (a) Cover the surface of the dishes completely with 0.1% gelatin solution (~ 5 ml for 10 cm plates).
   (b) Leave the dishes for approx. 2 min at room temperature.
   (c) Aspirate the gelatin solution and allow to dry before using the plates.
2. Thaw a vial of SNL STO cells and expand by passage and growth at 37°C, 5% CO_2 to the required number of 150 mm gelatinized dishes (see below).
3. Remove half of the medium (12.5 ml) from the confluent cultures and add 30 μl of mitomycin C (1 mg/ml stock) to each plate. Swirl each plate gently to mix the medium and to make sure the whole surface is covered.
4. Incubate the cells for about 2 h (37°C, 5% CO_2).
5. Follow *Protocol 2*, steps 8–12.
6. Count the cells and dilute to a concentration of 2.5–3.5 × 10^5 cells/ml.
7. Plate the feeder cell suspension on to the size and number of gelatinized plates needed (see *Table 1* for recommended cell volumes).

We passage and prepare mitomycin C treated feeder layers twice per week. In a typical schedule for EMFI cells, a vial of cells is thawed on Tuesday onto five 150 mm plates, one or two of these plates are passaged on Friday at a 1:5 dilution, and the medium is changed on the remaining plates. On the following Monday, feeder layers are made from the cells passaged the previous Friday and the remaining EMFI plates passaged (1:5). The cells passaged on Monday are then used on Thursday (after three days) to prepare feeder layers. SNL STO cells are typically passaged on Mondays, onto 150 mm dishes (8.75 × 10^5 cells/plate) for mitomycin C treatment on Thursday, and onto 100 mm dishes (3.5 × 10^5 cells/plate) for further expansion. On Thursday, the cells on the 100 mm dishes are passaged onto 150 mm dishes (8.75 × 10^6 cell/plate) for making feeder layers on Monday, and onto 100 mm dishes (3.5 × 10^6 cells/plate) for further passaging.

2.4 Culturing ES cells

In order to maintain the totipotency of ES cells, stringent culture conditions are required. Variant ES cell clones can arise spontaneously during culture. Some, but not all, of these variants will have an obvious abnormal karyotype and most will not likely contribute to the germline. Some variants will have a growth advantage by having a shorter doubling time or better plating efficiency, and thus could quickly outgrow the normal cells and become the predominant cells in the culture if stringent culture conditions are not used. We therefore recommend that in general you should follow precisely the ES cell culture protocols, observe cell morphology under a microscope, use a rich medium, and passage the cells frequently (every two days) with a minimum dilution of cells (1:5 to 1:7), depending on their growth rate. In general, ES cells are passaged when they have just reached subconfluence (i.e. individual clumps of cells are nearly touching one another) at a dilution which will permit them to be cultured for up to 48 h until this density is again reached (*Figure 2A*). If the cells require more frequent passage or a higher dilution, or their morphology suddenly changes, then an abnormal variant might have taken over the culture. Karyotype analysis of ES cells (the parental line and subclones) can be used to monitor gross chromosomal changes, but a

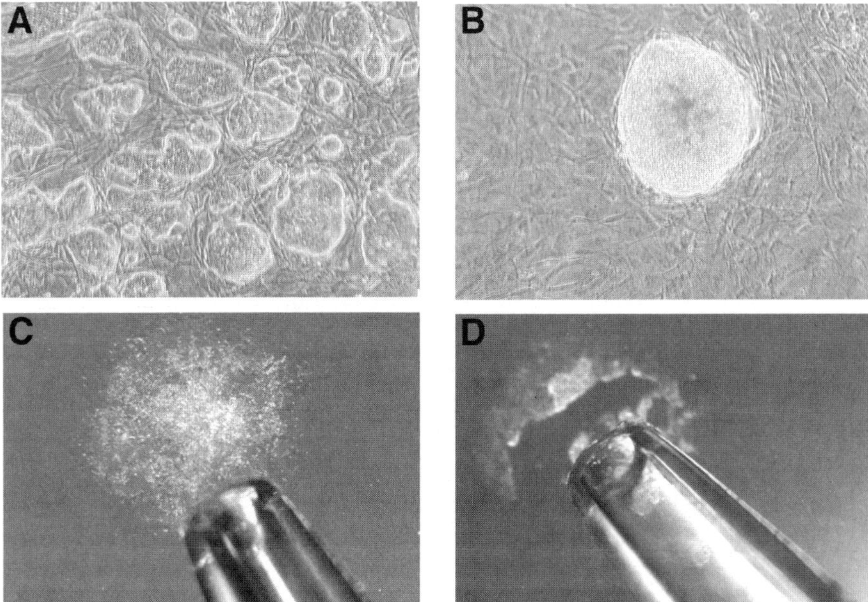

Figure 2. ES cells in culture. (A) Subconfluent cells on EMFI feeders. Culture is ready to be passaged when individual colonies grow close together. (B) ES cell morphology on feeder layers. (C) ES cell clone on gelatin prior to picking. Note size relative to plastic pipette tip. (D) Picking ES cell clone by scraping from bottom of dish with pipette tip.

seemingly normal karyotype does not guarantee germline transmission. Do not use ES cells that have gone through a 'crisis'; i.e. have suddenly changed their morphology or begun to differentiate extensively (see *Figure 3*).

Another important factor to keep in mind is that all cells grown in culture can become infected with mycoplasma. Such infections do not always have an obvious effect on cell growth or morphology. However, mycoplasma can cause chromosome damage and reduce the efficiency of obtaining ES cell chimeras. Therefore, all ES cell lines and sublines as well as STO and EMFI feeder stocks should be tested for mycoplasma after two weeks growth in the absence of antibiotics (see Chapter 4, Section 2.1.1 for further discussion). If you are planning to make your chimeras in an SPF (specific pathogen-free) facility, ES cells also should be mouse antibody production (MAP) tested (available from Charles River Lab, 800–522–7287) as well.

Protocol 5. Growth of ES cells on feeders or gelatin-coated plates

Equipment and reagents

- 100 mm dishes containing feeder layers (*Protocol 2* or *3*)
- ES cell medium briefly pre-warmed to 37 °C
- 0.05% trypsin in saline/EDTA
- PBS without Ca^{2+} and Mg^{2+}
- Gelatin, 0.1% solution in water, autoclaved

Method

1. Quickly thaw one vial of frozen ES cells (as prepared in *Protocol 7*) by warming in your hand or at 37 °C, and transfer cells to a 12 ml tube containing 10 ml of ES cell medium before all the ice has disappeared.
2. Centrifuge at 270 *g* for 5 min.
3. Aspirate supernatant and resuspend pellet in 10 ml ES cell medium and plate on a 100 mm dish with a feeder layer.
4. Change the medium the next day by swirling the medium in dish to collect debris, then aspirate. Add fresh medium gently to the side of the plate so that the feeder layer is not disturbed.
5. On the second day (cells should be just subconfluent, see *Figure 2A*) wash cells twice with PBS and add 2 ml trypsin/EDTA.
6. Incubate for approx. 5 min at 37 °C until cells begin to come off the plate.
7. Gently agitate the plate and observe under a microscope. When trypsinization is complete, cells should detach as small clumps, not as a single sheet or single cells.
8. Add 5 ml ES cell medium and gently pipette the cells up and down to break cell clumps. If the cells are sticky, gently pipette before adding medium.

9. Transfer to a sterile 12 ml tube and pellet cells in centrifuge (5 min, 270 *g*).
10. Aspirate supernatant and gently resuspend cell pellet in 5–7 ml medium (depending on desired dilution).
11. Add 1 ml of the cell suspension (about 2–5 $\times$ 10^6 cells) to a fresh 100 mm feeder layer dish containing 9 ml ES cell medium. Disperse ES cells evenly by pipetting gently and rocking the plate prior to incubating at 37 °C, 5% CO_2.
12. Change the medium the next day and pass cells every second day as described above.

For the propagation of ES cells on gelatinized plates (not recommended for newly thawed ES cells), follow the same protocol, substituting gelatin-coated plates for plates with EMFI cells (see *Protocol 4*, step 1). It should be noted that ES cells grown in LIF on gelatin have a more flattened morphology than cells grown on feeders (compare *Figure 2B* to 2C, *3A*). However, we have found that when the cells are returned to feeder layers they revert to their original, round morphology after one passage.

2.5 Testing serum batches

To test serum batches, three concentrations of FBS (10%, 15%, 30%) for each test lot should be compared to a serum batch which is known to give good results (with respect to plating efficiency, colony morphology, and the toxicity to ES cell culture). The plating efficiency [(number of established colonies/number of cells seeded) $\times$ 100] of ES cells plated at a low density (10^3 to 10^4 cells/100 mm dish) should be at least 15%, and the colonies should retain an undifferentiated morphology for a few days of growth once the colonies are visible under a microscope about four days after plating. The plating efficiency should increase with the concentration of FBS, although generally there is a larger difference between 10% and 15% than between 15% and 30% FBS.

A serum batch is judged suitable if the plating efficiency (PE) in 15% FBS is 15–20% and slightly higher at 30%. If the PE is lower in 30%, this indicates toxicity. The cell morphology should also be very good when the colonies become visible and for the following three to four days. The serum can be further checked by growing ES cells in ES cell medium (15% FBS) for at least five passages (check the morphology and cell density after each passage), and by testing that the plating efficiency is similar (10–15%) on gelatinized plates with LIF-containing medium.

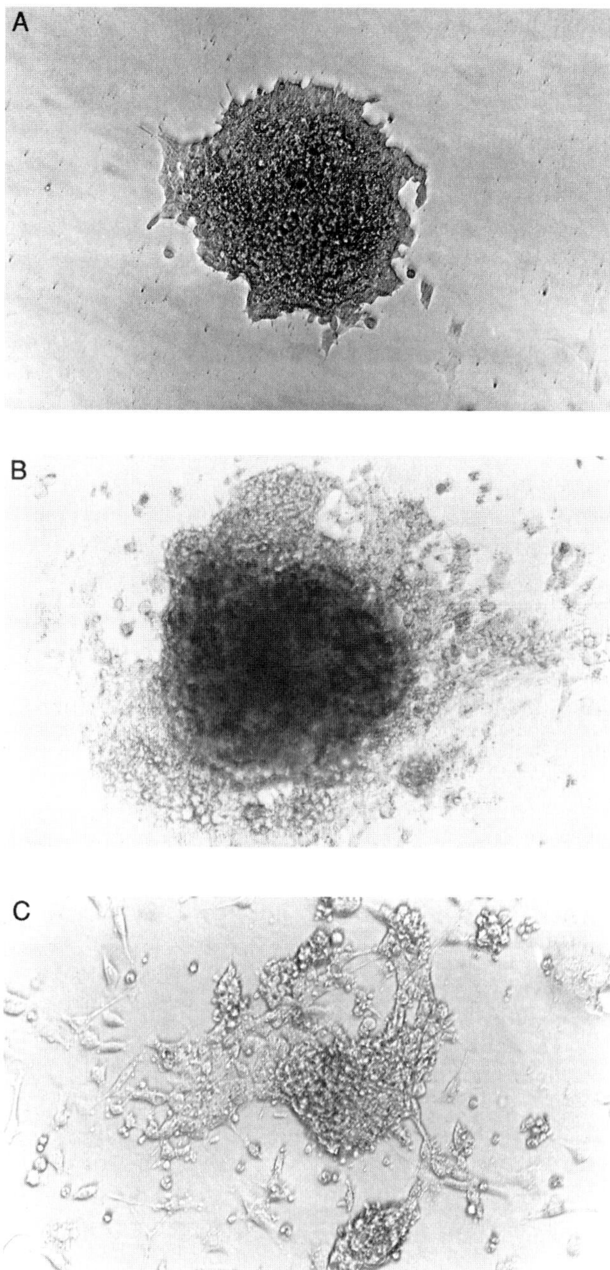

Figure 3. Cell morphology of ES cell colonies growing on a gelatinized plate in LIF. R1 cells were electroporated as in *Protocol 8* and grown in selective medium for eight days. Colonies have a number of different morphologies due to differentiation. (A) Undifferentiated ideal ES cell colony (see also *Figure 2C*). (B, C) Various differentiated colonies.

Protocol 6. Screening batches of serum for optimal ES cell growth

Equipment and reagents

- FBS (control and test batches) (we have had good results with HyClone and Gemini serum)
- Methylene blue stain: 3.3 g methylene blue, 1.1 g basic fuchsin in 1 litre of methanol
- 60 mm dishes containing feeder layers (*Protocol 2* or *3*) in ES medium (Section 2.2) with 10%, 15%, and 30% FBS
- PBS without Ca^{2+} and Mg^{2+}

Method

1. Plate 10^3 single ES cells on each 60 mm dish containing a feeder layer in 10%, 15%, and 30% FBS medium. Prepare duplicates for each serum concentration.
2. Let the colonies grow and change medium every second day. Note cell morphology.
3. After seven to ten days of culture, rinse the plate with PBS.
4. Add 2 ml methylene blue to each plate for 5 min at room temperature (RT). Rinse the plates several times with water.
5. Examine the plates with respect to colony size, morphology, and number. Note that the morphology is clearer before staining.

2.6 Freezing, storage, and thawing of ES cell lines

ES cells can be frozen, like other tissue culture cells, in freezing medium containing 25% FBS and 10% DMSO as a cryoprotectant. As a general rule, freeze cells slowly and thaw them quickly. For long-term storage (indefinitely), cells should be kept under liquid nitrogen and for short-term storage (up to six months) they can be kept in a –70°C freezer. ES cells freeze best at a cell density of 5–10 × 10^6 cells/ml of 1 × freezing medium. This corresponds to approximately three tubes of 1 ml each per confluent 100 mm dish. Cells can be frozen from different plate sizes using less freezing medium (60 mm dish 1 ml, 35 mm dish 0.5 ml, and 12 mm dish 0.25 ml). It is important to thaw the cells rapidly and remove the DMSO-containing medium as soon as possible (see *Protocol 5*). Also, the cells should be re-plated onto a dish that is one size larger than the one they were cultured on prior to freezing. Freezing plus thawing is counted as one passage.

Protocol 7. Long-term freezing of ES cell stocks

Equipment and reagents

- Round-bottom freezing vials (1–2 ml, e.g. Nunc)
- Styrofoam containers, pre-cooled to –70 °C (e.g. shipping containers from Boehringer Mannheim)
- Liquid nitrogen tank
- –70 °C freezer
- 1 × freezing medium to be made up fresh (final concentrations: 25% FBS, 10% DMSO in DMEM) and kept on ice
- ES cell medium (Section 2.2)
- Slow-cool freezing container (optional) pre-cooled to 4 °C (Nalgene, Cat. No. 5100–0001)

Method

1. Trypsinize cells (see *Protocol 5*) from a 100 mm dish (about 2 or 4 × 10^7 RI or W4 cells respectively).
2. Pellet cells by centrifugation (270 *g*, 5 min) and resuspend in 3 ml pre-cooled freezing medium.
3. Immediately aliquot 1 ml of the cell suspension into freezing vials on ice.
4. Immediately transfer vial to a Styrofoam box pre-cooled to –70 °C, or slow-cool container, and then into a –70 °C freezer. It is important to work quickly.
5. After 24 h transfer the tubes on dry ice to liquid nitrogen.

3. Electroporation of DNA into ES cells

The most common method for introducing DNA into ES cells is by electroporation. This technique has the advantage that it is technically relatively simple. However, one disadvantage is that the stable transformation efficiency is low, ranging from 10^{-3} to 10^{-4}, and thus necessitates the use of selectable marker genes in the targeting vectors (see Chapter 1 for details). However, advances in selection schemes and the simplicity of the electroporation procedure currently make it the method of choice.

3.1 Standard electroporation conditions

DNA can be electroporated into ES cells by application of a high voltage electrical pulse to a suspension of cells and DNA (16). After application of this pulse, the DNA passes through pores in the cell membrane. This procedure results in the death of about 50% of the cells with conditions that give optimal transfection efficiency. The parameters that influence the efficiency and cell viability are the voltage, the ion concentration, the DNA concentration, and the cell concentration. Ideally these parameters should be determined in advance by measuring cell viability and transfection efficiency. The following protocol is based on our experience with the W4 and R1 ES cell lines, and gives a stable transfection efficiency of 10^4 *neo*^{r} colonies/cells electroporated with a PGK *neo* vector. Before starting an electroporation with a new gene targeting experiment, it is advisable to test that your screening strategy works well with DNA isolated from ES cells in 96-well plates (*Protocol 11*). In particular, be sure that the screening can be done with a probe taken from the genomic region outside the vector arms of homology (see Chapter 1).

After electroporation, ES cells are plated onto gelatinized plates for drug selection. *Neo*-resistant feeders have also been used for this purpose. Such feeders can be made from β-Microglobulin knock-out mice purchased from Jackson Labs. However, since selection usually takes place over an eight to ten day period, EMFI cells can become unhealthy and detach from the plate, possibly bringing ES cells with them. Therefore, it is best to gelatinize the feeder plates, or preferably use gelatinized plates without feeders for post-electroporation selection with W4 and R1 cells. In addition, the G418 concentration required to select cells on a feeder layer is higher than on plates without feeders.

Each electroporation cuvette of cells can be plated onto a single 100 mm plate, or split onto 1.5 (R1) or 2 (W4) plates for selection. This is empirically determined, and is based on the transfection efficiency with the goal being to achieve the maximum number of well spaced *neo*-resistant clones on each plate, usually ~ 50 colonies.

Protocol 8. Standard electroporation of ES cells

Equipment and reagents

- Electroporation apparatus: Bio-Rad Gene Pulser (Cat. No.165–2105) and Capacitance Extender (165–2107)
- Electroporation cuvettes (Bio-Rad, Cat. No. 165–2088)
- ES cell medium (Section 2.2)
- PBS without Ca^{2+} and Mg^{2+}
- 100 mm gelatinized plates (*Protocol 4*)
- Geneticin (G418) (e.g. Gibco, Cat. No. 860–18111J)
- Gancyclovir (Cytovene, Syntex) (optional)
- Plates of growing, low passage (8–12) ES cells
- 0.05% trypsin in saline/EDTA

Method

A large scale preparation of gene targeting vector DNA should be prepared (at least ~ 200 μg for five cuvettes) and purified over a CsCl gradient. We have found that DNA isolated using commercial columns can be toxic to ES cells. After restriction enzyme digestion to linearize the gene targeting vector, the DNA should be extracted twice with phenol:chloroform, ethanol precipitated by adding 2 vol. 100% ethanol, washed twice in a large volume (0.5–1 ml) of 70% ethanol, and resuspended in sterile PBS (at a concentration of 1 μg/μl) in a laminar flow hood.

Protocol 8. *Continued*

1. On the second day ($\leqslant$ 36 h) after passaging recently thawed ES cells (cultured on EMFI for two to three passages), trypsinize cells, as in *Protocol 4*, but trypsinize for longer to obtain single cells.
2. After pipetting the cells gently in 3 ml trypsin (for a 100 mm plate) to break up the clumps, add 7 ml medium and pipette gently up and down again.
3. Pre-plate the cells by incubating them in the dish for 15–20 min (37 °C, 5% CO_2) to allow feeder cells to re-attach to the plate.
4. Harvest the ES cells by carefully mixing and withdrawing medium, and transfer cells to a 50 ml Falcon tube, combining the cells from two to five plates.
5. Pellet the cells for 5 min at 270 *g*.
6. Aspirate the supernatant and resuspend the cells in a minimal volume of PBS (~ 1 ml/100 mm plate starting culture). Keep cells on ice.
7. Count the ES cells using a haemocytometer and adjust cell concentration to 7–10 $\times$ 10^6 cells/ml with sterile PBS (the yield should be approx. 1 or 2.5 $\times$ 10^7 R1 or W4 cells/100 mm dish, respectively, at confluence).
8. Mix 0.8 ml of the cell suspension with 25–40 μg of linearized vector DNA (per cuvette), and transfer into an electroporation cuvette which has been pre-cooled on ice.
9. Set up the electroporation conditions in advance: 240 V, 500 μF for the Bio-Rad Gene Pulser using the Capacitance Extender.
10. Transfer each cuvette into the cuvette holder with electrodes facing the output leads. Deliver the electric pulse.
11. Remove each cuvette from the cuvette holder and place on ice for 20 min.
12. Remove the cells from each cuvette and dilute in appropriate volume of ES cell medium (15–20 ml/cuvette). Cells from several cuvettes can be combined.
13. Transfer 10 ml resuspended ES cells (mixed in total volume needed) onto each 100 mm gelatinized plate (using 1.5–2 plates per cuvette at final dilution). Disperse cells evenly by rocking the plate.
14. Change medium the next morning.
15. Two days after the electroporation, begin drug selection (for positive-negative selection, add G418 to 150 (for R1) or 200 (for W4) μg/ml[a] and fresh gancyclovir to 2 mM. Gancyclovir can be delayed for 24 h).
16. Change the medium every day ($\leqslant$ 24 h between changes) if working with gancyclovir, since it can break down and become toxic to the cells, otherwise every two days when using G418 selection only.

Widespread cell death should be apparent after two to three days of drug selection.

17. After about six to eight days of selection, individual drug-resistant colonies should have appeared (see *Figure 2C, 3A*) and be large enough to pick and subclone for screening by Southern blot analysis (Section 4.1) or by PCR (Section 4.2) and to freeze for storage.

[a] For different ES cell lines or for cells growing on G418^r feeder layers, the appropriate concentration of G418 must be determined by performing kill curves as follows. Plate half the number of cells used per plate after electroporation and add G418 after one day at different concentrations. Identify the lowest concentration that kills all the cells within ten days. The final G418 concentration will depend on the specific cell lines, batch of G418, and whether feeder layers or gelatinized plates are used.

4. Screening single colonies for homologous targeting events

4.1 Screening for homologous targeting events by Southern blot analysis

The use of isogenic DNA vectors has greatly increased the gene targeting frequency (17, 18). Using this type of vector, targeting frequencies from 1/10 to 1/100 seem to be quite common using a positive-negative selection scheme. With such high targeting frequencies and with improved methods in ES cell culture techniques and DNA isolation methods, screening by Southern blot analysis has become a straightforward procedure (see *Figure 4*).

On the 10–12th day after electroporation, ES cell clones should be reaching picking size (1.5–2 mm) (see *Figure 2C*). Picking usually occurs over a two to three day period. If positive-negative selection is used, gancyclovir selection is complete at the beginning of this time; cells can be grown in G418-containing medium only during picking. See *Figure 4* for schematic of procedure.

Protocol 9. Growing drug resistant clones (modified from ref. 19)

Equipment and reagents

- Dissecting microscope
- 96-well tissue culture plates with flat (Nunc, Cat. No. 167008) and V-shaped (Costar, Cat. No. 3894) bottom
- p20, p200 Pipetmen, sterile tips
- ES cell medium
- Multichannel pipettor and sterile reservoirs for 96-well plates
- 0.05% trypsin in saline/EDTA
- 0.1% gelatin (autoclaved)
- PBS without Ca^{2+} and Mg^{2+}

Method

1. Prepare two sets of 96-well plates, one containing 35 μl of trypsin/EDTA and one gelatin coated.

Protocol 8. *Continued*

2. Circle all visible ES cell colonies that will be picked on the bottom of each plate with a marker by holding plate up to light or using an inverted microscope.
3. Wash ES cell containing plates twice with PBS. Leave cells in PBS during picking. After picking, replace PBS with ES medium and return plate to incubator if additional, smaller colonies will be picked on subsequent days.
4. Pick the 1.5–2 mm drug-resistant ES cell clones that form after about eight days of selection with a drawn-out Pasteur pipette or a yellow Gilson tip under a dissecting microscope. Pick colonies of similar size, or take an equivalent portion of large colonies (*Figure 2D*).
5. Transfer each colony in a minimal volume of PBS into one well of a V-shaped 96-well plate containing 35 μl of 0.05% trypsin in saline/EDTA at room temperature. Usually every other row is used and 48 colonies picked at one time (*Figure 4*).
6. After picking 48 clones, incubate plate for approx. 10 min at 37 °C, 5% CO_2, until cell clumps break up.
7. Stop the reaction by adding 100 μl of ES medium containing G418 to each well using multichannel pipettor.
8. Mix cells gently by pipetting up and down and transfer to gelatinized 96-well tissue culture plate.
9. Wash each V-well from the trypsin plate with another 100 μl of medium per well and add to same well of the 96-well gelatinized plate (you should end with 235 μl per well).
10. Change the medium the next day, and then daily thereafter, using multichannel pipettor.
11. If the colonies have not grown to ~ 80% confluence in two to three days, 'tryplate' the cells using the following procedure:
    (a) Aspirate the medium and wash the cells with PBS.
    (b) Add 35 μl of trypsin/EDTA and incubate for 5 min at 37 °C, or until the cells lift off the dish.
    (c) Add 200 μl ES cell medium and break colonies up by gently pipetting up and down.
    (d) Change the medium 12–24 h later.
12. Repeat steps 1–11 on second and third picking days, using original plates of electroporated cells.
13. When the colonies in 96-well plates have reached at least 80% confluence you should passage them and freeze your clones and prepare replica plates for DNA analysis.

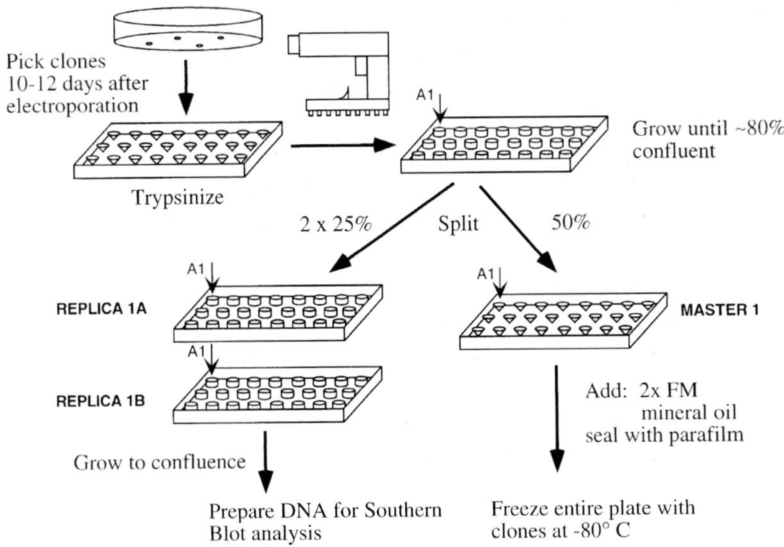

Figure 4. Schematic illustration of 96-well method for expanding and freezing resistant clones for Southern blot analysis. Repeated for each round of picking.

Protocol 10. Freezing drug resistant clones and preparation of replica plates for Southern blot analysis

Equipment and reagents

- 96-well tissue culture plates with flat and V-shaped bottom
- Multichannel pipettor and sterile reservoirs
- ES cell medium
- 0.05% trypsin in saline/EDTA
- PBS without Ca^{2+} and Mg^{2+}
- 2 × freezing medium: 20% DMSO, 50% FBS, DMEM
- Mineral oil, embryo tested (Sigma, Cat. No. M-8410)
- 0.1% gelatin (autoclaved)
- Styrofoam box, large enough to hold four to eight 96-well plates
- Aluminium foil
- Parafilm

Method

1. Prepare the required number of sets of two 96-well gelatinized flat-bottom plates (replica A and B) with 200 μl ES cell medium, and one 96-well V-bottom plate (master) containing 50 μl cold 2 × freezing medium, kept on ice (*Figure 4*).
2. Remove medium from each well of the 96-well plate containing growing ES cell clones using a multichannel pipettor.
3. Wash cells twice with 200 μl PBS.
4. Add 50 μl trypsin and incubate for 10 min at 37 °C, 5% CO_2.

Protocol 10. *Continued*

5. Observe cells under an inverted microscope. If all cells have detached, stop trypsinization by adding 50 μl of ES cell medium, then mix well by pipetting.
6. Working quickly, transfer 50 μl of cell suspension to the 96 V-well master plate, which contains 50 μl of pre-cooled 2 × freezing medium in each well, mix by pipetting up and down.
7. Layer 100 μl of cold mineral oil on the top of each well containing ES cells in freezing medium.
8. Seal plate by wrapping Parafilm around the outside edge where the top and bottom meet, wrap plate in aluminium foil, and transfer in a pre-cooled Styrofoam box (large enough to fit 96-well plates wrapped in foil) to –70 °C freezer.
9. Mix the remaining cell suspension and transfer 25 μl into each of the two (A and B) gelatinized 96-well replica plates containing 200 μl of medium per well. Return replica plates to incubator.
10. Change the medium daily on replica plates. Perform DNA extraction when cells are overly confluent (cell differentiation or poor morphology is not a concern on these plates). The goal is to obtain the maximum number of cells.

Protocol 11. DNA isolation from ES cell colonies (modified from ref. 19)

Equipment and reagents

- Sealing tape (Wallac, Cat. No. 1450–461)
- Paper towels
- Genomic tips (Bio-Rad, Cat. No. 223–9312) (with wide orifice to prevent shearing of DNA)
- PBS without Ca^{2+} and Mg^{2+}
- Lysis buffer: 10 mM Tris–HCl pH 7.5, 10 mM EDTA, 10 mM NaCl, 1 mg/ml proteinase K, 0.5% Sarcosyl or SDS
- NaCl/ethanol mix: 15 μl of 5 M NaCl per 1 ml of cold 100% ethanol (made fresh)
- 70% ethanol
- Restriction enzyme digestion cocktail (per well): 1 × restriction enzyme reaction buffer (appropriate for enzyme), 1 mM spermidine, 100 μg/ml BSA (nuclease-free), 50 μg/ml RNase A, ~ 20 U enzyme

Method

1. When cells are confluent, aspirate the medium from each well on the two replica plates and wash twice with PBS.
2. Add 50 μl of lysis buffer to each well.
3. Seal each plate with sealing tape, making sure tape seals well over wells with cells. Incubate overnight at 55°C.
4. Remove tape. Using a multichannel pipette, add 100 μl of freshly

prepared cold (on ice) NaCl/ethanol mix to each well and mix slowly. Mixture will turn cloudy.

5. Incubate at room temperature ⩾ 1 h, avoiding vibration. DNA precipitate should be visible against a black background, or you might see web-like DNA fibres under a low power microscope.
6. Remove lid, and put several layers of paper towels on plate over open wells. Invert plate together with paper towels. Lift the plate, blot away liquid. Precipitated DNA will remain attached to bottom of well. Trust us!
7. Wash three times with ~ 200 μl of 70% ethanol, each time blotting alcohol by inversion on paper towels. At this point, DNA can be stored in 70% ethanol in parafilm sealed plate at –20 °C (for four to five weeks) until proceeding with diagnostic restriction enzyme digestion and Southern blotting, or continue directly to step 8.
8. Air dry plate for 5 min.
9. Add 35 μl of restriction enzyme digestion cocktail per well, mix with DNA using genomic tips. Seal the plate with sealing tape. Incubate overnight at 37 °C.
10. Load all the DNA from a well into one or two wells of an agarose gel, and run the gel overnight for analysis by Southern blot.

This procedure produces a crude DNA preparation and not all restriction enzymes work well under these conditions. Therefore, you should check digestion efficiency on DNA prepared from normal ES cells before starting your gene targeting experiment. We have found the following enzymes work well with this procedure: *Eco*RI, *Eco*RV, *Hin*dIII.

An alternative to steps 6–9 that we have found produces cleaner DNA is to spool the DNA out of the 70% ethanol from the 96-well plate, into an Eppendorf tube, rinse twice in 70% ethanol, and resuspend in ~ 40 μl of TE. Restriction enzyme digestion is then carried out in these tubes.

Protocol 12. Expansion and freezing targeted ES cell clones

Equipment and reagents

- p200 Pipetman and yellow tips (sterile)
- 70% ethanol
- 35 mm and 100 mm plates with feeders
- ES cell medium
- 0.05% trypsin in saline/EDTA

Method

1. Remove the 96-well plate containing positive targeted ES cell clones from the –70 °C freezer.

Protocol 12. *Continued*

2. Unwrap, remove sealing tape, and warm quickly by placing in a 37 °C incubator.
3. When the ice crystals disappear (~ 5 min), sterilize the outside of the plate by cleaning it with a tissue moistened with 70% ethanol.
4. When mineral oil has thawed, add 100 μl of pre-warmed ES medium under the oil to each well containing a targeted clone, and transfer medium with cells to a 35 mm plate with feeders in 10 ml of ES cell medium.
5. Rinse original 96-well with 200 μl of ES cell medium under the oil and transfer to same 35 mm plate.
6. Incubate 35 mm plate at 37 °C, 5% CO_2.
7. Change the medium after at least 6 h, and then daily.
8. When cells reach 30–50% confluence (usually on the second or third day), passage all the cells to a larger (100 mm) plate containing feeders.
9. If cells have not reached 30–50% confluence after two days, tryplate them (*Protocol 9*, step 11).
10. Continue to passage 1/5 to 1/6 of the cells every two days onto a fresh 100 mm plate with feeders (*Protocol 5*). At each passage, freeze the remainder of the cells at –70 °C. Note passage number for future reference.

4.2 Screening for homologous targeting events using PCR

The basic strategy is to design PCR primers that will amplify a novel junction fragment created by the correct homologous recombination event, but not wild-type DNA (see Chapter 1 Section 2.1). This can be done most easily with replacement-type vectors by having one of the homologous arms approximately 600–1200 base pairs in length (short arm). The primers for the PCR reaction are then designed in such a way that one primer binds to the *neo* cassette, and the second to a region just past the short arm of the targeting vector within the endogenous locus. To test for PCR conditions that allow for the identification of a homologous recombination event, we make control vectors which include the two primer annealing sites and establish ES cell lines containing one copy of this control vector. It is advisable to make the sequences amplified in the control vector distinguishable from the expected targeted vector. We then perform a series of control experiments with these cell lines to optimize the PCR reaction conditions for detecting single homologous targeting events in pools of colonies with one targeted colony and many negative cells. Approximately half of a colony of control ES cells is mixed with different amounts of wild-type ES cells to determine the pool size under which the rare homologous events can be detected.

Protocol 13. Screening procedure using PCR (modified from ref. 20)

Equipment and reagents

- Pasteur pipettes
- Dissecting microscope
- Mouth pipette (Chapter 4) or pipetman
- PCR reaction tubes (500 μl) (GeneMate, Cat. No. T-3034)
- PBS without Ca^{2+} and Mg^{2+}
- Proteinase K (10 mg/ml stock, stored frozen in aliquots)
- Mineral oil
- dNTPs frozen at –70 °C: stock solution containing 10 mM each of dATP, dCTP, dGTP, and dTTP
- 10 × PCR buffer: 100 mM Tris–HCl pH 8.3, 500 mM KCl, 15 mM $MgCl_2$,[a] 0.1% gelatin
- *Taq* polymerase[b] (Cetus)

Method

1. After about eight days of drug selection (*Protocol 8*), circle the colonies on the bottom side of the plate by holding the plate up to the light.
2. Replace the medium with PBS.
3. Pick half of each colony to be included in a pool (or approximately equal cell volumes from each colony) with a drawn-out Pasteur pipette (heat Pasteur pipette with a Bunsen burner, pull one end quickly, and break pipette at appropriate diameter), and return cells to incubator. See *Figure 5*.
4. Pool between 20–50 colonies, depending on the sensitivity of the PCR reaction, in one reaction tube.
5. Centrifuge the cells for 15 sec in a microcentrifuge, and then aspirate the PBS, leaving approx. 5 μl.
6. Add 35 μl of dH_2O, resuspend cell pellet by vortexing.
7. Put the tubes on dry ice for 5 min.
8. Heat the tubes for 10 min at 95 °C.
9. Add 50 μl of mineral oil.
10. Cool the reaction tubes to 50 °C.
11. Add 1 μl of proteinase K stock (10 mg/ml) under oil. Incubate for 90 min at 50 °C.
12. Heat inactivate the proteinase K reaction at 94 °C for 10 min.
13. Add to each lysate under oil at 4 °C a PCR reaction cocktail containing:
    - 5 μl of 10 × PCR buffer
    - 1 μl dNTPs

Protocol 13. *Continued*

- 1 μl of each primer (1 μg/μl)
- 0.5–1 μl *Taq* DNA polymerase
- H_2O to a final reaction volume of 50 μl

15. Run the PCR reaction under predetermined optimal conditions. Sample conditions:
    (a) Initial denaturation: 94 °C/2 min.
    (b) Denaturation: 94 °C/1 min.
    (c) Annealing: 50–62 °C/1 min.
    (d) Elongation: 60–72 °C/1 min.
    (e) Repeat steps b–d for 30–40 cycles.
    (f) Final elongation 72 °C/10 min.
16. Analyse 10 μl of the sample on an agarose gel (for details see ref. 21).
17. Blot DNA onto a membrane.
18. Hybridize the blot with a probe spanning the synthesized DNA fragment.
19. Wash the membrane and expose for 5 min to1 h (control DNA bands and products of homologous recombination events should be visible after 15–30 min exposure).
20. Pick half of each colony from the pools which are positive on the Southern blot and repeat the PCR reaction (steps 5–19) on each single half colony. Transfer the other half of each colony to one well of 24-well plate containing EMFI feeder cells and 500 μl of ES cell medium.
21. When the positive clone is identified by PCR on single colonies, expand it after tryplating (*Protocol 9,* step 11) once or twice, depending on cell density, and analyse by Southern blot analysis.
22. Freeze and store at least one vial of cells in liquid nitrogen (*Protocol 7*) as soon as possible. Cells can be used for making chimeras (Chapters 4 and 5) once a confluent 35 mm plate is obtained.

[a] The optimal Mg^{2+} concentration should be determined.
[b] We found that not all batches of *Taq* polymerase work well on single colonies of targeted cells. Optimal batches must be identified before the experiment.

4.3 Selecting ES cells homozygous for the targeted allele

Although targeting strategies in ES cells are primarily designed to modify one of two alleles at a genetic locus, it is sometimes beneficial or necessary to have a cell line which is homozygous for a given mutation. For example, if a mutation results in haplo-insufficiency and is incompatible with postnatal

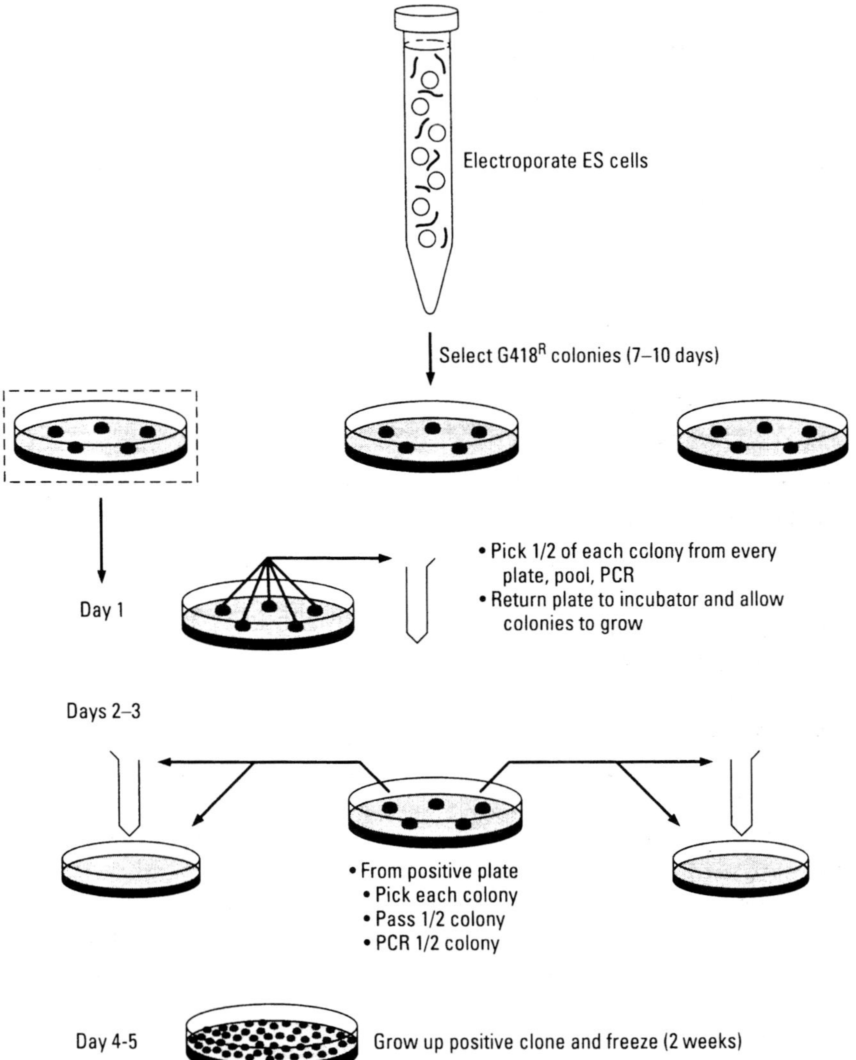

Figure 5. Screening for targeted colonies using PCR. A schematic diagram showing the steps taken to identify single targeted clones.

survival, or prevents breeding or transmission of the mutation through the germline. In these cases it is possible to consider generating a homozygous cell line to study the phenotype in chimeric embryos produced by aggregation with wild-type or tetraploid morulae (Chapter 5). In addition, it is often beneficial to have a homozygous mutant cell line for biochemical studies.

There are three approaches to obtain ES cells homozygous for an introduced mutation:

(a) By targeting the second allele with a construct containing a different selection marker (for example *puromycin* instead of *neomycin* resistance gene).

(b) By excising the selectable marker at one allele using Flp or Cre recombinase (if the targeting strategy was so designed with Frt or *loxP* sites surrounding drug resistance marker) and re-targeting the second allele with the same vector (see Chapter 2).

(c) By increasing the concentration of the selection agent to select cells which have two copies of the mutant gene.

The first approach is the least likely to be used since it requires the creation of two targeting constructs. Multiple rounds of electroporation and selection, required in the second approach, could be deleterious to ES cell pluripotency, thus it can be a limitation if making chimeras that will transmit the mutation through the germline is the ultimate goal. Furthermore, a second round of targeting may be difficult if the homologous recombination frequency is low at a given locus. The last method requires only one additional selection step and there is growing evidence that it does not alter the ability of ES cells to contribute to chimeras. This method appears to depend on the use of a mutant *neo* gene that codes for an attenuated protein (22). This mutant *neo* allows for low G418 concentrations to be used to select cells carrying one copy of *neo*, which also permits homozygous cell lines to be selected at higher G418 concentrations (23). The mechanism by which homozygosity is achieved is poorly understood, but evidence suggests that animals derived from these cell lines are similar to homozygous mutant animals generated by heterozygous intercrossing.

Protocol 14. Selection of ES cells homozygous for *neo* in a high concentration of G418

Equipment and reagents

- Heterozygous *neo*-resistant ES cell clone/s
- Gelatinized 100 mm and 96-well plates
- ES cell medium
- G418

Method

1. Subculture near-confluent ES cells from a 60 mm plate into two gelatinized 100 mm plates in regular ES cell medium. Make approximately four 100 mm plates.
2. Next day change regular medium to selection medium. Each plate should contain a different concentration of G418 between 600–1000 μg/ml.

3. Change medium daily at the beginning when there are a lot of dead cells, and every second day thereafter.
4. After ~ 12 days of selection, identify plates with only 10–100 colonies on them and pick clones into gelatinized 96-well plates (*Protocol 9*).
5. Expand cells for isolating DNA to determine the genotype by Southern blot analysis, and freeze the cells (as described in *Protocol 7*).

5. Establishing new ES cell lines

It is not necessary for each laboratory to develop their own ES lines for standard gene targeting experiments. This process requires previous ES cell culture experience and is lengthy since each new line must be tested for its ability to be transmitted through the germline. A number of different male ES cell lines (mostly from different substrains of 129 mice) are available (see Chapter 4, *Table 1*). Male ES cell lines have a number of practical advantages over female lines (discussed in Chapter 4) and their karyotype remains more stable. With chimera studies being used more frequently to study the function of genes (see Chapter 5 for discussion), it has become important to be able to establish ES cell lines from homozygous mutant embryos. The establishment of ES cell lines has been covered in this series by Robertson (24). In this chapter we describe our modified method using 96-well plates. We have used this method to successfully establish new ES cell lines, W4 and WB6, from 129SvEvTac and C57BL/6 mice, respectively.

Protocol 15. Establishing new ES cell lines

Equipment and reagents

- 3.5 dpc pregnant mouse of appropriate strain
- Tissue culture plates with EMFI feeder layers (96-well, 4-well, 35 mm, and 60 mm)
- PBS without Ca^{2+} and Mg^{2+}
- 0.05% trypsin in saline/EDTA
- M2 medium (Sigma, Cat. No. M7167)
- ES cell medium
- 0.1% gelatin, autoclaved in H_2O, and stored at 4°C
- 1 × freezing medium: 10% DMSO, 25% FBS in DMEM

Method

1. Flush blastocyst stage embryos from uterine horns of 3.5 dpc females in M2 medium (see Chapter 4). Wash in ES cell medium.
2. Plate each embryo individually in one well of a 96-well tissue culture plate containing EMFI feeders in ES cell medium containing 25% FBS and 2000 U LIF.[a] Embryos should hatch out of the zona pellucida and attach within two days. At this point the trophectoderm cells migrate out and flatten on the feeder cells and the ICM should start to expand. The embryos should be observed daily.

Protocol 15. *Continued*

3. Two days later disaggregate the ICM-derived clumps with trypsinization using a multichannel pipette. Remove the medium and add 35–50 μl of trypsin per well and incubate at 37 °C for at least 5 min. Observe this process under a microscope. When ICM-derived clumps detach, but before they disaggregate into single cells, stop reaction by adding 200 μl of medium and gently pipette up and down. Let all cells re-attach in the same well. The EMFI cells will form a new feeder layer and the ICM cells should begin to proliferate in clumps as ES cells. Some non-ES cells might also proliferate for a few days.
4. Change the medium after overnight incubation.
5. Two to three days after the first trypsinization, trypsinize again as in step 3, leaving the cells in the same plate or transfer them into a fresh 96-well plate containing fresh EMFI feeder cells. If transferring cells to a new well, stop trypsinization with 100 μl of medium. Transfer to a new plate, and wash the original well with another 100 μl of medium so as not to lose any cells. Combine with initial 135 μl. Change the medium the following day.
6. Two to three days later, in each well you should see small colonies of undifferentiated cells, as well as several types of differentiated ones. To separate undifferentiated cells from differentiated ones and from fibroblasts, trypsinize cells briefly (3–5 min), then transfer cells that have detached, to a gelatinized 96-well plate (plate A) containing 200 μl ES cell medium/well. The feeder cells will round up but should not detach as a layer at this stage. Add trypsin to original plate, incubate at 37 °C for 5–10 more min, stop trypsinization by adding 200 μl medium/well, and transfer cell suspension to a second gelatinized 96-well plate (plate B).
7. Change medium next day on both plates and observe each well under microscope. You should expect to find some ES cells on plate A. If any are also found on plate B, you will need to repeat step 6 to plate B after two to three days when colonies are visible, and then follow steps 6–10.
8. In one to two days, trypsinize and transfer undifferentiated cells in plate A to a 4-well plate with feeders, and in another two to three days passage cells to a 35 mm plate with feeders if cells are near confluence. Clones that are not confluent after two to three days on 4-well or 35 mm plates should be trypsinized and re-plated onto the same plate. Do not keep cells longer than three days without trypsinization because they will differentiate. Generally, clones that grow faster and are ready for trypsinization every two days produce the best ES cell lines, while more slowly growing cells tend to have some chromosome changes.

9. When trypsinizing confluent cells from the first 35 mm dish, split cells into two portions, spin down and freeze half of the cells in 0.4 ml of 1 × freezing medium (*Protocol 7*), and seed remaining cells on a 60 mm dish. Count this as the first passage.
10. Change medium and monitor cells daily, trypsinize and expand when 80% confluent (two to three days). Every second passage, reduce the serum and LIF concentration by 5% and 500 U, respectively, until 15% FBS and 1000 U LIF is reached. Passage 1/5 to 1/6 of cells, depending on density, and freeze remainder of the cells at all early passages (*Protocol 7*). We check cells for their ability to produce chimeras at passage 4 and later (~ 8 and 15). Each line should be karyotyped and analysed using the Southern blot technique using a Y chromosome-specific probe or PCR with Y-specific primers (25) at passage 4, and karyotyped after making a large stock.

[a] Alternatively several blastocysts can be plated in one well of a 4-well plate and later individual ICM clumps transferred to single wells of a 96-well plate with feeders using a drawn-out Pasteur pipette to pick the clumps under trypsin.

Acknowledgements

We would like to thank Wolfgang Wurst for his contribution to the previous version of this chapter (26). We also thank Anna Auerbach and Natalia Shalts for their support of our ES cell facility and comments on the protocols described. A. Joyner is an investigator of the HHMI.

References

1. Martin, G. (1981). *Proc. Natl. Acad. Sci. USA*, **78**, 7634.
2. Evans, M. J. and Kaufman, M. H. (1981). *Nature*, **292**, 154.
3. Joyner, A. L. (1994). In *Targeted mutagenesis in mice: a video guide* (ed. R. A. Pederson, V. Papaiannou, A. Joyner, and J. Rossant). Cold Spring Harbor Laboratory Press.
4. Nagy, A., Rossant, J., Nagy, R., Abramow-Newerly, W., and Roder, J. C. (1993). *Proc. Natl. Acad. Sci. USA*, **90**, 8424.
5. Smith, A. G., Heath, J. K., Donaldson, D. D., Wong, G. G., Moreau, J., Stahl, M., and Rogers, D. (1988). *Nature*, **336**, 688.
6. Williams, R. L. Hilton, D. J., Pease, S., Willson, T. A., Stewart, C. L., Gearing, D. P., Wagner, E. F., Metcalf, D., Nicola, N. A., and Gough, N. M. (1988). *Nature*, **336**, 684.
7. Moreau, J.-F., Donaldson, D. D., Bennett, F., Witek-Giannotti, J., Clark, S. C., and Wong, G. G. (1988). *Nature*, **336**, 690.
8. Gough, N. M., Gearing, D. P., King, J. A., Willson, T. A., Hilton, D. J., Nicola, N. A., and Metcalf, D. (1988). *Proc. Natl. Acad. Sci. USA*, **85**, 2623.

9. Nichols, J., Evans, E. P., and Smith, A. G. (1990). *Development*, **110**, 1341.
10. Mereau, A., Grey, L., Piquet-Pellorce, C., and Heath, J. K. (1993). *J. Cell Biol.*, **122**, 713.
11. Doetschman, T. C., Eistetter, H., Kutz, M., Schmidt, W., and Kemier, R. (1985). *J. Embryol. Exp. Morphol.*, **87**, 27.
12. Martin, G. R. and Evans, M. J. (1975). In *Teratomas and differentiation* (ed. M. I. Sherman and D. Solter), pp. 169–87. Academic Press, New York.
13. Ware, L. M. and Axelrad, A. A. (1972). *Virology*, **50**, 339.
14. Rudnicki, M. A. and McBurney, M. W. (1987). In *Teratocarcinomas and embryonic stem cells: a practical approach* (ed. E. J. Robertson). IRL Press, Oxford.
15. McMahon, A. P. and Bradley, A. (1990). *Cell*, **62**, 1073.
16. Potter, H., Weir, L., and Leder, P. (1984). *Proc. Natl. Acad. Sci. USA*, **81**, 7161.
17. te Riele, H., Maandag, E. R., and Berns, A. (1992). *Proc. Natl. Acad. Sci. USA*, **89**, 5128.
18. Wurst, W., Auerbach, A. B., and Joyner, A. L. (1994). *Development*, **120**, 2065.
19. Ramirez-Solis, R., Davis, A. C., and Bradley, A. (1993). *Methods in Enzymology* **225**, 855–78.
20. Joyner, A. L., Skarnes, W., and Rossant, J. (1989). *Nature*, **338**, 153.
21. Sambrook, J., Fritsch, E. F., and Maniatis, T. (ed.) (1989). *Molecular cloning: a laboratory manual.* Cold Spring Harbor Press, NY.
22. Yenofsky, R. L., Fine, M., and Fellow, J. W. (1990). *Proc. Natl. Acad. Sci. USA*, **87**, 3435.
23. Mortensen, R. M., Connner, D. A., Chao, S., Geisterfer-Lowrance, A. A. T., and Seidman, J. G. (1992). *Mol. Cell. Biol.*, **12**, 2391.
24. Robertson, E. J. (1987). In *Teratocarcinomas and embryonic stem cells: a practical approach* (ed. E. J. Robertson). IRL Press, Oxford.
25. Nadeau, J. H., Davisson, M. T., Doolittle, D. P., Grant, P., Hillyard, A. L., Kosowsky, M., and Roderick, T. H. (1991). *Mamm. Genome*, **1**, 461.
26. Wurst, W. and Joyner, A. L. (1993). In *Gene targeting: a practical approach* (ed. A. L. Joyner), p. 33. IRL Press.

4

Production of chimeras by blastocyst and morula injection of targeted ES cells

VIRGINIA PAPAIOANNOU and RANDALL JOHNSON

1. Introduction

The ability of mammalian embryos to incorporate foreign cells and develop as chimeras has been exploited for a variety of purposes including the elucidation of cell lineages, the investigation of cell potential, the perpetuation of mutations produced in embryonic stem (ES) cells by gene targeting, and the subsequent analysis of these mutations. The extent of contribution of the foreign cells depends on their developmental synchrony with the host embryo and their mitotic and developmental potential, which may be severely restricted if the cells bear mutations. If the goal in making chimeras is the transmission of a mutation produced by gene targeting to the next generation, the mutant ES cells must have the capacity to undergo meiosis and gametogenesis.

Cells from two different mammalian embryos were first combined experimentally to produce a composite animal, dubbed a chimera, nearly four decades ago. Pairs of cleaving, pre-implantation embryos were mechanically associated *in vitro* until they aggregated together to make single large morulae; these in turn resulted in chimeric offspring (1). Genetic markers were used to distinguish the contributions of the two embryos in these animals. Since then, various methods for making chimeras have been explored to address different types of questions (2). In 1972 it was reported that highly asynchronous embryonic cells, which had been cultured *in vitro*, could contribute to chimeras upon re-introduction into pre-implantation embryos (3). Not long afterward, several groups working with teratocarcinomas, tumours derived from germ cells of the gonad, discovered that stem cells from these tumours, known as embryonal carcinoma cells, could contribute to an embryo if introduced into pre-implantation stages (4–6). It appeared that the undifferentiated stem cells of the tumour had enough features in common with early embryonic cells that they could respond to the embryonic environment, differentiating in a normal manner, even after long periods *in vitro*. Their embryonic potential was limited, however, and many teratocarcinoma cell lines made only meagre contributions to the developing fetus or even produced tumours in chimeras (7). Either their derivation from tumours or their extended sojourn *in vitro* rendered these cells so dissimilar from early embryonic cells that they rarely, if ever, had full embryonic potential. Although specific genetic mutations could be selected in these cell lines, none was ever propagated to the next generation through a chimera.

All of this began to change dramatically in 1981 when cell lines were derived directly from pre-implantation embryos and maintained *in vitro* in an undifferentiated state (8, 9). These primary cell lines, called embryonic stem (ES) cell lines, correspond closely to cells of the inner cell mass of the blastocyst. Indeed, even after extended periods of culture, ES cell lines can remain multipotential and participate in the formation of all tissues of chimeras, including the gametes. The past decade has seen the rapid exploitation of ES cells to propagate mutations created by gene targeting *in vitro*, taking advantage of the possibility of transmitting the altered gene through the germline of a chimera (e.g. see the knock-out databases `http://www.biomednet.com/db/mkmd`, `http://tbase.jax.org/`, and `http://www.bioscience.org/knockout/knochome.htm`).

In this chapter we will detail two methods, suitable for use with ES cells, that utilize injection of cells into pre-implantation embryos for producing chimeric mice. Chapter 5 describes another method involving aggregation of ES cells with morula stage embryos to make chimeras. For the most part, the injection methods are identical to procedures developed for and used extensively in experimental embryology (10, 11) and their application to ES cell work has been described earlier in this series by Bradley (12) and others (13, 14). Commonly used markers of chimerism are discussed and test breeding procedures will be outlined. We also describe simple breeding schemes to propagate mutant alleles in generations subsequent to the founder generation and discuss special problems associated with mutants of unknown or unpredictable phenotype. For additional information on mouse husbandry and handling see Hetherington (15). As the use of gene targeting techniques has proliferated, not all laboratories have met with uncomplicated success. We will discuss and evaluate some of the factors that theoretically and practically might enhance the successful germline transmission of the ES cell genotype from chimeras.

2. The starting material

2.1 ES cells

The derivation of ES cell lines directly from embryos (Chapter 3, *Protocol 15*) was an extension of a large body of work on the stem cells of teratocarcinomas.

Table 1. Allelic differences in marker genes for some strains and substrains commonly used in ES cell experiments, and the ES cell lines derived from them (see *Table 3* for an explanation of the effects of the coat colour alleles)

Substrain	ES cell lines	Agouti[a]	Albino (*Tyr*)[a]	Locus Pink-eyed dilution[a]	Gpi-1	References
129/J	mEMS32	A^w	Tyr^c, $Tyr^{c\text{-}ch}$	*p*	$Gpi\text{-}1^a$	18
129/SvJ	PJ5	A^w	Tyr^c, $Tyr^{c\text{-}ch}$	*p*	$Gpi\text{-}1^a$	23
129/Sv	D3, CJ7	A^w	$+^{Tyr\text{-}c}$	$+^p$	$Gpi\text{-}1^a$	24, 25
129/OlaHsd	E14TG2a	A^w	$Tyr^{c\text{-}ch}$	*p*	$Gpi\text{-}1^a$	26
129/SvEvBrd	AB1	A^w, *A*	$+^{Tyr\text{-}c}$	$+^p$	$Gpi\text{-}1^c$	27, *b*
129/SvEv	CP-1	A^w, *A*	$+^{Tyr\text{-}c}$	$+^p$	$Gpi\text{-}1^a$	28
129/SvEv	CCE, CC1.2	A^w, *A*	$+^{Tyr\text{-}c}$	$+^p$	$Gpi\text{-}1^c$	28, 29
129/SvEvTac	W4	A^w	$+^{Tyr\text{-}c}$	$+^p$	$Gpi\text{-}1^c$	*c*
(129/Sv × 129/J)F_1	R1	A^w	$Tyr^{c\text{-}ch}/+^{Tyr\text{-}c}$	$p/+^p$	$Gpi\text{-}1^a$	18, 30
C57BL/6J	ES632	*a*	$+^{Tyr\text{-}c}$	$+^p$	$Gpi\text{-}1^b$	24

[a] Two alleles separated by a comma means the mouse strain is segregating for these alleles, whereas two alleles separated by a slash means the ES cell line is heterozygous.
[b] A. Bradley, personal communication.
[c] Joyner and Auerbach, personal communication; see Chapter 3.

These rare germ cell tumours are most prevalent in the testes of 129 mice and in the ovaries of LT mice (16). In the attempt to derive embryonic stem cell lines directly from embryos, these strains were the obvious choices and some of the first ES cell lines were produced from 129 embryos (8). Although the methods have been extended to embryos of other inbred and random-bred strains (e.g. refs 9 and 17), 129 ES cell lines remain the most widely available and commonly used. However, the 129 strain has had a complicated history of outcrossing and development of divergent substrains (18) and ES cell lines have been derived from several 129 substrains differing for known genetic markers (*Table 1*). This is important for several reasons: first, the genotype of the ES cell line will determine what markers are available to detect chimerism; secondly, it is known that efficient gene targeting is dependent on the use of isogenic DNA so that discrepancies between the genetic background of the ES cells and the targeting DNA can greatly influence targeting success (19, 20). An added complication is the recent determination that 129/SvJ, which is the source of several commonly used genomic libraries, is a genetically contaminated substrain containing genomic regions of non-129 origin (21) (see Chapter 7 for a further discussion of the genetics of 129 strains of mice).

2.1.1 The parent cell line

Dutiful tissue culture is critical to maintaining an ES cell line and preserving its capacity to contribute to the germline. Cells should be kept at high density, passaged every second or third day in order to minimize overt differentiation. It is important to avoid passaging cells in large clumps, since these clumps will differentiate readily, especially on their outer borders. This differentiation, usually into endoderm, can often be seen as a ring of rounded, diffractile cells around the edges of large ES cell colonies. ES cell lines or clones which have a large percentage of differentiated cells generally contribute poorly to chimeras following embryo injection. However, we have found that clones which have already differentiated extensively can, in some cases, be rescued by selecting colonies of undifferentiated cells from the differentiated culture. This is done much as though one were picking clones in a selection experiment, but the undifferentiated cells are pooled for subsequent culture.

ES cell culture should also include monitoring for the presence of mycoplasma, which can be an insidious spoiler of chimera formation and germline transmission. Mycoplasma are common contaminants of tissue culture cells, and can be transmitted from culture to culture by contaminated tissue culture facilities and/or poor sterile technique. Even the most fastidious culturist may pick up mycoplasma from contaminated reagents (such as fetal calf serum or trypsin), so a monthly check for the presence of mycoplasma is good practice. There are several ways to detect the various species of mycobacterium. For routine monitoring, we use the Gen-Probe kit, which utilizes a labelled probe for the presence of mycobacterium-specific ribosomal RNA in tissue culture medium. As an alternative or for confirmation of the presence of mycoplasma contamination, we also use a Hoechst stain on fixed cells, which allows visualization of mycoplasma under a fluorescence microscope (see ref. 22 for detailed protocol); they appear as a thousand points of light around the stained nucleus. Once a mycoplasma contamination is detected, it may be tempting to rescue the contaminated cells using antibiotics, such as gentamicin, effective against mycoplasma. In general, however, while it is possible to cure an ES cell line, the uncertainty about the effect of the mycoplasma or treatment on the cells' pluripotency makes the best solution an immediate disposal of all contaminated cultures and frozen cells. This rule should hold true for all contamination with bacteria or yeast, as it is usually safest to simply start over once a culture has been invaded. It is for this reason that we routinely culture without antibiotics, since they can mask a contamination by keeping it at a low level. Antibiotics can be added to ES cell cultures which are deemed irreplaceable, such as initial expansions of targeted clones; good sterile technique should make them unnecessary for routine culture.

Clones from all of the ES cell lines listed in *Table 1* have been used to make germline chimeras. It is nevertheless important, when starting from a low passage number (less than 12) frozen aliquot obtained from another laboratory or when working with an untested ES cell line, to ensure that the parent cell line to be used for experiments can contribute to the germline. Chimera production with the parent line is thus the first experiment that should be undertaken by a laboratory beginning to do gene targeting. Positive results will indicate a good cell line as well as adequate cell culture and embryo manipulation techniques.

2.1.2 Feeder cells and media supplements

As discussed in Chapter 3, the routine culture of ES cells requires either monolayers of inactivated feeder cells and/or media supplements to prevent differentiation. The most commonly used media supplement is leukaemia inhibitory factor (LIF), sold commercially as a purified, bacterially-expressed recombinant protein (Gibco BRL, ESGRO). The feeder layers can be one of two types: primary murine embryonic fibroblasts (EMFI) or STO cells (a transformed murine embryonic fibroblast line). The main disadvantage of EMFIs is that they will not grow indefinitely and thus must be re-derived periodically. However, they are very effective in maintaining ES cell pluripotency. Neomycin-resistant EMFIs can be obtained from mouse strains carrying targeted mutations with the *neomycin* gene inserted into the locus to make subsequent selection of ES cell clones easier. The STO cell line is easily available and easy to grow and can in principle be expanded indefinitely. STOs can also be transfected with the neomycin resistance gene and the LIF gene, making media supplements unnecessary. A variety of lines are commercially available (Genome Systems, Inc; Lexicon Genetics). For further information on feeders and supplements, see Chapter 3.

2.1.3 Drug selection and cloning

Selection protocols vary with the selectable marker employed (see Chapter 1). Germline transmission from selected clones is certainly possible but this is not to say that ES cells are completely unaffected by selection agents. Care should be taken to ascertain that the selection protocol uses the minimum amount of drug necessary to select transfectants. Once they are subcloned, cells should not be maintained under selection conditions. A careful initial isolation will prevent non-clonal isolation of non-resistant cells and allow the selected cells to be cultured without drugs. After clones are characterized, they should only be expanded enough to allow freezing of a reference stock in addition to a seed stock for making chimeras. This minimal expansion will help limit the total number of passages of the clone. The reference stock should include one or two aliquots of 10^6 to 10^7 cells. The aliquots for making chimeras should have enough cells to thaw directly into a single well of a 24-well plate (see Chapter 3 for details). When confluent, the well will contain enough cells for a day of making chimeras, as well as enough to pass into another well for later experiments.

2.2 Mice: setting up for embryo recovery and transfer

A major consideration is the health of the animals for all aspects of a gene targeting experiment, from the mice used for producing chimeras to expansion and maintenance of the mutant strains. Ideally, the animal colony should be free of all specific pathogens (a specific pathogen-free or SPF facility). With time, most animal facilities that do transgenic and knock-out research are being raised to this standard. Although the additional techniques needed in maintaining such SPF mice can be somewhat cumbersome, the advantage is that fertility, embryo production, and viability will be optimized.

C57BL/6J mice are widely used for blastocyst injection, are commercially available, and differ from 129 mice in coat colour and at other genetic loci which are useful as markers (*Table 1*). Their embryos proved early on to be compatible hosts for 129 ES cells; chimera formation and most importantly germline transmission could be efficiently obtained. They are by no means the only possible choice of host embryo for ES cell chimeras, and considerations of availability, background genotype, and ease of obtaining large numbers of embryos may also come into play. Although random-bred mice, such as MF-1, CD-1, or Black Swiss, can be used, genetically defined host embryos with several genetic markers that can be used to detect chimerism are usually preferable. In addition, a common experience is that random-bred host embryonic cells tend to out-compete the ES cells for population of the germline when chimeras are made by blastocyst injection.

In choosing a strain for embryo transfer recipients, a primary consideration should be its reproductive performance, including maternal behaviour. For this reason, commercially available F_1 mice, such as B6D2, or less expensive, random-bred mice that have high fecundity and good maternal behaviour make good embryo transfer hosts. Likewise, a random-bred strain is suitable for the vasectomized males used to produce pseudopregnancy in the embryo recipients. Any vigorous strain will do but it is useful to incorporate a coat colour marker that would distinguish offspring of the vasectomized male from the experimental offspring, in the unlikely event of a reanastomosis of the vas deferens following vasectomy.

2.2.1 Selection for oestrus

The oestrous cycle in mice is three to four days long with ovulation occurring at approximately the midpoint of the dark period of a light/dark cycle, although this can be affected by crowding, male pheromones, and by exogenous hormone treatment. The timing of mating can be controlled by altering the light/dark cycle of the mouse room to provide embryos at a given stage at a convenient time. For most strains, a year-round 14 h light/10 h dark cycle, with midnight the middle of the dark period, will provide blastocysts suitable for injection in the late morning of the fourth day. The most straightforward way of obtaining sufficient embryos of known age is to determine the stage of the oestrous cycle and mate animals on the night ovulation is predicted. In practical terms, this will require a bank of proven fertile, stud males, set up in individual cages, and a stock of females that can be selected for overnight mating when they are in oestrus or pro-oestrus. External vaginal changes are reliable signs of progression through the oestrous cycle in female

Table 2. Changing appearance of the vaginal epithelium during different stages of the oestrous cycle in mice

Stage of cycle			
Pro-oestrus	**Oestrus**	**Metoestrus 1and 2**	**Dioestrus**
Moist	Dry	Dry	Wet
Pink/red	Pink	White	Bluish-red
Folded	Folded	Flaky	Smooth
Swollen	Swollen	Less swollen	Not swollen

mice (*Table 2*) (31). The actual size of the opening is not a good criterion, since this varies considerably from mouse to mouse, but the following external vaginal epithelial characteristics should be chosen to select females for mating:

- dry but not flaky
- pink rather than bluish red or white
- swollen so that the tissue bulges out
- wrinkled or corrugated epithelium on both upper and lower vaginal lips

Females caged together without a male tend to cycle together which simplifies oestrus selection. The morphological criteria for selection are more obvious in some mice than others, for example, vaginal colour changes are more evident in albino than in pigmented animals. In addition, inbred animals generally breed less well than outbred mice and this will be reflected in a larger proportion of non-cycling females in a given population.

The morning following pairing of oestrous females with stud males, the vagina of the females must be checked to ascertain whether mating took place. This will be indicated by the presence of a white or yellow vaginal plug, which is composed of male secretions from the vesicular and coagulating glands. This plug can appear loose in the vagina or may seal it shut. It may be at the surface or out of sight deep within the vagina. A narrow, stainless steel dental spatula (3 mm wide) is a convenient probe for detecting plugs, which feel harder than the surrounding tissue (*Figure 1*). Vaginal plugs usually persist for 12 h after mating, and may even remain for more than 24 h. Since this is variable, it is wise to check for plugs during the morning after mating. Even with careful selection, one rarely obtains 100% mating, but with experience, the number of animals that will need to be selected for a given number of matings can be determined for each strain being used. In our experience, good stud males are capable of more than one fertile mating per night and of mating three or more times per week, but there is individual and strain variation. A mating record of each stud should be kept to ensure that each male is sexually active. Characteristically, the use of natural matings will entail the maintenance of a relatively large stock of mice.

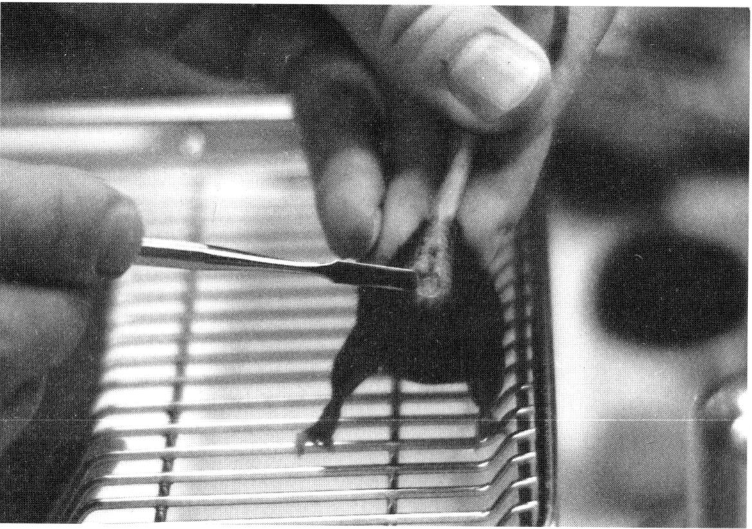

Figure 1. Vaginal plug in a C57BL/6J mouse the morning after mating. A stainless steel dental spatula is used as a probe.

2.2.2 Superovulation

An alternative to oestrus selection is the hormonal regulation of ovulation with exogenous hormones. In this procedure, a large cohort of immature follicles are induced to mature by an injection of pregnant mare serum gonadotropin (PMSG, Sigma) which has follicle stimulating hormone activity. This is followed 44–48 h later by an injection of human chorionic gonadotropin (HCG, Sigma) which has luteinizing hormone (LH) activity, thus mimicking the LH surge that normally brings about ovulation. Ovulation will occur approximately 12 h after HCG administration and should be timed to coincide with the midpoint of the dark cycle. The procedure for superovulation is as follows:

(a) Inject 2–10 IU PMSG intraperitoneally between noon and 16.00 h on day one.

(b) Inject 2–10 IU HCG intraperitoneally at noon on day three.

(c) Put females with fertile stud males.

(d) Check for vaginal plugs before noon on day four (this is the first day of pregnancy).

Although superovulation would seem the perfect answer to timing experiments and increasing the number of embryos obtained per female used, the theory is far from practical reality. Many variables contribute to inconsistency in the response of the females. Strain and age are important factors, and in some strains immature females (three to five weeks) respond best. Although the number of embryos ovulated should theoretically be related to the dose of hormone, this too is strain- and age-dependent. Usually only a proportion of the hormonally treated animals mate, although more of them may ovulate. Economy would suggest that the unmated animals be reused, but this has only limited success. A recovery period of one to two weeks is necessary for recruitment of another wave of follicles and many females appear refractory to a second dose of hormones, possibly due to the production of antibodies to the foreign hormones.

In spite of these possible frustrations, superovulation is a method well worth trying, especially in circumstances where only a small numbers of animals are available or large numbers of embryos are needed on a specific day. Even if the number of embryos recovered per animal used is not higher than natural mating, superovulation can synchronize different groups of animals and can, on occasion, result in a bounty of 60–80 embryos from a single mated female.

2.2.3 Embryo transfer hosts

Successful embryo transfer depends on the quality of the embryos and also the suitability of the host maternal environment. Mice are spontaneous ovulators and can be rendered pseudopregnant by mating with sterile males during oestrus, that is, they display the hormonal profile of a pregnant female when the stimulus of mating occurs during oestrus. Embryo transfer at the appropriate time then provides the embryonic signal that leads to the maintenance of the corpora lutea, preventing a return to cyclicity.

Because the window of uterine receptivity to embryo implantation is narrow, and because the uterine environment is hostile to embryos in the first day of pseudopregnancy or pregnancy (when the embryos are normally in the oviduct), the timing and placement of embryos is critical. In general, embryos can be transferred orthotopically to a synchronous host, but since *in vitro* culture and manipulation have the effect of delaying embryonic development, the more efficient method is to transfer embryos to hosts that are one day asynchronous. Thus, fourth day blastocysts are transferred to the uterus of females in the third day of pseudopregnancy (the day of the plug is the first day). Even greater asynchrony can be tolerated, provided it is the embryos that are more advanced and that they are placed in the oviducts. Blastocysts transferred to the oviducts on the first day of pseudopregnancy will delay their development during transport to the uterus, and most will still be capable of implantation when the uterus becomes receptive. Thus, embryos at any stage between the one cell and blastocyst can be transferred to the oviducts of host females in the first day of pseudopregnancy and they will implant according to the host's schedule on the fifth day. With this flexibility, host females can be selected for oestrus and set up with sterile males a day or more after the mice providing the embryos.

2.2.4 Vasectomy

The final animal component is a stock of sterile males to mate with the embryo transfer hosts to produce pseudopregnancy. Vasectomized males can be purchased commercially from animal suppliers such as Taconic or Charles

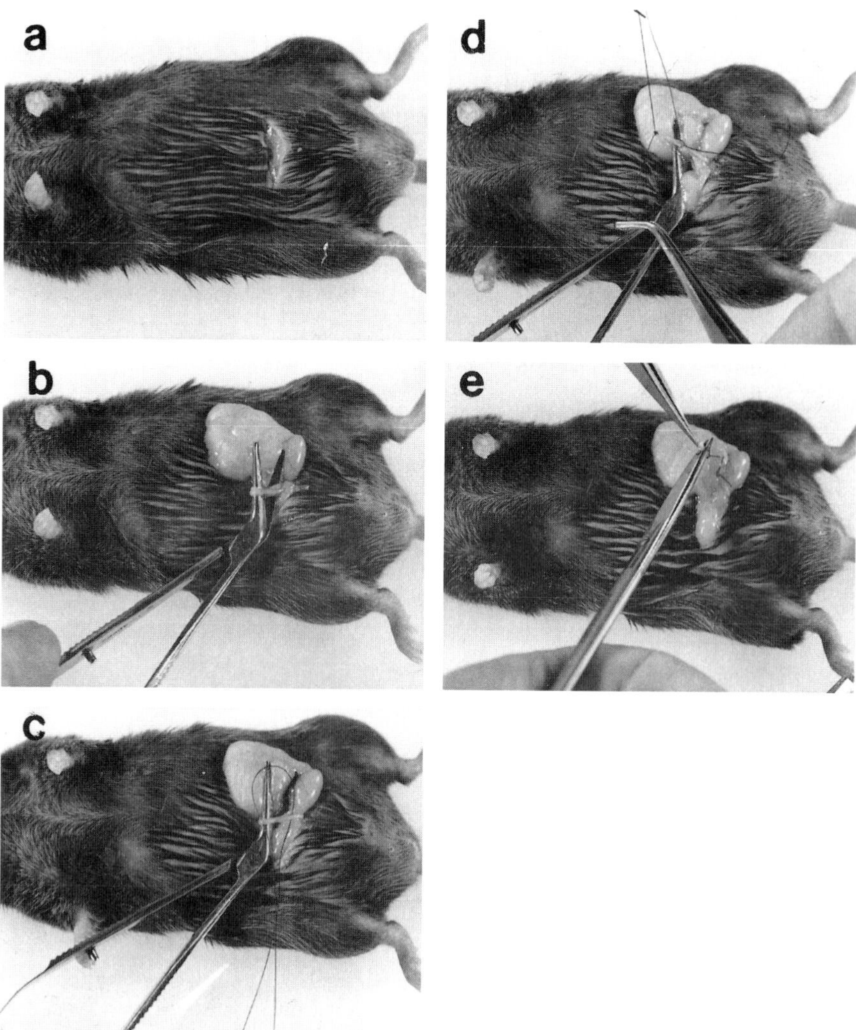

Figure 2. Vasectomy of a male mouse. (a) A transverse incision is made ventrally in the abdominal skin. (b) A corresponding incision is made in the peritoneum and the testicular fat pad and testes is exteriorized. Iridectomy scissors are used to isolate a short length of the vas deferens. (c) A loop of suture is passed under the vas deferens. (d) After the loop is cut, the two lengths of suture are used to tie off the vas deferens in two places, 2 mm apart. (e) The intervening length of vas deferens is cut out.

River, or produced in the laboratory. For vasectomy and other mouse surgeries detailed in *Protocols 7* and *8*, a thorough knowledge of the anatomy of the male and female reproductive tracts is essential (see refs 14 and 32). Tribromoethanol anaesthesia (*Protocol 2*) is recommended as a fast-acting, safe anaesthesia for all mouse surgeries (33). The following high quality surgical instruments are necessary for surgery:

- small, sharp scissors
- fine, blunt, curved forceps
- two or three pairs watchmaker's forceps (No. 5)
- iridectomy scissors
- wound clips and applicator (9 mm; Autoclip, Clay Adams)

Because mice are extremely resistant to septicaemia, surgery can be done on an open bench, although antiseptic technique should be observed to minimize the possibility of infection. Instruments should be sterilized by soaking in 95% alcohol followed by flaming or sterilized in a hot glass bead well (Germinater). Depth of anaesthesia should be checked by loss of the palpebral reflex and failure to respond to a digital tail pinch.

Protocol 1. Vasectomy (*Figure 2*)

Equipment and reagents

- Sterile surgical instruments
- Surgical silk suture (5.0) and needle
- Tribromoethanol anaesthesia (see *Protocol 2*)
- 70% alcohol and cotton

Method

1. Weigh and anaesthetize a mouse and swab the abdomen with alcohol.
2. Make a 1 cm, ventral, midline, transverse skin incision 1 cm rostral to the penis by lifting the skin free of the peritoneum (*Figure 2a*).
3. Make a corresponding incision through the peritoneum.
4. With blunt forceps, probe laterally and caudally into the peritoneum and grasp the testicular fat pad, which lies ventral to the intestine and is attached to the testes. Pull it out through the incision, bringing the testis with it.
5. Locate the vas deferens, being careful to distinguish it from the corpus epididymis, and separate it from the mesentery by inserting the closed tips of the iridectomy scissors through the mesentery. Open the scissors and leave them in this position, supporting and isolating a length of the vas deferens (*Figure 2b*).
6. Insert a doubled length of suture under the vas deferens (*Figure 2c*) and cut it, leaving two separate pieces.

Protocol 1. *Continued*

7. Tie off the vas deferens in two places 2 mm apart and cut out the intervening length (*Figures 2d* and *2e*).
8. Return the testis to the peritoneum and repeat the procedure on the other side.
9. Suture the peritoneal wall and clip the skin incision with one or two small wound clips.
10. Following surgery, leave vasectomized males for at least ten days to clear the tract of viable sperm and then, before use in experiments, test mate the male with a fertile female to ensure that he will mate but is sterile.

Protocol 2. Recipe for tribromoethanol anaesthesia

Reagents

- 2.5 g 2,2,2-tribromoethanol[a] (Aldrich)
- 200 ml distilled water
- 5 ml 2-methyl-2-butanol (tertiary amyl alcohol; Aldrich)

Method

1. Add tribromoethanol to butanol and dissolve by heating (approx. 50°C) and stirring.
2. Add distilled water and continue to stir until butanol is totally dispersed.
3. Aliquot into brown bottles or foil covered tubes (10 ml) and store in the dark at 4°C.[b]
4. Warm to 37°C and shake well before use. The dose for mice is 0.2 ml per 10 g body weight although up to 1.5 × this dose can be given safely, particularly to fat mice.

[a] Tribromoethanol should be discarded if crystals become dark or discoloured.
[b] Stored properly, this solution is stable for several months. Decomposition to dibromoacetic aldehyde and hydrobromic acid can result from improper storage in light or at high temperatures (above 40°C). If this occurs, the solution is toxic and will cause death within 24 h of injection. pH can be used as an indicator of decomposition (a drop of Congo red will turn the solution purple at pH < 5), but this test will only be valid if the original solution was pH > 5, which can vary considerably with the water source used (33).

2.3 Recovering embryos

2.3.1 Culture media and conditions

Making chimeras necessitates a period of embryo culture *ex vivo*, during which embryos will be in less than optimal conditions. During embryo collection and ES cell injection, when the embryos are in room atmosphere, a phosphate or Hepes-buffered medium should be used. If the embryos are to be incubated for any period of time, they should be placed in bicarbonate-buffered medium and kept in 5% CO_2 in air. It should be noted that optimal culture conditions for pre-implantation embryos and for ES cells are not the same so that in making chimeras there is a compromise while the cells are being introduced into the embryo. Suitable media for morulae recovery and culture are M2 and M16, respectively, or the optimized KSOM media which is also suitable for blastocysts (34) (Speciality Media, Inc.). We routinely use PB-1 + 10% serum (*Protocol 3*) for recovery and handling of blastocysts and DMEM/Hepes (Sigma) + 20% serum for ES cell injection, although the blastocysts can equally well be recovered and injected in the latter. Keeping embryos cool (10°C) during manipulations by means of a cooling stage has no adverse effects of their subsequent viability and may facilitate handling of the ES cells during injection.

Protocol 3. Modified PB-1 medium with 10% fetal calf serum for embryo recovery and handling

Reagents and Method

1. Make up the following stock solutions and mix the indicated volumes:

Reagent	Stock solutions (g/100 ml)	Volume (ml)
• NaCl	0.9	68.96
• KCl	1.148	1.84
• $Na_2HPO_4.12H_2O$	5.5101	5.44
• KH_2PO_4	2.096	0.96
• $CaCl_2.2H_2O$	1.1617	0.88
• $MgCl_2$	3.131	0.32
• Na pyruvate	0.020 (in stock NaCl)	22.40

2. Add the following to make up 104 ml of medium:
   - penicillin 6.2 mg
   - glucose 104 mg
   - distilled water 2.16 ml
   - phenol red (1%) 0.1 ml
3. To this medium add heat inactivated fetal calf serum (56°C for 30 min) at a final concentration of 10% (e.g. 10 ml/90 ml medium).
4. Filter sterilize with a 0.2 μm filter, aliquot into sterile containers, and refrigerate until use.

2.3.2 Blastocyst recovery (*Protocol 4*)

Around midday on the third day of pregnancy, embryos traverse the utero-tubal junction. They are at the 8–16 cell stage and will be forming compact morulae (*Figure 3*). Recovery of morulae from the reproductive tract is covered in Chapter 5 (*Protocol 5*). By the middle of the fourth day blastocysts will be present in the uterine lumen and can be recovered by removal and flushing of the uterine horns with the culture media just described.

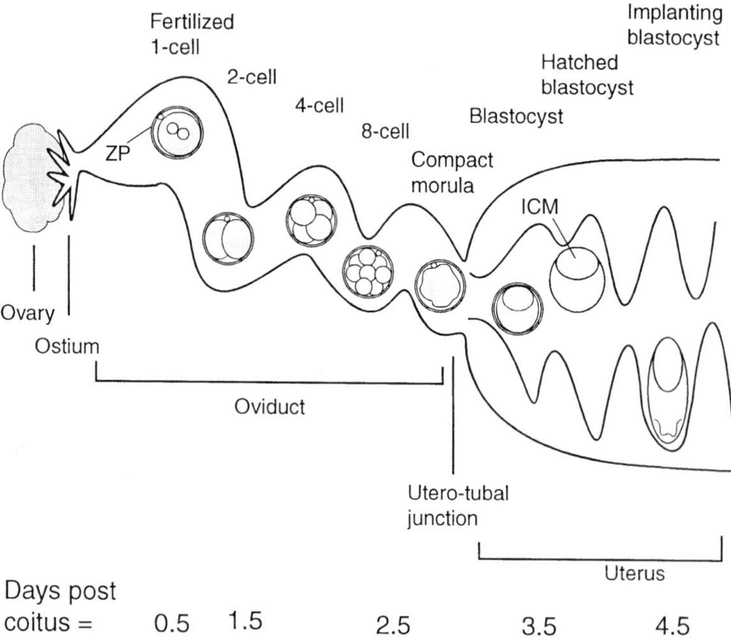

Figure 3. Representation of the pre-implantation stages of mouse development and the location of different stage embryos in the reproductive tract at different days post coitus. ICM, inner cell mass; ZP, zona pellucida.

Protocol 4. Recovery of blastocysts (*Figure 4*)

Equipment and reagents

- 70% alcohol and cotton
- Surgical instruments
- Pasteur pipettes and bulbs
- Sterile Petri dishes
- Sterile watch glasses (optional)

Method

1. Kill the pregnant animal on the fourth day of pregnancy (three and a half days post-coitus).

2. Swab the abdomen with 70% alcohol and make a small, transverse incision midventrally. Wipe away the cut hairs with alcohol-soaked cotton.
3. Gripping the skin anterior and posterior to the incision, tear the skin, pulling it back to the front and hind limbs respectively (*Figure 4a*). Tearing obviates the problem of cut hairs in the incision.
4. Open the peritoneum with a large transverse incision to expose the abdominal cavity.
5. Locate the dorso-caudal reproductive tract by pushing the intestines craniad. Grasp the point of bifurcation of the uterus, which is near the base of the bladder, with blunt forceps (*Figure 4b*).
6. Lift the uterus, cut across the cervix and continue to lift the uterus, clipping away the mesometrium, the mesentery supporting the uterus, on both sides.
7. Make a cut through the bursa between the oviduct and the ovary or through the utero-tubal junction on each side, freeing the entire tract (*Figure 4c*).
8. Now trim the tract prior to flushing using a low power dissecting microscope:
   (a) Spread the tract out in a dry sterile Petri dish and trim away any remaining mesentery and fat.
   (b) Cut off the oviduct at the utero-tubal junction (if this was not already done) and make a 0.5 cm longitudinal cut in the end of each horn.
   (c) Cut across the cervix to expose the entrances to both uterine horns.
9. Rinse with a few drops of medium and place the tract into a sterile Petri dish or watch glass. Using a syringe 30G needle or a rubber bulb on a Pasteur pipette which has been pulled out and broken off, flush approx. 0.5 ml of medium (*Protocol 3*) through each horn by inserting the needle or pipette tip into the cervical end. The uterus will balloon slightly and medium should flow freely through the tract.

Embryos will settle to the bottom of the flush dish very quickly. It is useful to have round-bottom watch glasses (Carolina Biological) to flush the uteri so that the embryos can be swirled to the centre of the dish, but a small Petri dish is a suitable alternative. The cleaner the dissection of the uterus, the less cellular debris, fat, and blood will obscure the visualization of the blastocysts (*Figure 5*). Embryos can be collected from the flush dish and transferred to a watch glass containing fresh medium prior to their introduction into the micromanipulation chamber or dish. A simple and convenient mouth-

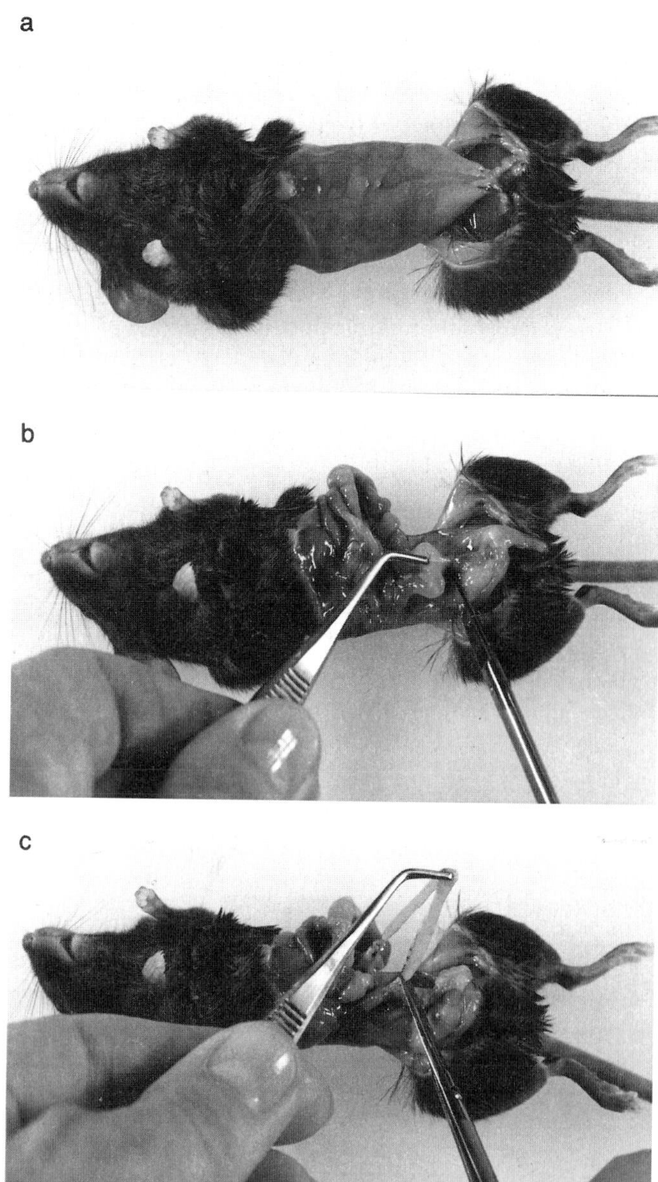

Figure 4. Blastocyst recovery at three and a half days post-coitus. (a) The peritoneum is exposed by tearing the skin. (b) After the peritoneum is opened and the intestines are pushed craniad, the bifurcation of the uterine horns is grasped and a cut is made across the cervix. (c) The uterus is lifted and each horn is freed by cutting the mesentery and finally by cutting at the utero-tubal junction, as shown here, or between the oviduct and ovary if the oviducts are to be recovered.

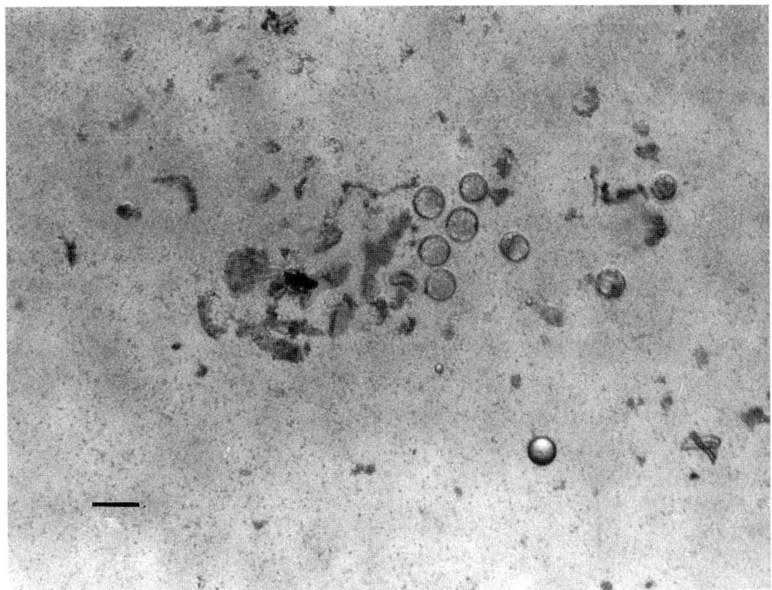

Figure 5. Eight embryos including seven blastocysts immediately after being flushed from the uterus. Blood, cellular debris, fat, and oil drops can make embryos difficult to see. Bar = 0.1 mm.

controlled transfer pipette can be made by drawing out a Pasteur pipette by hand over a flame, breaking it off at a diameter slightly larger than a blastocyst, and attaching it to rubber tubing with a mouth piece.

3. Injection of ES cells into embryos

3.1 Micromanipulation apparatus

3.1.1 Microscopes

ES cell injections can be done under phase, bright-field, or differential interference contrast optics. The microscope, which is best located on an anti-vibration table, should have at least one low and one higher power objective giving a magnification range of approximately × 63 to × 200. Another very useful feature is a fixed stage so that the embryos and microinstruments remain in the same focal plane relative to each other when the microscope focus is adjusted. The convenience of this feature is well worth the effort of modifying a microscope if a fixed stage is not available. However, if stage-mounted micromanipulators are used, this is not a concern.

Another useful addition to the microscope is a cooling stage. Its main advantage is to prevent stickiness during cell injections and to impart a degree of rigidity to cells and embryos. Any cooling device that will allow free access

a

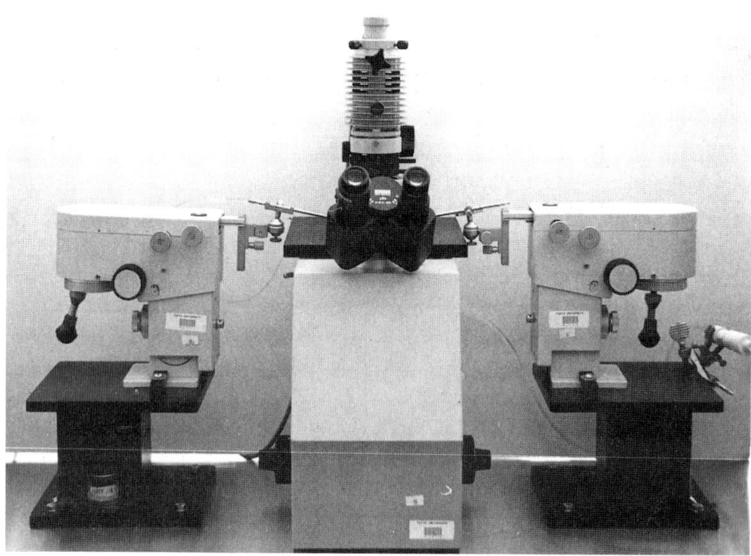

b

Figure 6. Micromanipulation set-up with Leitz micromanipulators and (a) an inverted microscope or (b) a standard microscope. Both are placed on pneumatic, anti-vibration tables. (a) For the inverted microscope, the manipulators are raised to the level of the stage on metal supports. A Gilmont micrometer syringe can be seen mounted on the right. Instrument holders are positioned for use in a culture dish. Note that the manipulators are level and the instrument holders are tilted at the ball and socket joint to attain the correct angle. (b) In the set-up with the standard microscope, the manipulators are mounted on a Leitz base plate. On the right is a Narishigi micrometer syringe used for the holding pipette, and on the left is the Stoelting micrometer for cell injection. The control box for the Physitemp cooling stage is on the left.

to the stage and will also keep the injection chamber at 10°C will suffice (13). We use a Physitemp cooling stage with a modified stage to accommodate either an inverted or standard microscope (*Figure 6*).

There are several microscope configurations that can be used for blastocyst injections. An inverted microscope (*Figure 6a*) requires that the injections be done on the bottom of a culture dish (or lid) or on a depression slide. The embryos and cells can either be placed in a large volume of culture medium or they can be placed in microdrops of medium covered with a layer of inert oil such as heavy liquid paraffin oil (HLP, Fisher or mineral oil, Sigma). Microdrops will allow for separation of different groups of embryos or cells and also obviates the problem of searching for the embryos in a large field. An alternative microscopic configuration is the standard microscope (*Figure 6b*) and the use of a micromanipulation chamber (*Figure 7*) in which the embryos and cells are kept in drops of medium hanging from a coverslip. This configuration requires a long working distance condenser and a fixed-stage microscope. The main difference in the two methods is in the shape of the instruments used for cell injection and the fact that in one case the embryos are resting on a the solid surface and in the other they are resting on an interface of oil and medium. Other than this, the choice may be dictated by the microscope and manipulators available, since both methods can be used quite successfully.

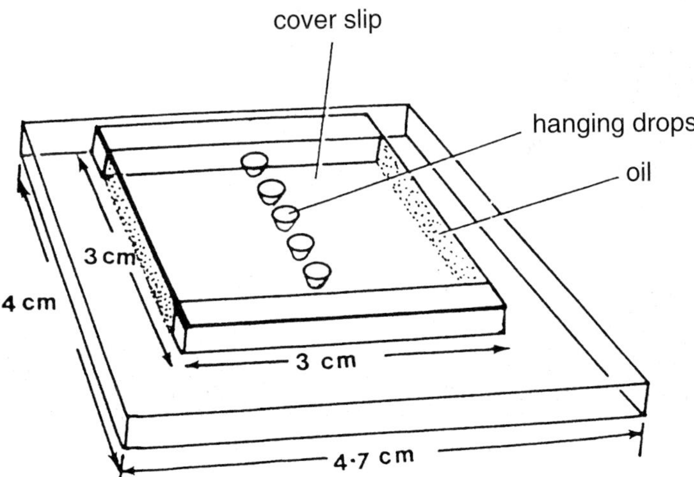

Figure 7. Diagram of a hanging-drop micromanipulation chamber (Leitz). The chamber is made of glass with two parallel supports on which a coverslip rests. To load the chamber, petroleum jelly is placed on the top of the supports. Drops of medium are placed on a row on an acid-alcohol cleaned, sterilized glass coverslip which is then inverted onto the supports. Finally, the space under the coverslip is filled with HLP. A chamber can be made by gluing the cut off ends of a Pasteur pipette onto a glass slide (reproduced from ref. 11).

3.1.2 Micromanipulators

Several types of micromanipulator are available that are suitable for ES cell injection. Cell injection into blastocysts is a fairly large-motion procedure so a wide range of movement, particularly in the coarse adjustments, is important. For this reason, stage-mounted manipulators are not as convenient as those that are free-standing. The precision and working range of movement needed for the holding pipette is less than for the cell injection pipette so that a simpler manipulator could be used for that instrument. However, we recommend two Leitz micromanipulators (one right- and one left-handed model; Leitz Instruments) which satisfy all the requirements for ES cell injection and can be used for virtually any other micromanipulation as well. The Leitz manipulator is a direct mechanical device with an XY joystick movement with a wide range of adjustment of the movement ratio. Movement in the third plane is by a screw device. All three planes of movement have coarse and fine adjustments and there is flexibility also in the positioning of the instrument holder in all directions.

3.1.3 Micrometer syringes

Mechanical suction-and-force syringe devices are used for both the holding and cell injection pipettes in conjunction with Leitz instrument holders. There is a wealth of choices of micrometer-driven syringe devices, but it is essential that they both push and pull the syringe plunger to create suction and force. It is a mistake to use too fine a device (one with very small displacement per

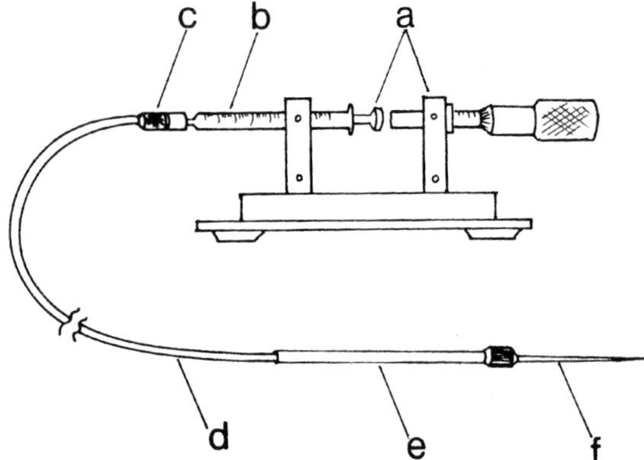

Figure 8. Diagram of a Stoelting micrometer syringe assembly with attached injection pipette. A rubber band connecting the two points indicated by (a) will allow the syringe to be used for suction as well as expelling. The assembly includes a Hamilton microlitre syringe (b) and an adapter (c). Flexible polythene tubing (d) is fitted to connect the syringe to a Leitz instrument holder (e) into which the injection pipette (f) is inserted. The syringe and tubing are filled with oil. It is useful to have a three-way valve (not shown) in the line (reproduced from ref. 35).

turn), especially for the holding pipette. Both devices are attached to instrument holders with polythene tubing filled with HLP or silicon oil. A three-way valve is placed in the line between the syringe and instrument for easier filling and instrument changing. For the holding pipette, we use either a Gilmont 2 ml micrometer syringe or a Narishige Model IM-58, with a 3 ml syringe, which has the added advantage of a magnetic base. More precision and finer control over suction is needed for the cell injection pipette. For this we either use a deFonbrune Suction and Force pump or a Stoelting micrometer syringe assembly (*Figure 8*). This comes with a 50 μl Hamilton syringe that should be changed to a larger size (e.g. 0.5–1 ml) for better control. A rubber band or spring clip will need to be added to allow retraction as well as pushing of the plunger. Alternatively, an Eppendorf CellTram Oil System (Cat. No. 5176 000.025) can be used for the injection needle. The syringes are placed on the opposite side of the manipulator as the instrument they control so that suction or force can be applied at the same time as the instrument is being moved with the joystick. As an inexpensive alternative, a somewhat less precise but adequate set of injection and holding syringes can be made using air-filled, 10 cc disposable plastic syringes connected to the instrument holders.

3.2 Microinstruments

3.2.1 Holding pipettes

Microinstruments for use in blastocyst injection are fashioned out of glass capillary that will fit the instrument holders (e.g. Drummond custom capillary, 1 mm OD, 0.75 mm ID). Only two instruments are necessary, a pipette to hold and stabilize the blastocyst and a pipette to pick up and inject the cells. The tips of these instruments are fashioned in the same way regardless of the microscope/manipulator configuration used; only the more proximal shaping of the shafts will differ depending on the manipulation set-up.

The overall tip diameter of a holding pipette should be broad enough to stabilize an embryo. The opening should be small enough to prevent gross deformation of the embryo or the zona pellucida (ZP) while still allowing sufficient area for suction to hold the embryo tightly . If a microforge is available, reproducible, precision instruments can be made (see *Protocol 5* for the method). However, by trial and error, holding pipettes can also be made by simply breaking off pulled capillary, selecting those with the appropriate diameter and shape, and passing the tips quickly through a flame to close the tip down to the correct diameter.

Protocol 5. Making an embryo holding pipette

Equipment

- Pipette puller (such as Kopf)
- Microforge (deFonbrune)
- Glass capillary
- Spirit lamp or Bunsen burner

Protocol 5. *Continued*

Method

1. Pull out a length of glass capillary either with a pipette puller or by hand over a small flame. We prefer to make this initial pull by hand over a low flame to give a taper of 2–3 cm, which confers better control over the suction than the shorter taper obtained with a pipette puller.
2. Break the pipette cleanly at an overall diameter of 60–90 μm (*Figure 9*) using a microforge.
   (a) Start with a bead of glass on the filament tip.
   (b) Focus on the pipette and bring the filament into the same focal plane.
   (c) Turn on the filament and touch the molten glass bead to the pipette at the desired diameter just until it sticks, taking care not to melt the pipette.
   (d) Turn off the filament and contraction will break the pipette.
3. Turn on the filament and bring it close to the tip of the pipette to fire-polish and close it down to a diameter of 12–20 μm. The pipette is now ready to use in a hanging-drop.
4. Place additional bends in the shaft of the pipette if depression slides or culture dishes are used for manipulation, so that the final orientation of the tip is horizontal when it enters the microscope field (*Figure 10*). Make these bends with the microforge in the narrow part of the shaft (*Figure 11*) or in the thick part of the shaft over the small flame of a microburner that can be made by connecting a 30G hypodermic needle to a gas supply.

3.2.2 Cell injection pipettes

Either blunt or bevelled pipettes can be used for cell injection. Blunt tips require some force to penetrate the trophectoderm layer of the blastocyst while bevelled tips can be inserted more deliberately. Both methods have their proponents and are equally effective for blastocyst injection, although making the bevelled tip requires more steps. Morula injection requires the use of a bevelled pipette to avoid damage to the blastomeres.

For both types of pipette, glass capillary is drawn out to a gradual 1 cm taper with a pipette puller. For a blunt pipette, the tip is broken off using a microforge (see *Protocol 5*, step 2 and *Figures 9a–c*) at an internal diameter of 12–15 μm, a diameter just large enough to allow the entry of single ES cells without deformation (see *Figure 13*). For a bevelled tip, the pipette is placed on a silicon pad under a dissecting microscope and a scalpel blade is used to

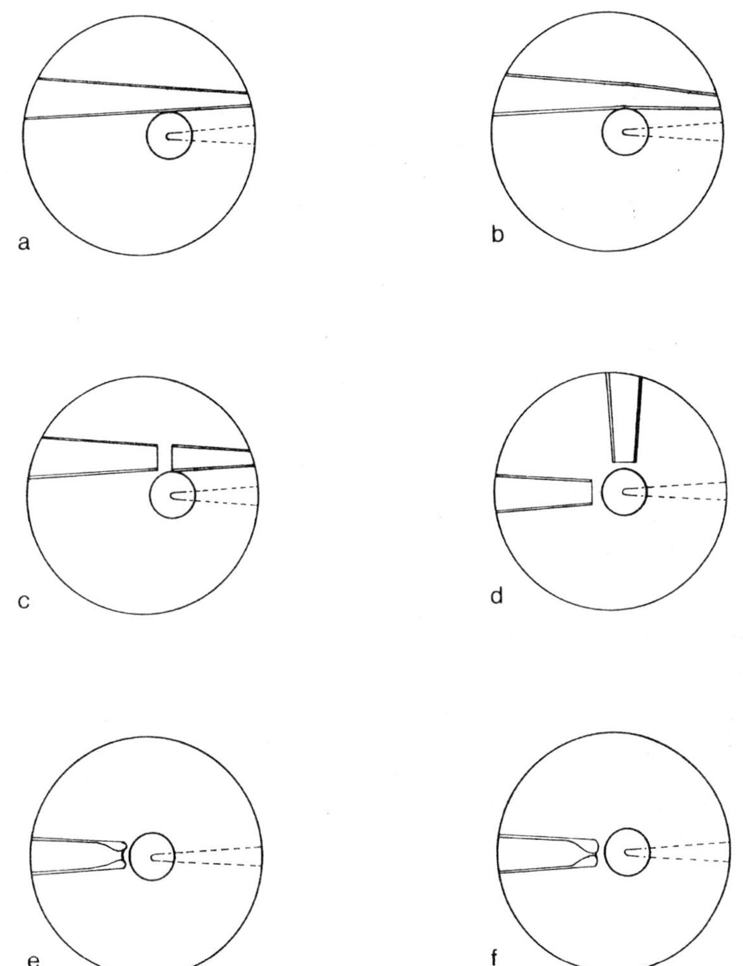

Figure 9. Making an embryo holding pipette. (a) A molten glass bead on the tip of the microforge filament is brought into contact with a pulled-out pipette at the appropriate diameter and (b) the bead fuses with the pipette. (c) The filament will contract when the element is switched off, breaking the pipette. (d–f) The pipette is closed down to a small opening by fire-polishing the tip with a hot filament (reproduced from ref. 12).

break off the tip. Pipettes of the correct diameter, with smooth, sharp bevels that are neither jagged nor too long (see *Figure 14*), are then selected for use (12). No fire-polishing of either type of injection pipette is necessary. For use in a culture dish or depression slide, bends must be made in the shaft of cell injection pipettes so the final working position of the tip is horizontal with respect to the microscope stage (see *Protocol 5*, step 4 and *Figures 10* and *11*).

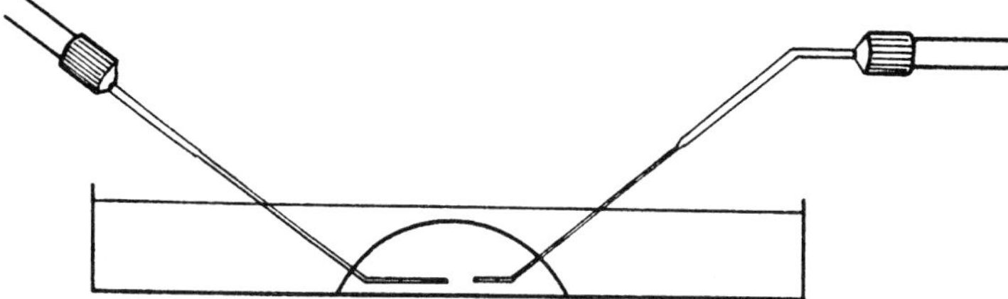

Figure 10. Positioning of instruments for use with an inverted microscope, with bends in the glass pipettes. The pipette on the right has two bends which allows the instrument holder to be positioned horizontally, while the one on the left has a single bend so that the instrument holder is at an angle. Which position is used will depend on the location of the manipulators with respect to the microscope stage.

3.2.3 Embryo transfer pipettes

Making the pipette for transferring embryos into a host female is very similar to making a holding pipette (*Protocol 5*). Capillary is drawn out by hand in the same way and the tip is broken off similarly using a microforge or by hand. The difference is that the overall diameter at the tip is larger so that the opening can easily accommodate embryos (110–130 μm internal diameter). For this same reason, fire-polishing of the tip is minimal, just enough to take off the sharp edges without closing down the opening. No bends are necessary (*Figure 12*).

3.3 Injection procedures

3.3.1 Blastocyst injection

On the day of an experiment, blastocysts are recovered from pregnant females in the morning. ES cells in an exponential growth phase are fed; one hour later they are trypsinized to single cells (Chapter 3, *Protocol 5*) and resuspended in DMEM/Hepes +20% FCS and maintained as a single cell suspension at 4 °C. This suspension can be used throughout the day.

Blastocysts and a single cell suspension of ES cells are introduced into drops of medium in the manipulation chamber using a drawn-out, mouth-controlled Pasteur pipette. This can be done under a dissecting microscope, for a hanging-drop chamber (*Figure 7*), or under the manipulation microscope for a dish or depression slide. When using a hanging-drop chamber, set up the microscope so that the embryos are in focus and then remove the chamber from the microscope. Now the injection instruments can be set up in the micromanipulators, without fear of breaking them on the chamber, as follows:

(a) Put each instrument into its holder and manipulator, fill with oil by positive pressure.

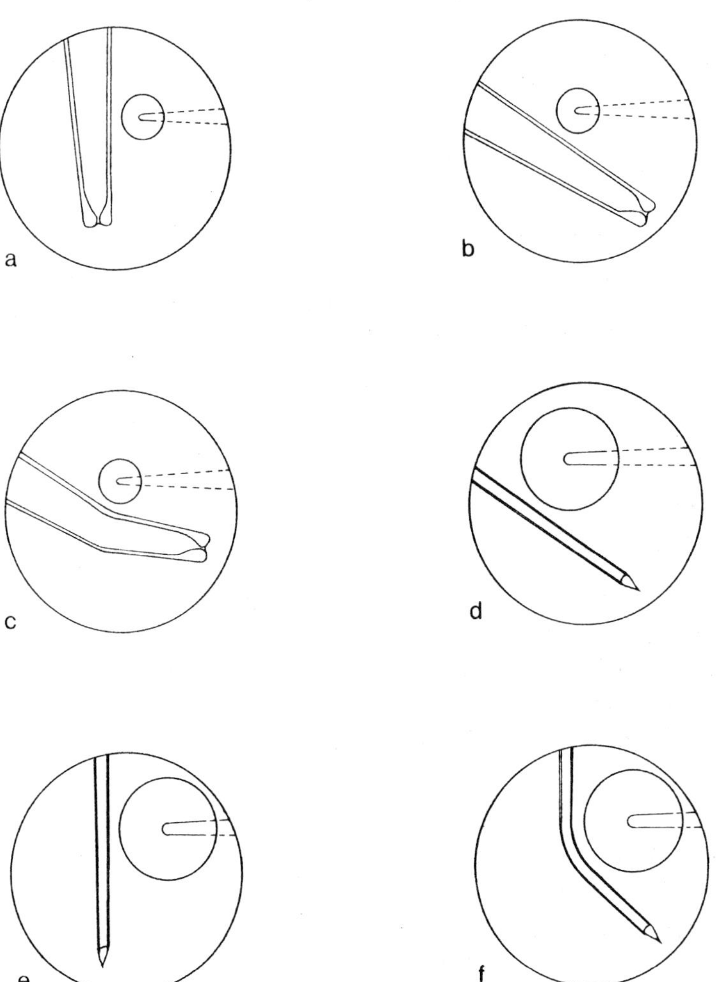

Figure 11. Making bends in a pipette with a microforge. (a–c) Holding pipette. The pipette is oriented as in either (a) or (b) and the filament is heated, causing the pipette to bend towards the filament (c). (d–f) Bevelled injection pipette. If the bend is made near the end of a fine pipette, a higher magnification and lower heat can be used. While making the bend, the bevel should be oriented so that its final position in the injection chamber will be as indicated in *Figure 13* (reproduced from ref. 12).

(b) Without adjusting the microscope focus, roughly position the tips parallel in the field using the coarse adjustments of the manipulator; put them into the same focal plane using the fine adjustments of the manipulators.

(c) Withdraw the instruments far enough to allow the introduction of the manipulation chamber onto the stage and restore the instruments to position.

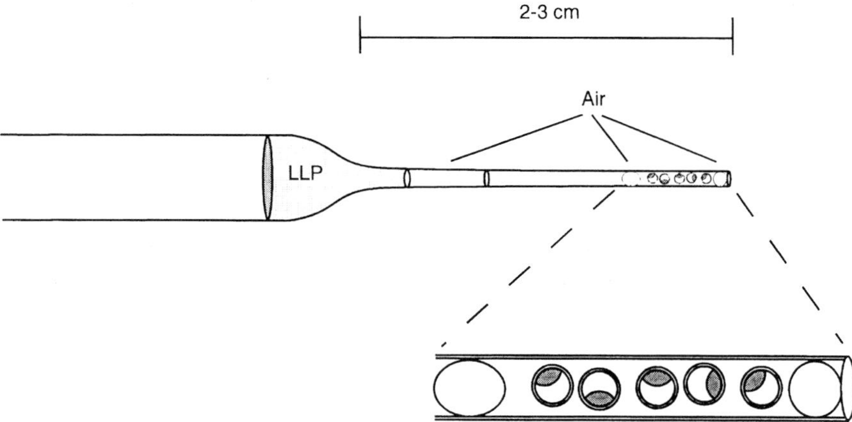

Figure 12. Diagram of embryo transfer pipette ready for embryo transfer. The narrow, pulled-out tip should measure 2–3 cm. In filling the pipette, light liquid paraffin oil is taken up to the thick region, then a column of air to separate oil from medium, then medium, then a small marker air bubble followed by the embryos (six blastocysts are shown here) and, finally, a small bubble at the tip.

A dish or depression slide can be left in place while setting up the micro-injection instruments. The blastocyst injection procedure is outlined in *Protocol 6*.

Protocol 6. Blastocyst injection

A. *Method using blunt-tipped pipettes (11, 36) (see Figure 13)*

1. Pick up 12–15 small, round ES cells in a tight column in the tip of the injection pipette using suction.
2. Pick up a blastocyst with the holding pipette by suction, positioning it so that the ICM can be seen in profile against the pipette (*Figure 13a*).
3. With the cell injection pipette in the centre of the field, and the blastocyst in focus but offset to the north or south, position the injection pipette so that it is level with the surface of the ICM when it is at the full extent of the range of movement of the manipulator joy stick (*Figure 13a*).
4. Retract the injection pipette with the joystick, reposition the blastocyst to the centre of the field, check that the equatorial plane of the blastocyst is in the same focal plane as the injection pipette, i.e. that the edge of the zona pellucida (ZP) is in sharp focus (*Figure 13b*), then apply force with the joystick to pop the injection pipette through the trophectoderm wall. The pre-set limit on the joystick movement will prevent damage to the ICM (*Figures 13c, 13d*).

5. Expel the ES cells into the blastocoelic cavity and remove the pipette (*Figure 13e*).
6. Place the injected embryo to one side of the chamber drop and repeat the procedure.

B. *Method using a bevelled pipette (12) (see Figure 14a)*

1. Follow part A, steps 1 and 2.
2. Put the injection pipette in the same focal plane and slowly penetrate the trophectoderm layer, trying to insert the tip of the pipette at a junction between two cells.
3. Follow part A, steps 5 and 6.

The microinstruments will only need changing if they become clogged or sticky. As a rule, we set up a fresh manipulation chamber with fresh ES cells after several hours or when the pH of the medium changes. When all of the embryos in a chamber have been injected with ES cells, they can be transferred to culture medium for a brief period of culture in a 5% CO_2 incubator prior to transfer to pseudopregnant hosts.

3.3.2 Morula injection

Introduction of ES cells into embryos at the morula stage is another means of producing chimeras (see Chapter 5, *Protocol 5* for morula recovery methods). The forceful method of blastocyst penetration cannot be used for injecting morulae without damage to the blastomeres. Instead, a bevelled pipette is used to penetrate the ZP and cells are inserted just under the ZP (*Figure 14*). Alternatively, the cells can be placed among the blastomeres in an embryo that has not yet compacted or one that has been decompacted by a brief culture in calcium-free medium. No study has been published that critically compares the success of morula versus blastocyst injection. Using morula injection we have produced ES cell chimeras, however, our experimental results indicate no clear advantage of this method over blastocyst injection.

4. Embryo transfer

Oviduct transfers (*Protocol 7*) and uterine transfers (*Protocol 8*) are similar although oviduct transfers require somewhat greater skill. Since the ostium of the oviduct is enlarged and fairly obvious during the day after ovulation, this is a convenient time for oviduct transfers. Uterine transfers should be done on the third or fourth day of pseudopregnancy (see Section 2.2.3 for details on setting up the embryo transfer hosts). As an alternative to tribromoethanol anaesthesia, inhalant veterinary anaesthetics (e.g. metafane) can be used provided an adequate gas scavenging system is available. These inhalants also require somewhat more experience on the part of the surgeon, as the window

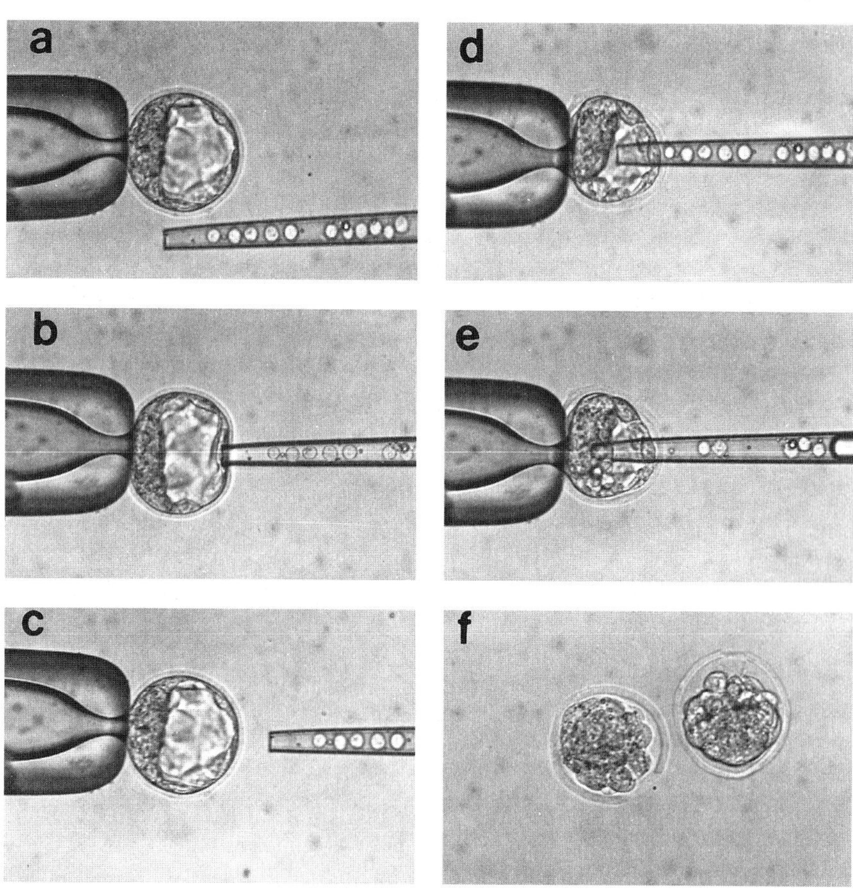

Figure 13. Blastocyst injection using a blunt pipette. (a) The limit of travel of the injection pipette is set when the pipette is in the position indicated. ES cells can be seen in the injection pipette and the blastocyst is held by suction on the holding pipette. (b) The focal planes of the injection pipette and blastocyst can be checked by gently touching the pipette to the embryo. (c) The injection pipette is repositioned with the joystick and (d) forced through the trophectoderm layer by a sudden movement, or bash, of the joystick. (e) Cells are then expelled into the blastocoelic cavity. (f) Injected embryos prior to re-expansion.

of complete anaesthesia is small without continuous administration of the drug.

Protocol 7. Embryo transfer to the oviduct

Equipment and reagents

- Sterile surgical instruments
- Embryo transfer pipette (*Figure 12*)
- Mouth pipette
- Tribromoethanol anaesthesia (see *Protocol 2*)
- 70% alcohol and cotton
- Light liquid paraffin oil (Fisher, Sigma)

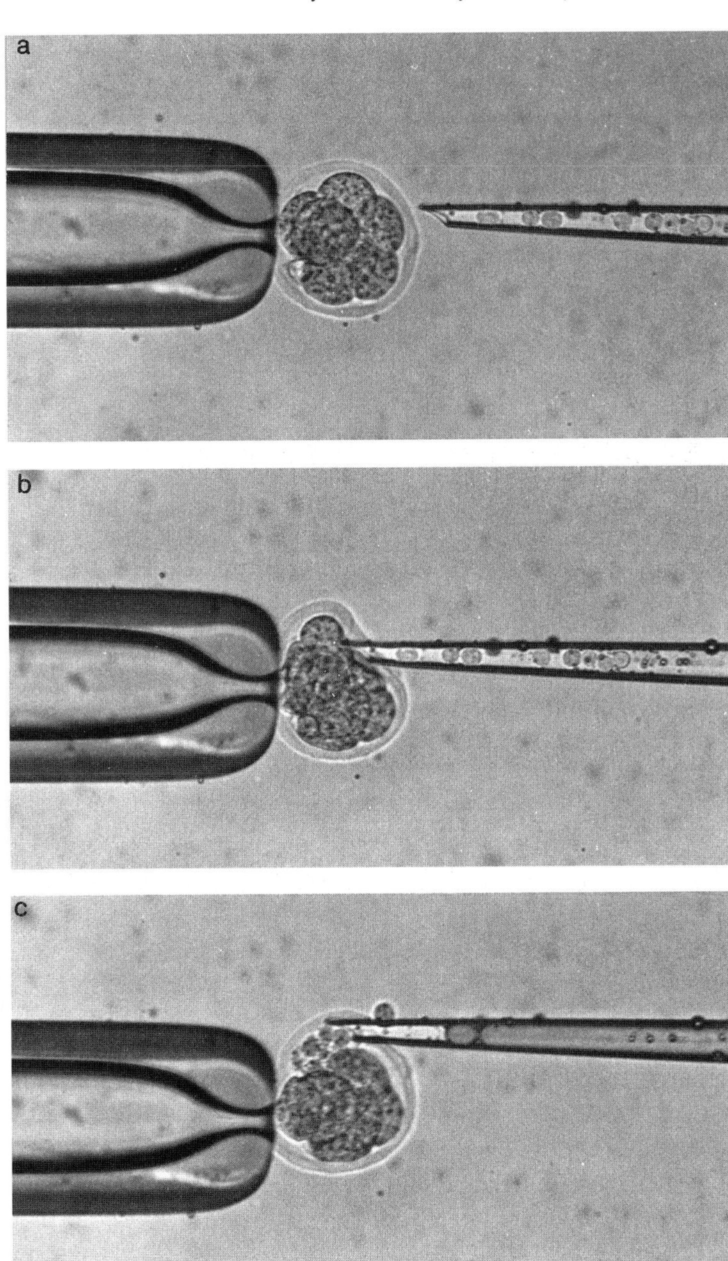

Figure 14. Injection of cells under the ZP of an 8-cell embryo. (a) The embryo is held by suction and ES cells are in the bevelled injection pipette. (b) The tip of the pipette is slowly inserted through the ZP and (c) the cells are expelled into the perivitelline space once the pipette has fully penetrated.

Protocol 7. *Continued*

Method

1. Weigh and anaesthetize a pseudopregnant female.
2. Prepare the embryo transfer pipette by connecting it to a mouth pipette and filling it with light liquid paraffin oil at least to the end of the taper. Take up a column of air, medium, a small air bubble, the embryos to be transferred, and finally a small air bubble at the tip (*Figure 12*). Set this aside while the mouse is prepared.
3. Clean the back of the mouse with alcohol and make a 1 cm transverse incision in the skin at the level of the first lumbar vertebra (*Figure 15a*). Wipe the hair away with cotton dampened with alcohol.
4. Slide the skin incision laterally until the right ovarian fat pad and ovary are visible through the peritoneal wall. Make a 3 mm incision through the peritoneum and grasp the fat pad with blunt forceps, pulling it and the ovary and oviduct out through the opening in the direction of the midline.
5. Position the mouse on a dissecting microscope stage with its head facing away from you at 11 o'clock (*Figure 15b*).
6. Stabilize the ovary outside the peritoneum by clamping with a small sepaphim clip or drape the fat pad over a small square of index card with a notch in one side, pulling the ovary, oviduct, and a short length of uterus through the notch (*Figure 15c*).
7. With a pair of fine watchmaker's forceps in each hand and the loaded transfer pipette in the right hand, locate the fimbria of the oviduct by looking in the space between the ovary and the oviduct. Using the forceps, tear a hole in the bursa, the vascular, transparent membrane enclosing the ovary. It may be necessary to remove fluid or blood with a tissue paper in order to visualize the ostium. Alternatively, a drop of epinephrine (Sigma) can be used to control bleeding (37).
8. Insert the tip of the pipette into the ostium of the oviduct and blow in the embryos along with the air bubbles on either side of them. The orientation of the oviduct opening is such that if the mouse is facing away from you at about 11 o'clock, and the fat pad is pulled toward the midline, the pipette can be inserted from the right, parallel to the backbone for either a right or left oviduct transfer (*Figure 16*).
9. The marker air bubbles will be visible in the first turn of the oviduct but the transfer pipette should also be checked to ensure that all embryos were expelled.
10. Carefully replace the tract into the peritoneum.
11. Repeat steps 4–10 on the left side for a bilateral transfer.
12. Provided the peritoneum incisions are small, they can be left open. Close the skin incision with a small wound clip.

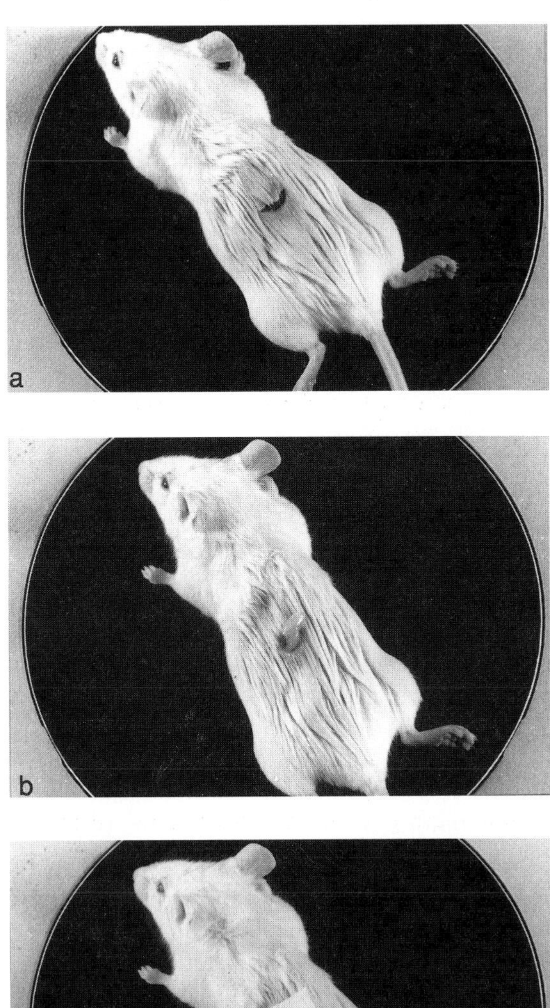

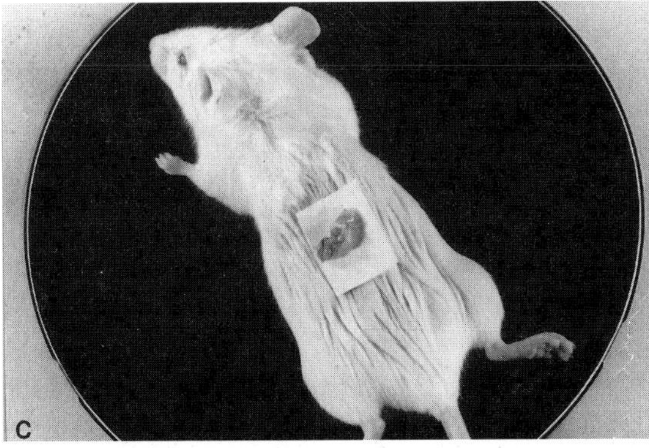

Figure 15. Preparation for embryo transfer. (a) A single dorsal transverse incision will serve for embryo transfer to both the left and the right. The mouse is in the correct position for transfer to either side. (b) The left ovary and ovarian fat have been exteriorized through a lateral peritoneal incision and the skin incision. With the exteriorization of a small, distal segment of the uterus, the mouse will be ready for embryo transfer to the uterus. (c) The fat pad, ovary, oviduct, and a small section of the uterus have been stabilized by anchoring them in a notch cut in the side of a square of card. Embryo transfer to the oviduct can now be done.

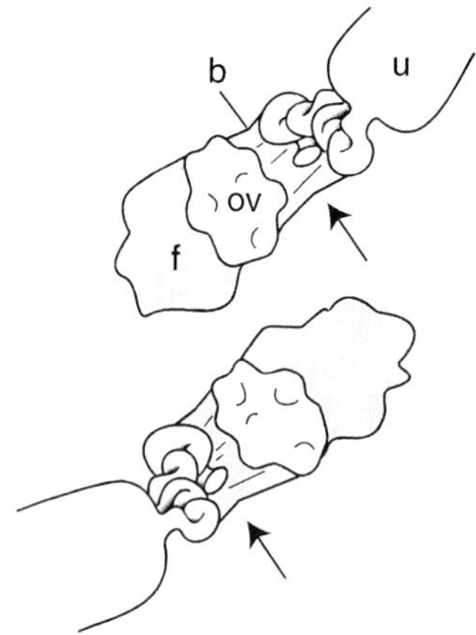

Figure 16. Diagram of the fat pad (f), ovary (ov), coiled oviduct, and utero-tubal junction connecting the oviduct to the distal end of the uterus (u). The diagram shows the position and orientation of the ostium when the mouse is prepared as illustrated in *Figure 15*. The embryo transfer pipette should be inserted into the ostium in the direction of the arrows once a hole has been opened in the bursa (b). The right ovary and oviduct is shown at the top, the left at the bottom.

Protocol 8. Embryo transfer to the uterus

1. Follow *Protocol 7*, steps 1–5 with the modification that a length of uterus is also exteriorized and the reproductive tract is directed away from the midline.
2. With the loaded embryo transfer pipette and a sewing needle (sharp) or a 5/8 inch 26G hypodermic needle in the right hand, and the uterus held firmly by the mesentery with watchmaker's forceps in the left hand, insert the needle through the muscle layers of the uterus, 2 mm from the utero-tubal junction. Direct the needle parallel to the long axis of the uterus so that the tip enters the lumen.
3. Remove the needle and without taking your eyes off of the hole, insert the tip of the transfer pipette into the lumen and expel the embryos and marker air bubbles. These will not be visible through the uterine wall but will serve as indicators of the expulsion of the embryos.
4. Check the transfer pipette to ensure that all embryos have been expelled.
5. Replace the tract into the peritoneum and repeat on the left side for a bilateral transfer.
6. Close the incision as for oviduct transfers.

5. Chimerism

5.1 Detection and quantification of chimerism (see Chapter 5 for additional markers of chimerism)

5.1.1 Coat colour

The most convenient and readily apparent genetic marker of chimerism is coat colour. The common coat colour alleles at variance in ES cell injection experiments are shown in *Table 1* and the cell types affected and phenotypic effects are shown in *Table 3*. Chimeric combinations of strains which differ at only one coat colour locus allow a simple visual appreciation of the degree of tissue contribution of each component in terms of the proportion of the coat that expresses the ES cell allele. This evaluation of chimeric animals is necessarily subjective but in general, the degree of coat colour chimerism of a particular animal correlates with the degree of germline contribution. In evaluating a given ES cell clone, its contribution to a number of chimeras should be considered. For example, if a clone consistently contributes more than 50–60% of the coat in a series of chimeras, then it will quite likely contribute to the germline at a similar level. Conversely, if it rarely contributes more than 5–10% of the coat to any chimera, the likelihood of germline transmission is correspondingly small. The decision of whether or not to test breed a particular chimera should thus depend on its level of chimerism as well as the behaviour of that ES cell clone in other chimeras.

5.1.2 GPI

Another method of detecting chimerism takes advantage of isozyme differences between strains, most commonly in the ubiquitous, dimeric enzyme,

Table 3. Some common coat colour alleles used as markers in chimera experiments and the cell type they affect

Locus	Alleles	Cell type affected and phenotypic effect
Agouti	A^w, A, a, a^e	Acts through hair follicle to affect the amount and distribution of yellow pigment; dominant wild-type (A^w or A) produces hair with a yellow sub-apical band.
Brown	B, b	Affects melanocytes; wild-type black (B) is dominant.
Albino	$+^{Tyr\text{-}c}$, Try^c, $Tyr^{c\text{-}ch}$	Affects melanocytes in hair and eye; wild-type full colour ($+^{Tyr\text{-}c}$) is dominant; homozygous albino (Try^c/Try^c) is epistatic over and will mask any other coat colour alleles; chinchilla ($Try^{c\text{-}ch}$) causes a reduced pigmentation and affects yellow pigment more than black; $Try^c/Try^{c\text{-}ch}$ mice are intermediate between albino (Try^c/Try^c) and chinchilla ($Try^{c\text{-}ch}/Try^{c\text{-}ch}$).
Pink-eyed dilution	$+^p$, p	Affects melanocytes in hair and eye; causes a reduced pigmentation but affects black pigment more than yellow; wild type ($+^p$) is dominant.

glucose phosphate isomerase (GPI-1) which has three variants (*Table 1*) distinguishable by electrophoretic mobility. Homozygous animals show a single, homodimer band while heterozygotes for any two alleles will have both homodimers and the corresponding heterodimer which runs intermediate to the corresponding homodimers. Since dimerization takes place intracellularly, chimeras between homozygotes for different alleles will show only homodimers, in a proportion that reflects the chimeric contributions of each cell type. The only exception is chimeric muscle which may show a heterodimer if the two cell types fuse during myogenesis.

The GPI electrophoretic assay is particularly useful since it can be used on any tissue at any stage of development and will provide a crude measure of the degree of ES cell contribution. About 5% contribution can be detected and the relative strength of bands indicates the proportion of tissue contribution. The enzyme is very hardy; even tissue from animals that have been dead for some hours can be typed successfully and samples can be stored frozen indefinitely. The one disadvantage is that the assay uses tissue homogenates, and so provides no information about the cell types derived from the ES cells. GPI isozymes can be separated by starch gel electrophoresis (38) or cellulose-acetate electrophoresis (Chapter 5, *Protocol 13*).

5.2 Phenotypic effects of chimerism

5.2.1 Sex ratio of chimeras

All of the ES cell lines in common use were derived from male embryos and thus contain a Y chromosome. There are several reasons for preferring to use male ES cell lines: male ES cells produce a higher proportion of phenotypic male chimeras, which can produce more offspring in a given period of time than females to test for germline transmission; female-derived XX cell lines are reputed to be unstable. This reputation for instability comes partly from work with female embryonal carcinoma cell lines, which were shown to lose one X after several passages to become XO. Female ES cells can be derived as readily as male ES cells, and there are no clear-cut reasons other than those mentioned above to avoid their use.

Sex bias among ES cell chimeras in favour of males is a common observation when male-derived, XY ES cells are used. In combination with a female embryo, male cells will often produce a fertile, phenotypic male chimera. This sex conversion presumably occurs when XY cells colonize sufficient portions of the various tissues which determine sex in the developing embryo. Sex conversion is not always complete in XX <-> XY chimeras, however, and occasionally results in an infertile hermaphrodite or gynandromorph. Hermaphroditism may be subtle, with the animal showing small testes and abnormalities of the urogenital system, or more obvious with the formation of ovotestes and components of male and female reproductive tracts in the same animal. Such anatomic curiosities are rarely fertile and should be discarded if germline transmission is the object. An exploratory laparotomy will quickly determine the status of an animal suspected of being a hermaphrodite either because of infertility or ambiguous external genitalia.

Fully sex-converted chimeras are advantageous for breeding since they will only transmit the ES cell (XY) genotype, due to the inability of XX cells to undergo spermatogenesis. On the other hand, transmission of the male-derived ES cell genotype has been reported to occur from female chimeras by a number of laboratories. Since XY cells do not normally undergo oogenesis in chimeras, this phenomena may be due to the loss of part or all of the Y chromosome, resulting in effectively XO cells which are capable of forming ova, as was shown in at least one case of transmission from a female chimera (39).

5.2.2 X-linked mutations and heterozygous effects

Chimeras with targeted ES cells contain heterozygous cells; the effects of heterozygosity, if any, may reveal themselves to a degree corresponding to the amount of ES cell contribution. In the case of targeted, X-linked genes in male ES cells, chimeric mice are hemizygous for the mutated allele and will demonstrate whatever phenotype is associated with the mutation dependent on the amount and type of chimeric tissue. For example, a mutation at the X-linked gene GATA-1, an erythroid-specific transcription factor, causes the death of those chimeras with the greatest contribution of mutant cells, due to failure of erythropoiesis (40). Similarly, lethality associated with increasing amounts of chimerism may arise in the case of an autosomal dominant mutation. These possibilities should be considered when designing targeting experiments, especially since they can define the end-point in cases where the chimeric phenotype is so severe as to prevent germline transmission.

In situations like this, a 'rescue' vector containing a functional copy of the targeted gene, integrated into another chromosome, may help the targeted allele through gametogenesis in the chimera. Segregation will then separate the mutant from the rescue vector insertion in subsequent generations. This will allow the mutant phenotype to be studied in a non-chimera, even if the effect is severe.

6. Maintaining a targeted mutation

6.1 Animal husbandry

Once chimeras have been produced, and throughout subsequent studies of the mutant phenotype, efficient breeding is crucial for an expedient analysis of the mutation. In addition to the time required to produce the targeted ES cell clone, the minimum time between injection of targeted cells into embryos and the birth of mice bred to homozygosity for the mutant allele is five months. Any deficiencies in the breeding program will increase this time considerably.

Thus, the highest standards of animal husbandry must be maintained and the breeding program of chimeric mice and their progeny must be scrupulously monitored. Monitoring for the presence of murine pathogens should be a routine feature of the health surveillance of the animal facility, whether it be SPF or conventional, since the presence of any pathogen has the potential to interfere with breeding or the analysis of mutant phenotypes. If an SPF facility is not available, a conventional mouse facility can be maintained in such a way as to minimize the possibility of such common pathogens as murine hepatitis virus (MHV) and Sendai virus. The routine use of micro-isolator cages and HEPE-filtered cage changing stations, along with strictly controlled import of animals, with judicious use of quarantine rooms, can keep a conventional facility virtually problem-free. As a final safeguard, cell lines to be injected into mice and obtained from outside sources should be murine antibody production (MAP) tested to ensure that they do not harbour pathogenic viruses.

6.2 Test breeding

Chimeric males (and sometimes females) are generally test bred to ascertain contribution of the ES cells to the germline. It is possible that other methods of determining germline contribution could be developed, such as molecular analysis of sperm DNA using PCR to detect the altered allele. Reasons for screening chimeras in this way, such as suspected effects of the mutation on gametogenesis in the chimera, may make the effort worthwhile, but for the most part, test breeding should reveal fairly quickly what secrets the germ cells conceal.

Chimeras should be mated to mice with genetic markers that will allow a distinction to be made between ES-derived and host blastocyst-derived gametes, keeping in mind that only half of the ES cell-derived gametes will carry the targeted allele. For example, a chimera between a 129/SvJ-derived ES cell, which has the albino and chinchilla alleles at the albino locus, and a non-albino blastocyst could be test mated with any albino mouse. Offspring derived from a host blastocyst-derived gametes would have black eyes, which are distinguishable at birth, while ES cell-derived gametes would produce progeny with pink eyes. We have used several different 129-derived ES cell lines, differing in coat and eye colour and GPI-1 alleles (*Tables 1* and *3*), in combination with C57BL/6J blastocysts to produce chimeras (*Figure 17*). A good test breeding scheme for this common combination is to breed chimeras to C57BL/6J mice. GPI-1 typing can be done at birth or animals can be distinguished on the basis of their agouti phenotype when yellow-banded hairs are evident at about seven days postnatally. Agouti animals are derived from the 129 ES cells and will be (129xC57)F_1; half of them will carry the targeted allele, provided there is not a dominant lethal effect of the mutation prior to birth, or a haploid effect on sperm function. Non-agouti animals will be pure C57BL/6J and can be raised and used for other experiments.

Figure 17. A chimeric mouse made by injecting 129/SvJ ES cells into a C57BL/6J blastocyst. This mouse has albino hairs, agouti hairs, and black hairs. Note the variegated eye pigmentation.

6.3 Breeding schemes to maintain targeted alleles and determine phenotype

Once a germline chimera has been identified, the first priority will be to obtain and maintain the targeted allele in living animals. The second priority will be to put the new allele onto a genetic background(s) that will allow its characterization. For many studies, a random or genetically undefined background may be adequate for the initial characterization, but for others, an inbred background may be preferable. See Chapter 7 for a further discussion of genetic analyses of targeted mutations.

The best breeding scheme to use will depend on the timing and severity of the effect of the mutant gene. Clearly, a dominant mutation cannot be maintained at all if it causes lethality before sexual maturity or if it adversely affects gamete function. In this case, the chimera itself will be the only transmitter of the allele and the mutant effects can only be studied in its offspring prior to their death (see Section 5.2.2), or, in specific cases, by making tetraploid chimeras (described in Chapter 5). Mutants with less severe dominant effects or recessive effects can be maintained by standard schemes for maintaining mutant genes (41). Mutations that are homozygous viable and fertile can be maintained by breeding homozygotes and thus, only the founder parent mice need to be genotyped. If a mutation is recessive and homozygous lethal, heterozygous mice can be intercrossed or crossed to wild-type mice. Resulting offspring, are then genotyped. In order to establish the time of death of a homozygous mutation, offspring of a heterozygous intercross are genotyped at weaning. If no homozygotes are present, mice at birth should be carefully monitored to see if any pups die, and if so they should be genotyped. If no homozygotes are present at birth, embryos at progressively earlier stages should be genotyped (see Chapter 7, *Figure 3*).

6.3.1 Segregating backgrounds

If test breeding is done as described above, the first available heterozygous animals are likely to be F_1 mice between two inbred strains (a 129 substrain and C57BL/6J). If the founder chimera is transmitting the ES cell genotype at a high frequency, it will be possible to produce a large number of F_1 offspring. These mice will be genetically uniform except for the targeted allele and any alleles that were segregating in the 129 substrain from which the ES cells were derived (18). Once they have been typed by PCR or Southern blot analysis of tail DNA, heterozygotes for the targeted allele can be mated together to produce F_2 litters with wild-type, heterozygote, and homozygous mutant animals. This breeding scheme takes advantage of the hybrid vigour of F_1 mice and is the fastest means of producing mutant homozygotes for analysis. The drawback is that the F_2 offspring will be segregating for all alleles that differ between the two parent strains and thus the effect of the mutant gene is being observed on a variable genetic background; each mouse will be genetically different from its sibs. The practical effect is that any variation in the mutant phenotype between animals might be attributable to unknown, segregating, modifier genes rather than the mutation alone. None the less, this scheme is very useful for a first characterization of genes with major effects. If the mutation is maintained on this mixed genetic background in a small, closed colony, the inevitable result will be inbreeding with its accompanying decline in reproductive fitness. Rigorous selection for high fecundity or occasional outcrossing to F_1 mice of the parent strains can help overcome this problem.

It is also worth mentioning that mutations unrelated to the targeted locus may have occurred fortuitously in the ES cells at any time during isolation, culture, or cloning. If this occurs, a distinctive phenotype may be detected, but this will segregate independently from the targeted allele and should be evident as a separate locus in one or two generations for all but the most tightly linked loci. Routine genotyping of animals will quickly distinguish between effects of a fortuitous mutation and effects of the targeted allele, assuming it is not closely linked to the targeted allele.

If the founder chimera is transmitting the targeted allele at a very low frequency, or if it ceases breeding after only a few carrier F_1 mice are produced, the mutant allele can be propagated by crossing heterozygotes to F_1 or even random-bred mice to ensure its survival. Transfer to an inbred background can then be done at any time independently of the founder chimera.

6.3.2 Inbred backgrounds

A uniform genetic background allows the most precise comparison to be made between mutant and non-mutant phenotypes. There are two ways to put targeted alleles on inbred backgrounds: by backcrossing and by mating the chimera with the strain from which the ES cells were derived. The former, which is described below, is straightforward and applicable to any inbred strain, but the latter may be more problematic than it seems if small genetic differences are important. Most commonly used ES cells were derived from 129 mice; however, some were derived many years ago from substrains that are not readily available today, or that might have diverged genetically from the ES cells in the intervening years (18, 21). An added problem is that some 129 substrains are poor breeders and are thus difficult to maintain. If it is desirable to have the mutant on an inbred background immediately, there may be no choice but to mate the chimera with the most closely related available 129 substrain and to accept the genetic variability that will be the inevitable result of crossing two substrains with genetic differences. In test breeding a chimera with 129 mice, all the offspring, whether derived from the host blastocyst or from the ES cells, will be agouti so that markers other than agouti must be used to distinguish germline transmission. If a chimera is chosen that only transmits ES cell-derived offspring, this will increase the percentage of heterozygous offsping to 50%. The advantage of this method is that the mutant gene will be immediately available for analysis in a relatively uniform genetic background.

Transferring the mutant allele to another inbred background requires backcrossing (see Chapter 7 for details). Heterozygous mice are identified by DNA analysis and crossed to mice of an inbred strain. Heterozygous progeny from this mating are then backcrossed to mice of the inbred strain. The more times this cycle is repeated, the more uniform the genetic background becomes. After seven or eight generations of backcrossing, 99% of loci not linked to the mutant allele will be homozygous, and the mice are considered congenic after ten generations, counting the first hybrid or F_1 generation as generation one (41). Backcrossing must continue for another 10–12 generations, however, to ensure genetic uniformity for closely linked loci.

In practice, backcrossing can be done very easily, with a minimum number of mice. Starting with an F_1 between the mutant-bearing stock and the inbred strain (generation one), we backcross a single F_1 heterozygous male with several inbred females. The first litter born (generation two) is typed and one or two heterozygous male offspring are mated with inbred females. All mice of the first generation can be discarded as soon as a pregnancy is detected in the next generation. In at least one backcross cycle, a female heterozygote is mated to an inbred male in order to perpetuate the inbred strain Y chromosome, otherwise it is more efficient to use male heterozygotes. As the stock comes closer to co-isogenicity, the breeding characteristics, along with all others traits, will approximate those of the inbred strain. Usually this will mean a gradual diminution of the hybrid vigour of the early backcrosses and it may be necessary to set up additional matings to ensure some of them will be fertile. Designation of the newly produced congenic strain should follow the rules for nomenclature as recommended by the Committee on Standardized Genetic Nomenclature (42) which is also available through the web site http://.informatics.jax.org/nomen/.

Once a suitable breeding scheme is in place for maintaining the mutation and the timing of any mutant effect is established, the big challenge, and the reason for making the mutant in the first place, is to determine the cause of phenotypic effects, whether they are mild or severe. Often producing double mutants by breeding with mice carrying mutations in related genes can uncover additional functions of a gene. The more that is known about the biochemical function of the gene and its expression pattern, the easier it will be to predict possible causes of any phenotypes. Histological analysis and marker gene analysis at embryonic or later stages prior to death can then be used to characterize progression of the defect and establish the cause.

Acknowledgements

We wish to thank James Spencer, John Anderson, and Barbara Van Lingen for technical assistance, Gerry Parker for photographic work, and Allan Bradley for permission to use published figures. This work was supported in part by NIH grant HD27295, HD33082, and the Raymond and Beverly Sackler Foundation.

References

1. Tarkowski, A. K. (1961). *Nature*, **190**, 857.
2. McLaren, A. (1976). *Mammalian chimeras*. Cambridge University Press, Cambridge.
3. Moustafa, L. A. and Brinster, R. L. (1972). *J. Exp. Zool.*, **181**, 193.
4. Brinster, R. L. (1974). *J. Exp. Med.*, **140**, 1049.
5. Papaioannou, V. E., McBurney, M. W., Gardner, R. L., and Evans, M. J. (1975). *Nature*, **258**, 70.
6. Mintz, B. and Illmensee, K. (1975). *Proc. Natl. Acad. Sci. USA*, **72**, 3585.
7. Papaioannou, V. E. and Rossant, J. (1983). *Cancer Surv.*, **2**, 165.
8. Evans, M. J. and Kaufman, M. H. (1981). *Nature*, **292**, 154.
9. Martin, G. (1981). *Proc. Natl. Acad. Sci. USA*, **78**, 7634.
10. Gardner, R. L. (1968). *Nature*, **220**, 596.
11. Papaioannou, V. E. (1981). In *Techniques in the life sciences, Vol. P1/1, techniques in cellular physiology, part 1* (ed. P. F. Baker), pp. 116/1. Elsevier/North Holland Biomedical Press, Amsterdam.
12. Bradley, A. (1987). In *Teratocarcinomas and embryonic stem cells: a practical approach* (ed. E. J. Robertson), p. 113. IRL Press, Oxford, Washington, DC.
13. Stewart, C. L. (1993). In *Methods in enzymology* (ed. B. M. Wasserman and M. L. DePamphilis), Vol. 225, p. 823. Academic Press, London.
14. Hogan, B., Beddington, R., Costantini, F., and Lacy, E. (1994). *Manipulating the mouse embryo, a laboratory manual*, 2nd edn. Cold Spring Harbor Laboratory Press, NY.
15. Hetherington, C. M. (1987). In *Mammalian development: a practical approach* (ed. M. Monk), p. 1. IRL Press, Oxford, Washington, DC.

16. Stevens, L. C. (1983). In *Teratocarcinoma stem cells* (ed. L. M. Silver, G. R. Martin, and S. Strickland). *Cold Spring Harbor Conferences on Cell Proliferation*, Vol. 10, p. 23. Cold Spring Harbor Laboratory Press, NY.
17. Robertson, E. J., Kaufman, M. H., Bradley, A., and Evans, M. J. (1983). In *Teratocarcinoma stem cells* (ed. L. M. Silver, G. R. Martin, and S. Strickland). *Cold Spring Harbor Conferences on Cell Proliferation*, Vol. 10, p. 647. Cold Spring Harbor Laboratory Press, NY.
18. Simpson, E. M., Linder, C. C., Sargent, E. E., Davisson, M. T., Mobraaten, L. E., and Sharp, J. J. (1997). *Nature Genet.*, **16**, 19.
19. te Riele, H., Maandag, E. R., and Berns, A. (1992). *Proc. Natl. Acad. Sci. USA*, **89**, 5128.
20. Deng, I. and Capecchi, M. R. (1992). *Mol. Cell Biol.*, **12**, 3365.
21. Threadgill, D. W., Yee, D., Matin, A., Nadeau, J. H., and Magnuson, T. (1997). *Mamm. Genome*, **8**, 390.
22. Hay, R. J., Macy, M. L., and Chen, T. R. (1989). *Nature*, **339**, 487.
23. Johnson, R. S., Speigelman, B. M., and Papaioannou, V. E. (1992). *Cell*, **71**, 577.
24. Doetschman, T. C., Eistetter, H., Katz, M., Schmidt, W., and Kemler, R. (1985). *J. Embryol. Exp. Morphol.*, **87**, 27.
25. Swiatek, P. J. and Gridley, T. (1993). *Genes Dev.*, **7**, 2071.
26. Handyside, A. H., O'Neill, G. T., Jones, M., and Hooper, M. L. (1989). *Roux. Arch. Dev. Biol.*, **198**, 48.
27. McMahon, A. P. and Bradley, A. (1990). *Cell*, **62**, 1073.
28. Bradley, A., Evans, M., Kaufman, M. H., and Robertson, E. (1984). *Nature*, **309**, 255.
29. Robertson, E., Bradley, A., Kuehn, M., and Evans, M. (1986). *Nature*, **323**, 445.
30. Nagy, A., Rossant, J., Nagy, R., Abramow-Newerly, W., and Roder, J. C. (1993). *Proc. Natl. Acad. Sci. USA*, **90**, 8424.
31. Champlin, A. K., Dorr, D. L., and Gates, A. H. (1973). *Biol. Reprod.*, **8**, 491.
32. Rugh, R. (1990). *The mouse: its reproduction and development.* Oxford Science Publications.
33. Papaioannou, V. E. and Fox, J. G. (1992). *Lab. Anim. Sci.*, **43**, 189.
34. Biggers, J. D., Summers, M. C., and McGinnis, L. K. (1997). *Hum. Reprod. Update*, **3**, 125.
35. Papaioannou, V. E. (1990). In *The post-implantation mammalian embryo: a practical approach* (ed. A. J. Copp and D. Cockcroft), p. 61. IRL Press, Oxford.
36. Babinet, C. (1980). *Exp. Cell. Res.*, **130**, 15.
37. Schmidt, G. and O'Sullivan, J. F. (1987). *Trends Genet.*, **3**, 332.
38. Papaioannou, V. and Johnson, R. (1993). In *Gene targeting: a practical approach* (ed. A. L. Joyner), p. 107. IRL Press, Oxford.
39. Kuehn, M. R., Bradley, A., Robertson, E. J., and Evans, M. J. (1987). *Nature*, **326**, 295.
40. Pevny, L., Simon, M. C., Robertson, E., Klein, W. H., Tsai, S.-F., D'Agati, V., *et al.* (1991). *Nature*, **349**, 257.
41. Green, E. L. (1966). In *Biology of the laboratory mouse* (ed. E. L. Green), p. 11. McGraw-Hill Book Co., NY.
42. Committee on Standardized Genetic Nomenclature for Mice. Chairperson: Davisson, M. T. (1996). In *Genetic variants and strains of the laboratory mouse* (ed. M. F. Lyon, S. Rastan, and D. M. Brown), 3rd edn, Vol. 1, p. 1. Oxford University Press, Oxford.

Appendix

Further details and suppliers of specialized equipment used in the authors' laboratories. These are not the only possible brands of equipment or suppliers. Requirements and additional information can be found in the text.

(a) Epinephrine chloride solution: Sigma or Parke Davis, Warner Lambert, Morris Plains, NJ, USA. Cat. No. N0071–4011013.

(b) Defonbrune microforge.

(c) Dental spatula for plug checking: Schein JE SPT 313, Dental Supply Co. of New England, 80 Fargo Street, Boston, MA 02210, USA.

(d) Electrostarch for GPI-1 electrophoresis: Otto Hiller Co., PO Box 1294, Madison, WI 53701, USA.

(e) Kopf Pipette Puller Model 700C: David Kopf Instruments, 7324 Elmo Street, PO Box 636, Tijunga, CA 91042–0636, USA.

(f) Micoplasma TC Detection Kit: Gen-Probe, San Diego, CA 92123, USA. Cat. No. 1004.

(g) Anti-vibration table: Micro-g, Technical Manufacturing Corporation, Woburn, MA 01801, USA.

Microscopes

(a) Dissecting scope for embryo recovery: Wild M5A, M8, or M10, Wild Heerbrugg Ltd.

Parco SM series, Parco Scientific Co., Instrument Division, 316 Youngstown-Kingsville Road, PO Box 189, Vienna, Ohio 44473, USA.

(b) Dissecting scope for embryo transfer: Reichart-Jung, Stereo Star Zoom (formerly AO Stereostar Zoom) model 570, .7 × to 4.2 x, 15 × widefield eyepieces.

(c) Microscope for embryo injection: Zeiss Standard 14 microscope or Zeiss IM inverted, Carl Zeiss.

(d) Stage cooling device: Physitemp TS-4 Controller, Sensortek Inc., 154 Huron Ave., Clifton, NJ 07013, USA.

Specialized equipment

(a) Syringe devices for cell injection:

(i) Beaudouin suction and force pump.

(ii) Stoelting Co. micrometer with Hamilton syringe.

(iii) Eppendorf CellTram Oil No. 5176 000.025.

(b) Syringe device for holding embryo: Narishige, Model No. IM-58.

(c) Surgical instruments: Roboz Surgical Instrument Co. Inc., 9210 Corporate Blvd., Ste. 220, Rockville, MD 20850, USA.

(d) Watch glasses: Carolina Biological.

(e) Wound clips: Autoclip 9 mm, Clay Adams, Division of Becton Dickinson & Co., Parsippany, NJ 07064, USA.

5

Production and analysis of ES cell aggregation chimeras

ANDRAS NAGY and JANET ROSSANT

1. Introduction

Embryonic stem (ES) cells behave like normal embryonic cells when returned to the embryonic environment after injection into a host blastocyst or after aggregation with earlier blastomere stage embryos. In such chimeras, ES cells behave like primitive ectoderm or epiblast cells (1), in that they contribute to all lineages of the resulting fetus itself, as well as to extraembryonic tissues derived from the gastrulating embryo, namely the yolk sac mesoderm, the amnion, and the allantois. However, even when aggregated with pre-blastocyst stage embryos, ES cells do not contribute to derivatives of the first two lineages to arise in development, namely, the extraembryonic lineages: trophoblast and primitive endoderm (2).

The pluripotency of ES cells within the embryonic lineages is critical to their use in introducing new genetic alterations into mice, because truly pluripotent ES cells can contribute to the germline of chimeras, as well as all somatic lineages. However, the ability of ES cells to co-mingle with host embryonic cells, specifically in the embryonic, but not the major extra-embryonic lineages, opens up a variety of possibilities for analysing gene function by genetic mosaics rather than by germline mutant analysis alone (3).

There are two basic methods for generating pre-implantation chimeras in mice, whether it be embryo ⟷ embryo or ES cell ⟷ embryo chimeras. Blastocyst injection, in which cells are introduced into the blastocoele cavity using microinjection pipettes and micromanipulators, has been the method of choice for most ES cell chimera work (see Chapter 4). However, the original method for generating chimeras in mice, embryo aggregation, is considerably simpler and cheaper to establish in the laboratory. Aggregation chimeras are made by aggregating cleavage stage embryos together, or inner cell mass (ICM) or ES cells with cleavage stage embryos, growing them in culture to the blastocyst stage, and then transferring them to the uterus of pseudopregnant recipients to complete development. This procedure can be performed very rapidly by hand under the dissecting microscope, thus making possible high throughput production with minimal technical skill (4).

In this chapter we describe some of the uses of pre-implantation chimeras, whether made by aggregation or blastocyst injection, but focus on the technical aspects of aggregation chimera generation. We also discuss the advantages and disadvantages of aggregation versus blastocyst injection for chimera production.

1.1 Types of aggregation chimeras

Blastomere stage cells, ICM cells, or ES cells can all aggregate with 8-cell embryos and be incorporated into resulting chimeras. However, more 'differentiated' cells like trophectoderm or primitive endoderm, or any later cell types, fail to attach and integrate after embryo aggregation. The ability of cells to incorporate into the embryo after aggregation probably depends largely on the compatibility of cell adhesion molecules on donor cells and host embryos. All cells that can aggregate with 8-cell stage embryos express high levels of the homotypic cell adhesion molecule, E-cadherin (5), which is also expressed by cleavage stage embryos. However, which later lineages are colonized by donor cells after aggregation depends on the intrinsic potential of the host and donor cells (*Figure 1*).

The original aggregation chimera was made by aggregating diploid cleavage stage embryos (pre-blastocyst), which are able to contribute to all later cell

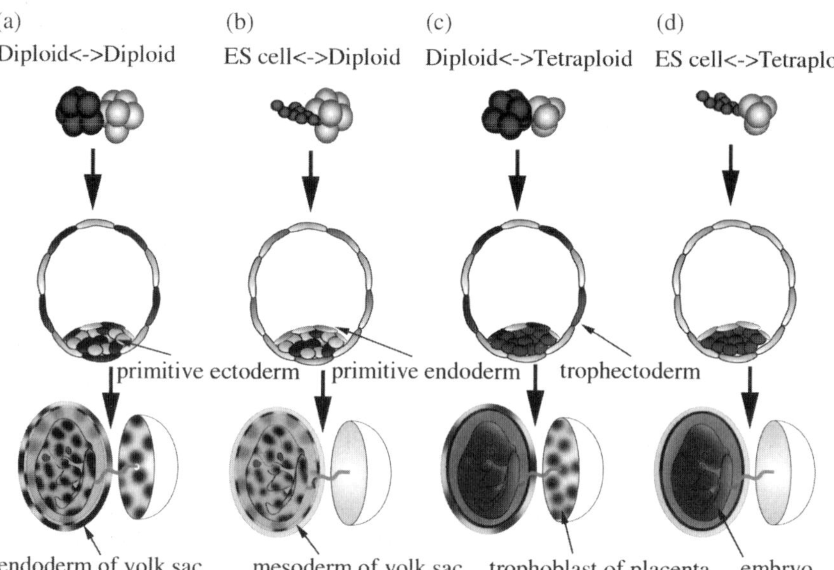

Figure 1. Diagram of lineage contributions in different kinds of aggregation chimeras. Solid colours indicate non-chimeric tissues, stripes indicate chimeric tissues.

lineages. Thus, such aggregates generate conceptuses with contributions from both embryos to all three pre-implantation lineages—primitive ectoderm or epiblast, primitive endoderm, and trophectoderm—and their later derivatives (*Figure 1a*). When ES cells are aggregated with 8-cell embryos, the result is rather different, because ES cells have limited potential to contribute to trophectoderm and primitive endoderm lineages (1, 2). ES cell ⟷ 8-cell chimeras produce ES cell contributions to the tissues derived from primitive ectoderm, i.e. the majority of the fetus itself, the amnion, the mesoderm of the yolk sac, and the allantois. However, the endoderm layers of the yolk sac and the trophoblast contribution to the placenta will be entirely derived from the host embryo (*Figure 1b*). These restrictions are valid for chimeras produced by blastocyst injection (Chapter 4) as well.

ES cell chimeras, therefore, provide a means of skewing ES cell contributions specifically to the embryonic lineages. A more extreme skewing of contributions can be obtained by generating chimeras between diploid and tetraploid embryos. Most tetraploid embryos die shortly after implantation (6, 7) although survival to late gestation has been reported (8, 9). However, when aggregated with diploid embryos, tetraploid cells could contribute quite well to the primitive endoderm and trophectoderm lineages but were generally excluded from the primitive ectoderm lineage (2, 7, 10). Such chimeras can be used to generate fetuses entirely of one genotype with extraembryonic membranes that are diploid ⟷ tetraploid mosaic (*Figure 1c*). An even more extreme skewing of contributions is produced when ES cells are aggregated with tetraploid embryos. Since ES cells do not readily form trophectoderm and primitive endoderm and since tetraploid cells fail to produce primitive ectoderm, the two components are complementary and generate conceptuses in which the entire fetus is ES cell-derived, but the primitive endoderm and trophectoderm components of the extraembryonic lineages are entirely host embryo-derived (2) (*Figure 1d*).

1.2 Uses of ES cell ⟷ diploid embryo aggregation

1.2.1 Germline transmission of genetic alterations

Chimeras generated by ES cell ⟷ embryo aggregation will transmit the ES cell genotype into the germline in the same manner as blastocyst injection chimeras and so can be a tool for generating knock-out mice. Successful germline transmission after ES cell ⟷ embryo aggregation depends on achieving a suitable balance between the donor ES cell line and host embryo cells.

Aggregation chimeras tend to have larger average ES cell contributions than blastocyst injection chimeras, presumably due to the longer period during which they can proliferate and contribute to the ICM after chimera formation (4). With most ES cell lines, there is an inverse correlation between the extent of ES cell contribution and chimera viability, which means that the aggregation technique may not be suitable for ES cells of less than optimal potential. For example, we have been unable to efficiently generate germline transmitting chimeras from D3 and AB1 ES cells by aggregation, although the same cells can still be transmitted through the germline after blastocyst injection. We routinely use early passage R1 cells, generated in-house (11), for all experiments requiring germline transmission. This is a cell line of high developmental potential, which efficiently generates germline chimeras after aggregation. Other laboratories have reported success with other ES cell lines, but the ability of any given ES cell line to function efficiently in this manner needs to be tested empirically. The W4 cell line described in Chapter 3 also contributes efficiently to the germline in aggregation chimeras (Auerbach and Joyner, personal communication).

The other important factor affecting the success of generating germline transmitting chimeras after embryo aggregation is the choice of host embryo strain. Given the high tissue contributions generated by ES cells after embryo aggregation, it is important to use a host embryo strain that produces chimeras with a good balance of host and donor contributions. We routinely use highly vigorous, outbred CD1 or ICR albino strains as 8-cell embryo donors to produce chimeras with a wide range of contributions from the 129-derived R1 donor ES cell line. There are some laboratories achieving successful germline transmission with outbred Swiss Webster and with different F_2 hybrid embryos, such as between C57 and CBA or C57 and DBA inbred strains. Use of pure inbred C57BL/6 or BALB/c strains as 8-cell donors is not recommended, as most embryos will then be almost totally ES-derived and of low viability. The use of outbred mice as host embryos is a major advantage of aggregation over blastocyst injection, because outbred mice are less expensive and respond well to superovulation.

Comparison of blastocyst injection and embryo aggregation chimeras in our laboratories shows that the efficiency of obtaining germline transmission from male chimeras, which are estimated by coat colour to be highly ES cell-derived (strong), is comparable between the two techniques. However, it takes more initial aggregations with CD1 8-cell embryos than injections into C57BL/6 blastocysts to generate the requisite five or so 'strong' male chimeras to achieve germline transmission of a given line (approximately 120 initial aggregation chimeras versus 40 chimeras made by blastocyst injection). However, the low cost and ease of obtaining 120 embryos from outbred mice and the fact that 120 aggregations can readily be performed under the dissecting microscope in under an hour, more than balances out this difference, in our estimation.

1.2.2 Chimeric analysis of mutant phenotypes

Chimeras are more than just an intermediate in transmitting mutations into mice; they can be a powerful tool in phenotypic analysis of genetic alterations. When a mutation results in a complex phenotype, analysis of the behaviour of mutant cells in combination with wild-type cells in a chimera can shed considerable light on the primary defects involved. Consistent exclusion of mutant cells from one particular tissue, for example, indicates that the gene is essential for that lineage and not others. Bypassing early lethal events by rescue with wild-type cells can also allow access to later phenotypic roles for the gene. In addition, mutations that give mild phenotypes alone may reveal more specific defects when placed in competition with wild-type cells (3).

Chimeras for this kind of analysis can be made by embryo ⟷ embryo or embryo ⟷ ES cell aggregation. However, for recessive lethal mutations, homozygous embryos will only be 1/4 of the offspring of heterozygous crosses and a means of distinguishing chimeras containing homozygous versus heterozygous cells is needed. Generation of chimeras with homozygous mutant ES cells greatly simplifies the generation of large numbers of chimeras for analysis. In addition, if aggregation rather than blastocyst injection is used, very large numbers of chimeras can readily be generated.

There are a variety of methods for producing homozygous mutant ES cells after targeting one allele. Increasing the G418 concentration on *neo*-containing cells can lead to loss of heterozygosity at the targeted locus (reviewed in ref. 12, and described in Chapter 3, *Protocol 14*). Alternatively, the second allele can be re-targeted using a different selectable marker. If the selectable marker in the first targeted event can be removed by Cre-mediated excision, the second allele can then be re-targeted using the original targeting vector. Finally, of course, it is always possible to derive new homozygous ES cells directly from mutant blastocysts (Chapter 3, *Protocol 15*).

The other important component for mosaic analysis of chimeras is a suitable independent marker to distinguish mutant and wild-type cells in all tissues. Electrophoretic analysis of GPI isozymes is a very sensitive and useful method (see *Protocol 13*) for ascertaining proportions of genotypes but it does not provide *in situ* spatial information on the distribution of mutant and wild-type cells. The current most popular *in situ* marker for mosaic analysis is the *E. coli* β-*galactosidase* gene product, which can be detected in whole embryos or in sections by a simple histochemical stain (see Chapter 6). Ubiquitous expression is needed and this can be achieved by use of the ROSA26 gene trap mouse line (13), which can be purchased from The Jackson Laboratory.

A human placental alkaline phosphate as a reporter transgene is gaining increasing popularity. This form of phosphatase is heat resistant, therefore the endogenous alkaline phosphatase activity can be eliminated with a brief heat treatment in a transgenic embryo or cell. The histochemical reaction to visualize expression is simple (*Protocol 14*) and allows at least lacZ equivalent, if not better, sensitivity of transgene expression at the cell level.

The green fluorescent protein and its derivatives from the jellyfish, *Aequorea victoria,* are becoming increasingly useful for mammalian studies, and ubiquitously GFP-expressing mice and ES cells have been produced (14, 15).

However, currently, GFP expression survives fixation and sectioning poorly, limiting its use to mosaic analysis in whole embryos or thick sections.

1.2.3 Separation of embryonic and extraembryonic defects in mutants

ES cell aggregation chimeras can be used to investigate whether embryonic phenotypes are primarily caused by gene action in the embryonic or extra-embryonic lineages, because ES cells fail to contribute to trophectoderm and primitive endoderm in chimeras. For example, if an early lethal phenotype is rescued when mutant ES cells are aggregated with wild-type embryos, even when the mutant cells make up the majority of the embryo, then this may indicate a primary role for the gene in extraembryonic lineages. To confirm this, the reverse experiment of aggregating wild-type ES cells with mutant embryos needs to be performed. If the defect is primarily extraembryonic, the mutant extraembryonic cells should impose the mutant phenotype on the wild-type embryonic cells. Identification of the one in four chimeras of the correct genotype can be made by PCR screening for homozygosity in either separated trophoblast or primitive endoderm cells from the chimeras.

1.3 Uses of ES cell ↔ tetraploid embryo aggregation

1.3.1 Rapid testing of developmental potential of ES cell lines

Although viable and fertile completely ES cell-derived mice have been successfully produced after ES cell ↔ tetraploid aggregation, the overall success rate is very low (11). The full reasons for the low success in obtaining viable completely ES cell-derived mice are not clear, but presumably include accumulated epigenetic and genetic alterations in ES cells in culture. Such mice are often born, but do not breathe and die shortly after birth. Thus, at least with current ES cell lines, tetraploid aggregation cannot be recommended as a means of efficiently producing ES cell-derived animals for germline transmission of genetic alterations. However, ES cells with good developmental potential should routinely be able to generate embryos up to late gestation stages after tetraploid aggregation. This method, therefore, can be used as a rapid test for the developmental potential of new ES cell lines or for choosing the ES cell clones most suited for undergoing the multiple rounds of transfections required in some current genetic manipulation strategies.

1.3.2 Separation of embryonic and extraembryonic defects in mutants

The clear complementary cell contributions of ES cells and tetraploid embryos in chimeras makes tetraploid embryo aggregation the technique of choice for assessing the embryonic or extraembryonic function of a genetic mutation. Unlike diploid embryo ↔ ES cell chimeras, where mixtures of mutant ES and wild-type cells occur in the embryonic lineage, in ES cell ⟷ tetraploid embryo aggregates, the tetraploid cells are usually totally excluded from the embryo proper by early post-implantation stages. Occasionally small contributions of tetraploid cells may persist and so it is important in these experiments, as in diploid chimera experiments, to include an *in situ* genetic marker to confirm the distribution of mutant ES and wild-type embryonic cells in the resulting chimeras.

1.3.3 Analysis of embryonic phenotypes without germline transmission

The ease of generating large numbers of ES cell-derived fetuses in early to mid-gestation after tetraploid aggregation allows analysis of embryonic lethal phenotypes without germline transmission. For dominant gain-of-function alterations which would be predicted to be lethal, such as ectopic over-expression of a transgene, tetraploid aggregation of ES cells containing the transgene allows analysis of the phenotypic effects without resorting to some kind of conditional or inducible expression strategy. For loss-of-function targeted alterations, generation of homozygous mutant ES cells and examination of tetraploid aggregation-derived fetuses allows examination of an embryonic lethal phenotype without waiting for germline transmission. This represents a considerable saving in time and expense and, indeed, for mutants which are haploinsufficient lethals, provides the only means of analysing the effects of homozygosity (16). There are two restrictions that currently preclude the use of this approach for all mutant analysis. First, it can only be used to define phenotypes, which are autonomous to the embryonic, not extra-embryonic lineages. Secondly, given the reduced viability of ES cell-derived postnatal animals, it can only be relied upon to assess embryonic phenotypes.

1.3.4 Generation of large numbers of mutant embryos for analysis

Although there are some caveats (detailed above) in the use of tetraploid aggregation as the primary means of assessing mutant phenotypes, it can be very useful as secondary tool. Once a lethal phenotype is defined and can be shown to be due to an embryonic, not extraembryonic defect, tetraploid aggregation can be used to generate large numbers of homozygous mutant embryos for more detailed phenotypic analysis. This removes the necessity to maintain a large stock of heterozygotes, from which only 1/4 of the offspring are the required genotype.

2. Embryonic stem cells

ES cell lines to be used to form aggregation chimeras should be maintained under the optimal conditions for the particular line used (see Chapter 3). Cells should be passed at least once after thawing, before using them for aggregation. The protocol given here was developed for R1 cells and our laboratory conditions and also works well for W4 cells. The goal is to produce clumps of 15–25 loosely connected cells just before aggregation.

Protocol 1. Preparation of ES cells for aggregation chimeras

Equipment and reagents

- Tissue culture plates with and without feeders (Chapter 3)
- Trypsin (Chapter 3)
- ES cell medium (Chapter 3)

Method

1. Day –4 (four days prior to aggregation). Thaw ES cells or passage a confluent plate of cells as described (Chapter 3, *Protocol 5*), and dilute cells 1 in 7, on feeders.
2. Change the medium the next day.
3. Day –2 (two days prior to aggregation). Trypsinize a subconfluent plate (60 mm) of ES cells to give a single cell suspension as described (Chapter 3, *Protocol 4*). After spinning down and resuspending the cells in 5 ml medium, let any large clumps settle in the tube for 5 min.
4. Plate cell suspension on gelatinized tissue culture plates in ES cell medium and LIF (Chapter 3, *Protocol 4*) without feeders. Plate at 1:50 dilution, to ensure small colonies.
5. Day 0 (day of aggregation) (after preparation of the embryos, see *Protocol 9*). Wash cells with PBS.
6. Quickly rinse the plate of ES cells with 3 ml of trypsin solution, then replace with 0.5 ml of fresh trypsin.
7. Trypsinize cells for 3 min at room temperature, or until gentle swirling lifts clumps of loosely connected cells (watch under the microscope). Then gently add 5 ml of ES cell medium. Do not pipette.
8. Keep the cell suspension in the plate and use to assemble aggregates within the next hour or two. Do not place in incubator, as clumps will stick and not be useful for aggregation.

3. Production of embryo partners

Embryos are recovered in M2 medium (17) (*Protocol 2*) which is buffered for normal atmospheric conditions, but cultured in either M16 (17) (*Protocol 2*) or KSOM medium (18) (*Protocol 4*) at 37 °C and 5% CO_2 in air. Since the aggregates are cultured for at least 24 hours, the quality of culture conditions is more critical than for injection chimeras, where the embryos are exposed to *in vitro* conditions for only a few hours. We have found KSOM medium to improve viability of aggregation chimeras after culture, as compared with M16.

General notes on media preparation are the following:

(a) Make up all media and stock solution with high quality water, at least triple glass distilled or MilliQ-filtered.

(b) Use plasticware if it is possible or the glassware should have been used for tissue culture only, and never exposed to detergent or organic solvents.

(c) All chemicals should be of the highest grade possible and reserved for media preparation only.

(d) The concentrated stock solutions can be stored at –75 °C for six months or –20 °C for one to two months. The final medium after filtration can be stored at 4 °C for two to three weeks.

Protocol 2. Preparation of stock solutions for M2 and M16

Reagents

Component	g/100 ml	Source	Cat. No.
• Stock A (10 ×)			
NaCl	5.534	BDH	B10241–34
KCl	0.356	BDH	B10198–34
KH_2PO_4	0.162	BDH	B10203–34
$MgSO_4.7H_2O$	0.293	BDH	B10151
Sodium lactate	2.610 or 4.349 g of 60% syrup	Fisher	S326–500
Glucose	1.000	Sigma	G8270
Penicillin G	0.060	Sigma	P3414
Streptomycin	0.050	Sigma	S6501
• Stock B (10 ×)			
$NaHCO_3$	2.101	BDH	B10247–34
Phenol Red	0.001	BDH	20090
• Stock C (100 ×)			
Sodium pyruvate	0.036	Sigma	P2256
• Stock D (100 ×)			
$CaCl_2.2H_2O$	0.252	BDH	B10070–34
• Stock E (10 ×)			
Hepes	5.958	Gibco	845–1344 IM
Phenol Red	0.001	BDH	20090

Method

1. Weigh solids into media bottles and add appropriate quantity of water.
2. For stock E add half of water volume, then adjust the pH to 7.4 with 1 M NaOH. Make up to volume using cylinder.
3. Filter stocks through 0.22 μm Millipore filter, aliquot into sterile tubes.

Protocol 3. Preparation of 100 ml of M2 and 100 ml of M16 from concentrated stocks

Reagents

- Stock solutions (see *Protocol 2*)
- Bovine serum albumin (BSA) (Sigma A4378 or A3311), embryo tested by the company

Stock	M2 ml	M16 ml
A (10 ×)	10.0	10.0
B (10 ×)	1.6	10.0
C (100 ×)	1.0	1.0
D (100 ×)	1.0	1.0
E (10 ×)	8.4	–
Water	78.0	78.0
BSA	400 mg	400 mg

Method

1. Rinse all pipettes and tubes used for aliquots with water and then transfer rinse into final flask. Add slightly less water (75 ml) to allow final volume adjustment. Add BSA, allow to dissolve, and gently swirl the medium without excessive frothing using magnetic stirrer.
2. M2. If necessary, readjust the pH to 7.2–7.4 with 1 M NaOH. Make up to volume. Filter sterilize and aliquot into polypropylene tubes. Store at +4 °C.
3. M16. Follow the procedure described in steps 1 and 2, except for pH adjustment to 7.4 by bubbling with 5% CO_2 in air; alternatively incubate the medium for several hours (or overnight) at 37 °C, 5% CO_2 with loose cap. The pH will not usually require adjustment, but it can vary with different batches of BSA. Make up to volume.
4. Filter sterilize and aliquot into polypropylene tubes. Store at +4 °C for no longer than two weeks.

Protocol 4. Preparation of KSOM medium

Reagents

Component	g/100 ml
• Stock A′ (10 ×)	
NaCl	5.55
KCl	0.186
KH_2PO_4	0.0476
$MgSO_4.7H_2O$	0.0493
Na lactate	1.87 of 60% syrup
D(+) glucose	0.036
Penicillin G	0.06
Streptomycin sulfate	0.05
Freeze in 10 ml aliquots.	

Component	g/100 ml
• Stock B′ (10 ×)	
$NaHCO_3$	2.10
Phenol Red 0.001 (or 1 ml of 1% solution)	
Freeze in 10 ml aliquots.	
• Stock C′ (100 ×)	
Na pyruvate	0.022
Freeze in 1 ml aliquots.	
• Stock D′ (100 ×)	
$CaCl_2.2H_2O$	0.25
Freeze in 1 ml aliquots.	

Component	g/100 ml	Component	g/100 ml
• Stock F′ (10 000 ×)		• Stock G′ (200 ×)	
$Na_2EDTA.2H_2O$	0.372 Freeze in 10 μl aliquots.	Glutamine	200 mM (in solution form) (Gibco, 320–1140AG) Freeze in 0.5 ml aliquots.
Or (100 ×)	0.0372/100 ml Freeze in 1 ml aliquots.		

Method

1. Preparation of 100 ml of KSOM from concentrated stocks:

Stock solution	ml
10 × stock A′	10
10 × stock B′	10
100 × stock C′	1
100 × stock D′	1
100 × stock F′	1
200 × stock G′	0.5
BSA	0.10 g
H_2O	90 ml

2. Follow instructions for M16 (*Protocol 3*).

3.1 Recovery of 8- and 2-cell stage embryos

The only point of note is that for aggregation chimeras non-compact 8-cell stage embryos are preferred. Flushing should therefore be performed in the morning of the day of aggregation, if the superovulating hormone injections are midday scheduled or the dark cycle is between 5 p.m. and 5 a.m. for the animals.

Protocol 5. Flushing of 8-cell stage embryos

Equipment and reagents

- Dissecting microscope
- No. 30 gauge needle (first, the sharp tip is cut off by a strong pair of scissors and then rounded using sharpening stone)
- 1 ml syringe
- No. 5 forceps (Dumont)
- Mouth pipette
- Dissecting instruments (fine pointed scissors, fine forceps)
- Pasteur pipettes
- Alcohol or Bunsen burner
- Sterile plastic Petri or tissue culture dishes
- Organ culture dishes (Falcon, 3037)
- M2 and KSOM medium

Method

1. Kill the superovulated (see *Protocol 6*, and Chapter 4, Section 2.2.2) or naturally mated females early on day 2.5 (day 0.5 is the day of the plug), in order to obtain non-compacted 8-cell embryos suitable for aggregation.

Protocol 5. *Continued*

2. Open the body cavity by making a small ventral incision in the skin across the midline at the level of the bladder, then pull the skin up over the chest and cut the body wall along both sides starting from the midline at the bladder.
3. Dissect the oviducts with the upper part of the uterus attached and place in drop of M2 in a Petri dish (see Chapter 4, *Figure 3*).
4. Transfer one of the oviducts into a small drop of M2.
5. Under dissecting microscope locate the infundibulum of the oviduct and insert the flushing needle attached to a 1 ml syringe filled with M2. Hold the needle in position with the No. 5 forceps and flush ~ 0.05 ml of medium through the oviduct (you should see the oviduct swell).
6. Repeat steps 4 and 5 with the remaining oviducts.
7. Collect embryos with mouth pipette attached to a drawn-out Pasteur pipette.
8. Wash the embryos through several M2 drops to get rid of the debris.
9. Wash them in KSOM and transfer them into an organ culture dish containing KSOM and place them into the incubator (at 37 °C, 5% CO_2 in air).

Recovery of 2-cell stage embryos which are used to produce tetraploid embryos is very similar to that of the 8-cell stage embryos (*Protocol 5*) but it must be done a day before assembling the aggregates using 1.5 days post-coitum (dpc) females. It is safer to recover late 2-cell stage embryos, to avoid the so-called 2-cell stage block. The presence of 10–15% 3/4-cell stage embryos among 2-cell stages at the time of flushing indicates appropriate timing. If the females are superovulated, the recommended time for flushing is 46–48 h post-hCG for most strains and mouse lines.

For a maximum yield of embryos after superovulation, young females (four weeks of age) are most favourable. However, oviduct flushing of such young females is difficult, since the utero-tubal junction (see Chapter 4, *Figure 3*) is usually very narrow. The majority of 2-cell stage embryos can be damaged unless the junction is ruptured with fine forceps before flushing.

Protocol 6. Production of 2-cell stage embryos for blastomere electrofusion

Equipment and reagents

- See *Protocol 5*
- Pregnant mare serum gonadotrophin (PMSG) (Equinex) 5000 units (DIN 330302 No. 5490)
- Human chorionic gonadotrophin (hCG) (APL) 10000 units (DIN 002291 No. 999)

Method

Day 0 is the day of assembling aggregates.

1. Day –5. Begin superovulation of young (four to six weeks of age is preferred) females by injecting 5 U PMS intraperitoneally at 2:00 p.m.
2. Day –3. Give 5 U hCG at 11:30 a.m. intraperitoneally and set up the females for mating with stud males.
3. Day –2. Check plugs.
4. Day –1. Start dissecting oviducts at 10 a.m. and flush oviducts as described for 8-cell stages (see *Protocol 5*). Before flushing, however, rupture the utero-tubal junction by tearing with two No. 5 forceps. Avoid long exposure to room temperature.

3.2 Production of tetraploid embryos

Several different methods have been used to produce tetraploid mouse embryos. These methods can be classified into two categories. One group of techniques utilizes the effect of agents like cytochalasin B (7, 8), that can prevent cytokinesis without influencing nuclear division. The techniques in the second category are based on the induction of fusion of two normal diploid embryonic cells, such as the blastomeres of 2-cell stage mouse embryos. Most of the procedures in this category suffer from variation in efficiency caused by the varying fusogenic activity of the agents used such as Sendai virus and polyethylene glycol (PEG). Electrofusion as devised by Kubiak and Tarkowski (19), is very simple and efficient, making it superior to other techniques. Here we describe this method in detail.

Fusion of the blastomeres of 2-cell stage mouse embryos occurs when an AC electric pulse is applied perpendicular to the plane of contact of the two cells. The precise parameters of the pulse will vary depending on the equipment used. The pulse shape of the CF-150B pulser (commercially available from Biochemical Laboratory Service, 31 Zselyi Aladar utca, Budapest, H-1165, Hungary, e-mail <bls@euroweb.hu>) that we use is a peak with a square shoulder. The effective field strength is 1–2 kV/cm. The equipment was designed specifically to meet the needs of blastomere fusion. The pulse amplitude and duration can be set within the range of 0–150 V and 0–150 μsec, respectively. The machine also provides an adjustable strength 1 MHz AC field to allow correct orientation of the 2-cell stage embryos between the two electrodes in non-electrolyte solution. A hand-held trigger provides a convenient means of initiating the pulse while working with embryos under the microscope. An electrode-chamber with a fixed distance (250 μm) is also provided. An alternative electrofusion set-up is the ElectroCell Manipulator (from BTX Instruments, California).

3.2.1 Electrofusion

In order to fuse the blastomeres the plane of contact between the two cells must be perpendicular to the electric field generated between the electrodes. In non-electrolyte (mannitol) the proper orientation of the embryos is provided by the 1 MHz AC field of the equipment (*Figure 2A*). Fusion can also be performed in electrolyte solution (i.e. culture medium) but the embryos must be oriented manually. As a consequence, in the latter case the embryos must be shocked individually, one at a time, whereas in non-electrolyte you can apply the electric pulse to a group of 20–40 embryos. We, therefore, recommend fusion in non-electrolyte.

The effective voltages and pulse duration for your particular fusion apparatus and electrode slide must be determined in a pilot experiment by following the steps of *Protocol 7* with fewer embryos in the groups. Choose the voltage and duration which causes no embryo lysis and still fuses at least 90% of the 2-cell embryos. This will be close to 40 V and 40 μsec for electrodes placed 250 μm apart with the CF-150B model. The earlier CF-150 model requires higher voltage for effective fusion. This model can be easily modified to reach the higher power level of the CF-150B model by following the instructions of the manufacturer. The effective orienting AC field (around 0.5 V) must be precisely determined as well. Too high an AC field also causes immediate lysis. With the BTX ElectroCell Manipulator 2001 use the electrodes set at 0.5 mm apart (Cat. No. BT450). The settings to use are 100 V (DC) for 30 μsec with two pulses. The AC field for orienting is 3 V.

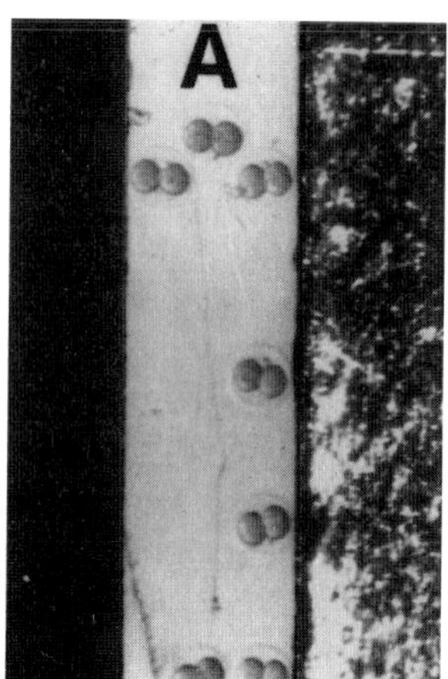

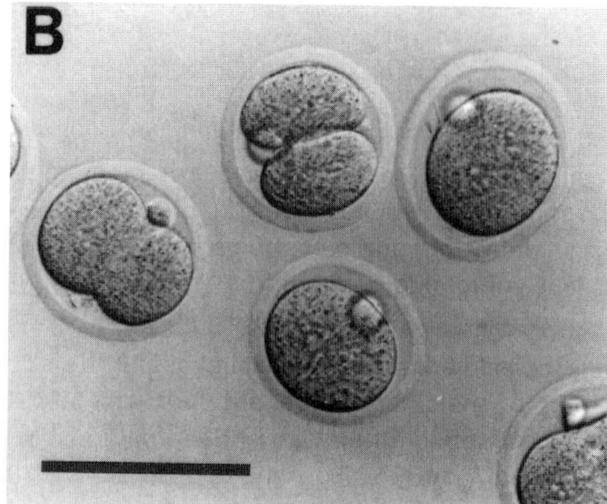

Figure 2. Electrofusion of 2-cell stage mouse embryos in non-electrolyte. (A) Embryos under orienting AC field. (B) Embryos during the fusion process (bar = 100 μm).

Protocol 7. Preparation of the fusion chamber and fusion in non-electrolyte

Equipment and reagents

- Two dissecting microscopes
- Electrode-chamber
- 0.3 M mannitol (Sigma, M4125): dissolve in ultrapure water with added 0.3% BSA (Sigma, A4378) and filter through 0.22 μm Millipore filter; store in aliquots at −20 °C; use fresh aliquot each experiment
- CF-150B pulse generator (or other similar machine)
- M2 medium (see *Protocol 3*)
- KSOM microdrops covered by paraffin oil (see *Protocol 8*) in 35 mm tissue culture dishes

Method

1. Set the pulse generator to the effective DC and zero AC voltages and the effective (~ 40 μsec) pulse duration.
2. Put a 100 mm Petri dish containing the electrode-chamber under a dissecting microscope and connect the cables to the pulse generator.
3. Place two large drops of M2 medium (drop A and D) and two drops of mannitol solution (drop B and C) in the dish as shown in *Figure 3*. Drop C should be eccentric with respect to the electrodes, to avoid blocking of vision by accidental air bubbles.
4. Place all the embryos in drop A.
5. Pass 15–20 embryos through drop B and place them in drop C between the electrodes. Slowly increase the AC field. This will properly orient (see *Figure 3*) most of the embryos. Pick up those which are not oriented and let them fall back again.
6. When all the embryos are properly oriented push the trigger to apply the fusion pulse.
7. Transfer the embryos into drop D.
8. Then wash the embryos in a large drop of KSOM and transfer the group of 15–20 embryos into a microdrop of KSOM under oil (prepared similarly to that described in *Protocol 8*) and put them into the incubator (37 °C, 5% CO_2).
9. Repeat steps 5–8 with new groups of embryos. The mannitol solution in drop C should be replaced with fresh solution every 15 min, as it rapidly evaporates.

3.2.2 Process of fusion, selection, culture

Fusion of the blastomeres should be complete in 30–60 min. It occurs at room temperature, but is faster and more effective at 37 °C. The boundary between the two cells completely disappears, the embryos round up, and they resemble

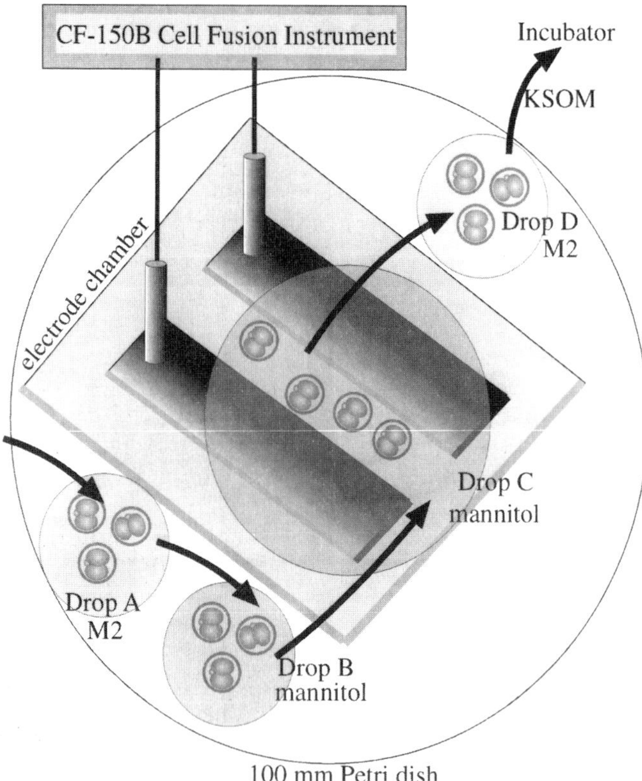

Figure 3. Set-up for electrofusion of 2-cell stage mouse embryos. The electrode chamber, the embryos, and the media drops are not proportional.

1-cell stage embryos (see *Figure 2B*). Embryos that fail to fuse can be subject to a second round of electrofusion, with some success.

Since embryos are recovered at the late 2-cell stage, the second mitotic division is expected soon after the fusion. Therefore it is important to select for the perfectly fused tetraploid embryos 30–60 min after application of the electric pulse. Occasionally one of the blastomeres lyses during fusion; such embryos will only contain one viable cell and could be confused with fused embryos. However, these embryos are readily recognized during the process of lysis right after applying the electric pulse or later by their size, which is half normal. It is safest to transfer the completely fused tetraploid embryos into a new culture dish or new KSOM microdrop and discard the unfused and lysed embryos to avoid any later confusion. Under optimal conditions the rate of unfused and lysed embryos does not exceed 5–10%. We place the fused embryos into KSOM medium in an organ cultured dish (Falcon 3037) for overnight.

4. Aggregating ES cells with cleavage stage embryos

4.1 Preparation of aggregation plate

The aggregation plate is where the actual assembly and overnight culture of aggregates takes place. The plate described below holds up to 60 aggregates. It has microdrops for the final selection of the ES cell clumps (see top and bottom rows of *Figure 4A*) and for the aggregates (middle two rows of *Figure 4A*). In the latter drops, six small depressions are made in the plastic with an aggregation needle that serve as cradles for the aggregates (*Figure 4B*). The walls of the depressions bring the embryos and ES cells into close proximity and therefore promote their aggregation. The same kind of plate can be used for embryo ⟷ embryo aggregations. The type of aggregation needle used is critical. They are commercially available from Biochemical Laboratory Service, 31 Zselyi Aladar utca, Budapest, H-1165, Hungary, e-mail: `<bls@euroweb.hu>`.

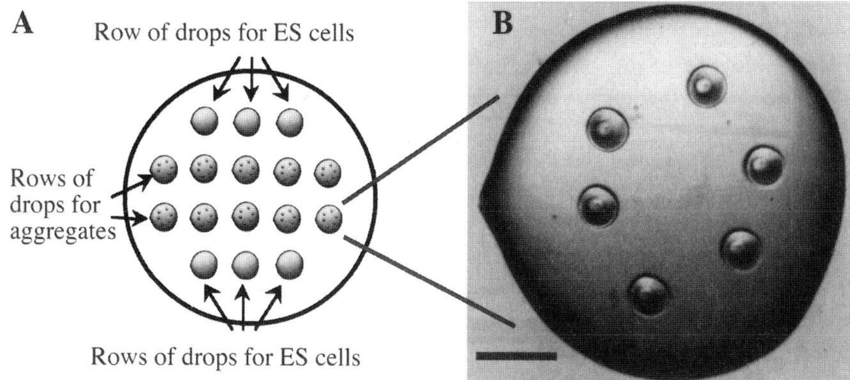

Figure 4. Arrangement of microdrops in aggregation plate (A). Note the depressions (B) made by aggregation in the microdrops of second and third rows (bar = 1 mm).

Protocol 8. Preparation of aggregation plate

Equipment and reagents

- 35 mm bacteriological Petri dish
- KSOM medium (*Protocol 4*) in 1 ml syringe with 26G needle
- Aggregation needle
- Paraffin oil (Sigma, Mineral Oil, M8410) embryo tested by the company

Method

1. Using the KSOM-filled syringe, put four rows of microdrops (roughly 3 mm in diameter) into a 35 mm tissue culture dish, three drops in the first and fourth, five drops in the second and third rows (*Figure 4A*).

Protocol 8. *Continued*

2. Cover the whole plate with paraffin oil.
3. Sterilize the aggregation needle by wiping it off with 70% ethanol.
4. Press the needle hard into the plastic through the paraffin oil and culture medium, while making a circular movement with the free end of the needle you are holding. This movement creates a tiny scoop of about 300 μm in diameter and depth with a clear smooth wall. To obtain the appropriate force to be applied to the needle, use *Figure 4B* as a reference.
5. Make six such depressions in a circle in each microdrop of the second and third rows (*Figure 4B*).
6. Return the plate to the incubator while you are preparing the embryos.

4.2 Preparation of diploid or tetraploid embryos for aggregation

4.2.1 Selection of embryos

As far as diploid embryos are concerned the selection is obvious. Choose only perfect 8-cell stage embryos with intact blastomeres for aggregation. After fusion the majority of tetraploid embryos (usually about 80–90%) cleave at the time of the normal second mitosis and reform the 2-cell stage. By noon of the day after fusion they have cleaved once more and reached the 4-cell stage. Since this stage is equivalent to the diploid 8-cell stage, tetraploid embryos start compacting at this stage. The remaining tetraploid embryos (about 10–20%) cleave slower or miss one or more cleavages. These embryos are still at the 2-cell stage at noon the day after fusion. You may discard the delayed embryos. Our experience shows that they are less efficient in supporting the development of ES cells in chimeras.

4.2.2 Zona removal

There are two different ways to remove the zona pellucida surrounding the embryos; one is enzymatic and uses Pronase, the other is dissolution and uses an acidified saline solution. Both are very effective and do not have any negative effects on the further development of the embryos. The only practical difference is the speed of the reactions. Zona removal by Pronase takes about 10–15 min, whereas acid-saline works in about 1 min. Therefore, we recommend the use of acid treatment. One potential problem encountered with this approach is that since the acid pH is critical for the reaction, transfer of too much medium along with the embryos to the acid drop can cause the pH to elevate. Conversely, transfer of too much acid solution with the embryos after zona dissolution can cause the pH of the medium to alter. To eliminate these problems always transfer minimal amounts of solutions with the embryos and use multiple washes.

Protocol 9. Removal of zona pellucida by acid Tyrode's solution

Equipment and reagents

- Bacteriological Petri dish
- Dissecting microscope
- Mouth pipette
- KSOM medium in 1 ml syringe with 26G needle
- Acid Tyrode's: room temperature or colder (can be purchased from Sigma, T1788)

Components	**g/100 ml**
NaCl	0.800
KCl	0.020
$CaCl_2.2H_2O$	0.024
$MgCl_2.6H_2O$	0.010
Glucose	0.100
Polyvinylpyrrolidone (PVP)	0.400

Adjust to pH 2.5 with 5 M HCl then filter sterilize.

Method

1. Put two drops of M2 and several drops of acid Tyrode's in the bacteriological Petri dish.
2. Transfer all of the embryos quickly into one of the M2 drops.
3. Pick up 20–50 embryos (the number depends on how quickly you can handle them) with as little medium as you can and wash through one drop of acid Tyrode's, then transfer to a fresh drop.
4. Agitate the embryos in the acid solution by moving them with the mouth pipette while observing zona dissolution.
5. Transfer the embryos into the fresh drop of M2 *immediately* after their zonas dissolve.
6. Repeat the procedure with the remaining embryos. Use new drops of acid Tyrode's for each group of embryos.
7. Wash the embryos through two to three drops of KSOM before putting them in the aggregation plate.

4.3 Assembly of aggregations

ES cell ⟷ diploid embryo aggregations are made by aggregating a small clump of ES cells with one 8-cell embryo. ES cell ⟷ tetraploid aggregates are made either the same way or by 'sandwiching' ES cells between two tetraploid embryos. The rationale for the sandwich-type aggregation is to increase the number of tetraploid cells in the aggregates, since one tetraploid embryo has only half the normal number of cells. Single tetraploid embryo aggregates also work well under optimal culture conditions. As far as the ES cell component is concerned, we routinely use 10–15 ES cells for each type of aggregate. This number was determined experimentally. Higher numbers of cells were found to interfere with normal development, and lower cell numbers were hard to manipulate. Loosely connected clumps of 10–15 cells can be handled easily, and did not impair development. However, each ES cell clone has its optimal number, which must be determined empirically.

Assembling aggregations as described in *Protocol 10* does not require as much technical skill as microinjecting cells into blastocysts. In the course of an hour, with practice, you can easily assemble more than 100 aggregates.

Protocol 10. Assembly of ES cell ↔ diploid embryo aggregation chimeras

Equipment

- Prepared aggregation plate (*Protocol 8*)
- Dissecting microscope
- Mouth pipette

Method

1. Directly after the removal of the zona pellucida, transfer groups of six 8-cell diploid embryos into each depression-containing microdrop of the aggregation plate (*Figure 5B*) and place them individually into each depression. Place in incubator.
2. Trypsinize ES cells as described in *Protocol 1*.
3. Choose a number of clumps of loosely connected ES cells (*Figure 5B*) from the plate under the dissecting microscope and transfer them into the first and fourth rows of the aggregation plate (there are no embryos in these microdrops). Make sure that the clumps are not too dense (not more than 80–100 clumps per drop), so you can easily make the final, individual selection among them.
4. Carefully select at least six clumps of ES cells of the required cell number (usually 10–15 cells) and transfer them to a drop containing embryos (*Figures 5B* and *5G*) and release them at a distance from the embryos.
5. Pick up one of the ES cell clumps and let it fall on the side of the embryo sitting in a depression (*Figure 5C*).
6. Assemble all the aggregates in this manner, check the plate, and then put the plate back into the incubator for overnight culture. Next morning most of the aggregates will be at the late morula/early blastocyst stage (*Figure 5D*).

To aggregate ES cells with a single tetraploid embryo, follow *Protocol 10*, but use 4-cell stage tetraploid embryos instead of 8-cell diploid. To produce sandwich-type ES cell ⟷ tetraploid embryo aggregation follow *Protocol 11*.

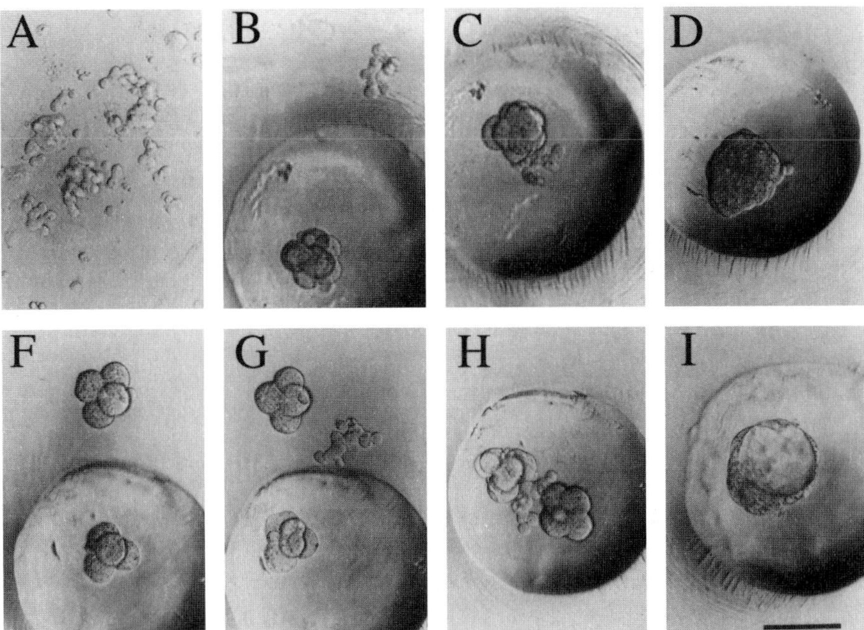

Figure 5. Steps of assembling aggregates (bar = 100 μm). (A) ES cells right after trypsin treatment. (B) The embryo is placed in the depression after removing its zona pellucida, then clumps of ES cells are placed beside for final selection for the right size. (C) The aggregate is assembled by putting the selected ES cell clump beside the embryo. (D) The aggregate forms compact morula after overnight culture. (F) Two tetraploid 4-cell stage embryos placed in and beside the depression. (G) The three components of 'sandwich'. (H) 'Sandwich', (I) after overnight culture.

Protocol 11. Assembly of ES cell ↔ tetraploid embryo aggregates

1. Assemble aggregation plate with clumps of ES cells and tetraploid embryos as described for diploid aggregates (*Protocol 10*, steps 1–4), except that groups of 12 4-cell tetraploid embryos should be transferred into each drop containing six depressions. In order to prevent accidental aggregation of embryos while you are preparing the drops, put one embryo into and one beside each depression (*Figure 5F*).
2. Add one clump of ES cells to the embryo sitting in the depression, pick up the embryo placed outside the depression, and place it on the free side of the ES cell clump (*Figure 5H*).
3. Assemble all aggregates in this manner, check and rearrange if necessary, then gently place the aggregation plate in the incubator for overnight culture. After 24 h, most aggregates will be at the early blastocyst stage (*Figure 5I*).

4.4 Culture and transfer of embryos

Usually more than 90% of the cultures aggregate overnight (*Figure 5D*) and form early blastocysts by the afternoon of the next day (*Figure 5I*). At this time they should be transferred into the uterus of pseudopregnant recipients (see Chapter 4, *Protocol 8*). We transfer a maximum of eight to ten embryos into each uterine horn. In a situation where there are too many embryos for the number of recipients, you can culture the aggregates for one more day and transfer into 2.5 day pseudopregnant recipients at this time.

4.5 Caesarian section

Depending on the nature of the experiment, the embryos derived from the aggregates can be recovered at different stages of development. The developmental stage of the embryos is determined by the recipient. Methods of dissection of many embryonic stages are found in Chapter 6. Therefore here we only describe the recovery of neonates by Caesarian section. Caesarian section can be helpful in recovering completely ES cell-derived or extensive ES cell chimeras at the newborn stage, since both of them experience severe survival problems at birth and are eaten by the mothers or are dead long before they are noticed. Caesarian section is also highly recommended if a recipient with chimeras is overdue and does not give birth on day 19.5 of pregnancy. If you are planning to foster the newborns that survive after Caesarian section, you must have foster mothers available by setting up normal matings that precede the pseudopregnant recipients day of delivery by one or two days.

Protocol 12. Caesarian section

Equipment

- Fine scissors
- Fine blunt-tipped forceps
- Small pieces of soft facial tissue paper
- Desk lamp

Method

1. Kill the recipient on day 18.5 of pregnancy.
2. Open the body cavity by making a small ventral incision in the skin across the midline at the level of the bladder, then pull the skin up over the chest, and cut the body wall along both sides starting from the midline at the bladder.
3. Locate the cervix and hold with a pair of forceps, cut across the vagina below the cervix.
4. Carefully cut the uterine wall beside one fetus and let the fetus pop out.

5. Cut the yolk sac and amnion and quickly wipe the nose and mouth of the newborn with tissue.
6. Recover the other newborns by repeating steps 4 and 5.
7. After the placental circulation has stopped, cut the umbilical cord close to the placenta.
8. It helps the recovery of the newborns if you keep them warm under a desk lamp and frequently wipe their noses and mouths until they start breathing regularly and turn pink.
9. If required, take a blood sample (e.g. for GPI assay to measure ES cell contribution) by cutting the tip of the tail and absorbing the little drop of blood at the wound with a small piece of tissue paper (4–5 mm^2). Put this tissue paper into an Eppendorf tube containing 20 μl of distilled water. Store at −20 °C. The yolk sac can be saved for PCR or Southern blot analysis if genotyping is required.
10. Remove the foster mothers from their cage and quickly take away half of their own pups. Mix the recovered newborns with the remaining pups and put the mothers back. It is recommended to have two foster mothers in the same cage.
11. After 2–3 h, when the mothers have accepted the new babies, you can remove the remaining original pups. Since the original pups are older, no special marker is necessary to distinguish the original and the Caesarian-derived pups.

5. Characterization of mosaicism in diploid or tetraploid chimeras

In any experiments using chimeras, one requires suitable reporters to follow the fate of the components. This is especially important in ES cell ⟷ tetraploid embryo chimeras, where the aim is to ensure that complete take over of the fetus by ES cells has occurred. For many experiments with ES cell ⟷ diploid embryo aggregation chimeras it will also be important to be able to follow the ES cell contribution in detail in different organs and tissues. For this reason it is advisable to use genetic markers other than simple coat colour differences in such experiments.

5.1 Marker systems

The most useful markers utilize natural or artificial genetic differences (genetic polymorphisms or transgenes) between the chimeric components. Here we describe four marker systems commonly used in connection with ES cell chimeras.

5.1.1 Glucose phosphate isomerase (GPI)

This is still a commonly used, traditional chimeric marker, based on electrophoretic polymorphisms of GPI. The enzyme is a dimer of the products encoded by the two alleles of the gene *Gpi1*. Therefore a mouse that is heterozygous for two different electrophoretic variants will show an electrophoretic pattern with three bands; the two homodimers and a heterodimer with intermediate mobility and double density. Three electrophoretically distinct alleles of the gene (*Gpi1a*, *Gpi1b*, and *Gpi1c*) have been described in laboratory mouse strains (see Chapter 4). The products of these genes are designated as GPI-A, GPI-B, and GPI-C, respectively.

Most 129-derived ES cells are GPI-A/A (see Chapter 4, *Table 1*). If these cell lines are used, GPI-B/B (C57xCBAF1 or C57BL) or GPI-A/B mice should be used as a diploid or tetraploid partner. In the case of ES cell ⟷ tetraploid embryo aggregation, the occasional presence of weak GPI-B/B (or GPI-A/B) band, indicates tetraploid cell contamination in ES cell-derived newborns. In chimeras with diploid embryos, where estimates of ES cell/host cell ratio are required the two components should be of different homozygous genotypes. With the GPI assay one can easily detect as low as 1–3% of contribution levels of the components. Different organs have different specific enzyme activity. *Table 1* shows suitable dilution factors or final volumes of the tissue homogenates of newborns to obtain equal activities for GPI cellulose-acetate electrophoresis.

Table 1. Newborn mouse tissue dilutions for GPI cellulose-acetate electrophoresis

Organs	Dilution factor
Brain	3 ×
Skin (0.5 cm^2)	2 ×
Adrenals	–
Kidneys	4 ×
Liver	8 ×
Stomach	2 ×
Pancreas	4 ×
Intestine	7 ×
Bladder	2 ×
Lungs	6 ×
Thyroid	2 ×
Spleen	2 ×
Testis	–
Muscle (deltoid)	2 ×
Heart	4 ×
Tongue	–
Thymus	–

Protocol 13. GPI electrophoresis using cellulose-acetate gel

Equipment and reagents

- Super Z Applicator Kit (Helena Laboratories, Cat. No. 4093): contains Super Z applicator (12 samples), Super Z aligning base, and Super Z well plate
- Titan III cellulose-acetate plates (Helena Laboratories, Cat. No. 3024)
- Disposable wicks (Helena Laboratories, Cat. No. 5081)
- Zip zone chamber (Helena Laboratories, Cat. No. 1283)
- Sample buffer: 50 mM Tris–HCl pH 8.0, 0.1% Triton X-100
- Running buffer: Supreheme buffer (Helena Laboratories, Cat. No. 5802); or 3 g Tris, 14.4 g glycine (Sigma, G7126) made up to 1 litre with distilled water
- Stock solutions for staining:

	Stock concentration
• Magnesium acetate 0.25 M	5.41 g/100 ml
• Fructose-6-phosphate (Sigma, F3627)	75 mg/ml
• Tetrazolium (MTT) (Sigma, M2128)	10 mg/ml
• Phenazine methosulfate (PMS) (Sigma, P9625) Store in light-proof container!	1.8 mg/ml
• Nicotinamide adenine dinucleotide phosphate (NADP) (Sigma, N0505)	10 mg/ml
• Glucose-6-phosphate dehydrogenase (G6PDH) (Sigma, G8878)	50 U/ml
• 1% agarose (1 g/100 ml) in 0.2 M Tris–HCl pH 8.0	2.42 g/100 ml

Store stocks in 200 μl aliquots at –20°C.

A. *Preparation of samples*

1. Freeze–thaw the samples at least twice in the appropriate volume of sample buffer. Then, if the tissue volume is large enough, standard homogenization techniques can be used. For small samples simply press the tissue against the wall of the tube with the tip of a pipette.
2. Spin the samples in microcentrifuge and transfer 8 μl of supernatant to the wells of the Super Z well plate.

B. *Electrophoresis*

1. Prepare the chamber by filling the buffer compartments with Supreheme buffer half-way and folding the wicks over the inner walls to electrically connect them to the buffer compartments.
2. Soak the cellulose-acetate gel in Supreheme buffer 15 min prior to applying the samples by lowering the gel slowly into the buffer to avoid bubble formation on the surface.
3. Take the gel out and remove excess buffer by blotting between paper towels. Mark the plastic backing with the necessary information, e.g. gel number, anode.
4. Press down the applicator into the samples in the Super Z wells and then blot it on tissue paper. Return the applicator to the sample wells and hold it down in samples for a few seconds. Apply samples to gel by holding the applicator for a few seconds on the gel placed on the aligning base.

Protocol 13. *Continued*

5. Transfer the gel (gel surface down) to the chamber and weight it down with a coin in the centre. Run at 300 V (from anodes to cathode) for 90 min.

C. *Staining*

1. While the gel is running boil 1% agarose in buffer and put it into a 55 °C water-bath.
2. Just before the electrophoresis is complete, place 200 μl of each stock solution of stain components 1–5 into a test-tube (> 12 ml) shielded from light with aluminium foil. Warm the mixture in the 55 °C water-bath, then add 10 ml of 55 °C agarose solution. Mix thoroughly.
3. Terminate gel run, blot the gel on tissue, and place the gel (gel side up) on a level surface.
4. Add 200 μl of staining stock 6 (G6PDH) to the stain mixture from part C, step 2. Mix quickly and pour it evenly over the gel. To decrease background work in dim light, since the stain is light-sensitive.
5. Stain for 10–15 min (depending on the intensity of the reaction) at 37 °C, then fix the gel in 1:3 acetic acid:glycerol for 10 min.
6. Read the gel and make a photographic record.

5.1.2 LacZ gene expression

The bacterial β-galactosidase enzyme encoded by the *lacZ* gene provides an easily detectable reporter for transgenic expression using simple histochemical staining procedures. This gene, under the control of suitable promoters, can be used either to mark the ES cell-derived or the embryo component of chimeras. It can be used in whole-mount embryos or in sections. The staining procedures for different embryonic stages are found in Chapter 6 (*Protocol 11*).

5.1.3 Human alkaline phosphatase (hAP) gene expression

Alkaline phosphatases are widely expressed during embryonic development and adult life. All the mouse enzymes are heat-sensitive, therefore a brief (10–15 min) 70 °C treatment of tissues destroys their activity. Human placental AP, however, is resistant to this treatment, and can be used as a reporter gene in the mouse system. Similarly to lacZ, hAP provides an excellent, easily detectable reporter to identify one of the components in chimeras after simple histochemical staining procedures. The stain does not penetrate into the tissue very well, so for internal tissues of embryos beyond 8.5 dpc the best results can be achieved with cryostat sections. The staining also works on paraffin sectioned specimens but it is less sensitive.

Protocol 14. Staining for human placental alkaline phosphatase in chimeric or transgenic mouse embryos or tissues

Reagents

- PBS with Ca and Mg
- PBS without Ca and Mg
- 1 M sodium phosphate pH 7.3
- 4% paraformaldehyde in 100 mM sodium phosphate pH 7.3 (aliquot and store at −20 °C)
- 1% sodium deoxycholate
- 2% Nonidet P-40 (NP-40)
- 1 M Tris–HCl pH 9.5
- 5 M NaCl
- 20% Tween 20
- BM Purple AP Substrate (Boehringer Mannheim, 1 442 074)
- 25% glutaraldehyde (Sigma, G-6257)
- 250 mM EGTA pH 7.3
- 1 M $MgCl_2$
- 30% sucrose in PBS
- O.C.T. compound (Tissue-Tek, 4583)
- Paraformaldehyde fix: 4% paraformaldehyde in100 mM sodium phosphate pH 7.3, 0.01% sodium deoxycholate, 0.02% NP-40
- Wash buffer: PBS without Ca and Mg, 0.01% sodium deoxycholate, 0.02% NP-40
- AP buffer: 100 mM Tris–HCl pH 9.5, 100 mM NaCl, 10 mM $MgCl_2$
- PTM: PBS without Ca and Mg, 0.1% Tween 20, 2 mM $MgCl_2$
- Glutaraldehyde fix: 100 mM sodium phosphate pH 7.3, 5 mM EGTA, 2 mM $MgCl_2$, 0.2% glutaraldehyde

A. *Preparation and staining of embryos (up to 9.0 dpc)*

1. Dissect embryos in PBS with Ca and Mg.
2. Fix in paraformaldehyde fix on ice for 1 h.
3. Rinse in wash buffer three times for 20–30 min at room temperature.
4. Inactivate endogenous alkaline phosphatase by heat treatment in PBS for 30 min at 70–75 °C.
5. Rinse in PBS, then wash in AP buffer for 10 min.
6. Stain with BM Purple AP Substrate for 0.5–36 h at 4–23 °C. Lower staining temperature reduces background but extends incubation time.
7. After staining, wash embryos extensively in PTM.

B. *Preparation of larger embryos (later than 9.0 dpc) and adult tissues*

1. Fix embryos/tissue for 30 min at room temperature in paraformaldehyde fix plus 0.2% glutaraldehyde. Cut larger tissues in half or smaller pieces. Continue fixing for an additional 30 min on ice.
2. Change fix to glutaraldehyde fix. Fix embryos for 1 h at room temperature; larger tissues for 4 h.
3. Wash embryos/tissue in wash buffer three times for 30 min at room temperature.
4. Prepare embryos/tissue for cryosectioning. Infiltrate with 30% sucrose overnight at 4 °C. Embed in O.C.T. compound, freeze at −70 °C, and section by cryostat. Heat inactivate endogenous alkaline phosphatase for 15 min and stain sections as above. The staining time is dependent on the expression level and temperature.

5.1.4 Green fluorescent protein (GFP)

GFP has been successfully used as a vital transgenic reporter system in the mouse to identify chimeric compartments (15), specific cell types, or the sex of the embryo as early as pre-implantation stages (19). This protein emits a 508 nm green light upon excitation with 488 nm blue light. With proper optical filter systems the excitation and the emission is separable and the protein positive cells become visible and easily distinguishable from the negative cells under simple microscopic observation or FACS sorting.

The current GFP variants, however, behave poorly in histology. They lose activity upon fixation, the protein leaks out from cryostat sections, and is not active after exposure to the organic solvents required for wax embedding and sectioning. Currently the best method to visualize GFP positive cells in tissues is to use fresh thick vibratome sections.

GFP also provides an opportunity to increase the germline transmission efficiency and decrease the cost of testing chimeras. An X chromosome-linked GFP transgenic has been recently established (19), which allows non-invasive pre-implantation sexing of embryos fathered by a transgenic male. All the female offspring are GFP positive while the males are negative. When male ES cells are aggregated with embryos, the resulting ES cell ⟷ embryo chimeras are either XY ⟷ XX or XY ⟷ XY in sex chromosome constitution. As discussed in Chapter 4, moderate or strong ES cell ⟷ embryo XY ⟷ XX chimeras are often phenotypically male and, if fertile, only transmit the XY, ES cell-derived gametes. Such chimeras give 100% germline transmission of the ES cell genotype and are much preferred for efficient generation of mutant mice over the XY ⟷ XY chimeras which usually give mixed genotype offspring. Pre-selecting the sex of host embryos by GFP expression prior to or after aggregation allows only XX embryos to be used in aggregates, biasing the outcome towards 100% germline transmitting chimeras.

Protocol 15. GFP expression and detection

A. *Source of GFP*

1. Wild-type GFP is not efficiently expressed in mammalian cells. However, several mutant forms of GFP are now available which exhibit greater thermostability and increased fluorescence (Clontech Inc.).
2. The EGFP variant of GFP (available from Clontech Inc.), which contains two amino acid substitutions in the vicinity of the chromophore and other alterations that improve the codon usage, can be successfully used in ES cell and transgenic applications.
3. Other colour variants of GFP are also available, including enhanced yellow and enhanced blue (EYFP, ECFP), but there are few data yet on their use in whole animal systems.

B. *Detection of fluorescence*

1. Observation of living embryos. Living embryos at all stages from pre- to post-implantation can be observed under appropriate illumination. Certain types of plasticware inhibit fluorescent observations and so embryos should be observed in glass dishes for optimum clarity. EGFP fluorescence can be detected under the dissecting microscope or inverted microscope equipped with suitable excitation sources and appropriate filters (e.g. Leica GFP plus fluorescence filter set, or a universal light source and filter sets from the fusion apparatus manufacturer, BLS, mentioned earlier).
2. Observation of living mice. Newborn and adult mice expressing GFP can be visualized macroscopically with a headset of light source and goggles manufactured by BLS (above).
3. GFP in sections. Tissue is fixed in 4% paraformaldehyde at 4°C for between 30 min to 4 h. 50 μm sections are cut in ice-cold PBS using a vibratome. Sections are mounted under coverslips in 1:1 glycerol:PBS and stored at 4°C for later GFP visualization.

6. Prospects and limitations

The ability to take cells from tissue culture and reconstitute an entire fetus or an animal is a remarkable testament to the stability and pluripotency of ES cells. We have tended to emphasize the difficulties in achieving viable fully ES cell-derived or extensively ES cell-derived live-born mice after embryo aggregation, in order to prepare the novice for some disappointments. However, even with existing established ES cell lines, it is possible to use the ES cell ⟷ tetraploid embryo aggregation approach outlined here to produce ES cell-derived fetuses. The usefulness of this tool now has been demonstrated in a variety of studies. It seems likely that, in the future, as we learn more about the optimal conditions for deriving and maintaining ES cells, the fully ES cell-derived animals will be more readily and reproducibly achieved. Whether it will be possible to maintain ES cell potency through extensive passages in culture, will depend on determining the nature of the changes that lead to loss of full potency, and then learning how to prevent them. In addition to this, ES cell ⟷ diploid embryo aggregation is an efficient, cost-effective, and easy-to-learn way of introducing ES cell-mediated mutations into mice and for performing informative chimera studies to dissect complex phenotypes.

Acknowledgements

We wish to thank many of our colleagues in the Samuel Lunenfeld Research Institute, Mount Sinai Hospital for their contributions to many aspects of developing the ES cell aggregation technologies and the BLS Ltd. for development of several instruments. This work was supported by grants from the Medical Research Council of Canada, and Bristol Myers Squibb Ltd. A. N. is an MRC/Bristol Myers Squibb scientist and J. R. is an MRC Distinguished Scientist and an International Scholar of the Howard Hughes Medical Institute.

References

1. Beddington, R. S. P. and Robertson, E. J. (1989). *Development*, **105**, 733.
2. Nagy, A., Gocza, E., Diaz, E. M., Prideaux, V. R., Ivanyi, E., Markkula, M., *et al.* (1990). *Development*, **110**, 815.
3. Rossant, J. and Spence, A. (1998). *Trends Genet.*, **14**, 358.
4. Wood, S. A., Allen, N. D., Rossant, J., Auerbach, A., and Nagy, A. (1993). *Nature*, **365**, 87.
5. Collins, J. E. and Fleming, T. P. (1995). *Trends Biochem. Sci.*, **20**, 307.
6. Snow, M. H. L. (1976). *J. Embryol. Exp. Morphol.*, **35**, 81.
7. Tarkowski, A. K., Witkowska, A., and Opas, J. (1977). *J. Embryol. Exp. Morphol.*, **41**, 47.
8. Snow, M. H. L. (1973). *Nature*, **244**, 513.
9. Kaufman, M. H. and Webb, S. (1990). *Development*, **110**, 1121.
10. James, R. M., Klerkx, A. H., Keighren, M., Flockhart, J. H., and West, J. D. (1995). *Dev. Biol.*, **167**, 213.
11. Nagy, A., Rossant, J., Nagy, R., Abramow-Newerly, W., and Roder, J. C. (1993). *Proc. Natl. Acad. Sci. USA*, **90**, 3424.
12. Nagy, A. and Rossant, J. (1996). *J. Clin. Invest.*, **97**, 1360.
13. Zambrowicz, B. P., Imamoto, A., Fiering, S., Herzenberg, L. A., Kerr, W. G., and Soriano, P. (1997). *Proc. Natl. Acad. Sci. USA*, **94**, 3789.
14. Okabe, M., Ikawa, M., Kominami, K., Nakanishi, T., and Nishimune, Y. (1997). *FEBS Lett.*, **407**, 313.
15. Hadjantonakis, A.-K., Gertsenstein, M., Ikawa, M., Okabe, M., and Nagy, A. (1998). *Mech. Dev.*, **76**, 71.
16. Carmeliet, P., Ferreira, V., Breier, G., Pollefeyt, S., Kieckens, L., Gertsenstein, M., *et al.* (1996). *Nature*, **380**, 435.
17. Whittingham, D. G. (1971). *J. Reprod. Fert. (suppl)*, **14**, 7.
18. Erbach, G. T., Lawitts, J. A., Papaioannou, V. E., and Biggers, J. D. (1994). *Biol. Reprod.*, **50**, 1027.
19. Hadjantonakis, A.-K., Gertsenstein, M., Ikawa, I., Okabe, M., and Nagy, A. (1998). *Nature Genet.*, **19**, 220.

6

Gene trap strategies in ES cells

WOLFGANG WURST and ACHIM GOSSLER

1. Introduction

Gene trap (GT) strategies in mouse embryonic stem (ES) cells are increasingly being used for detecting patterns of gene expression (1–4), isolating and mutating endogenous genes (5–7), and identifying targets of signalling molecules and transcription factors (3, 8–10). The general term gene trap refers to the random integration of a reporter gene construct (called entrapment vector) (11, 12) into the genome such that 'productive' integration events bring the reporter gene under the transcriptional regulation of an endogenous gene. In some cases this also simultaneously generates an insertional mutation. Entrapment vectors were originally developed in bacteria (13), and applied in *Drosophila* to identify novel developmental genes and/or regulatory sequences (14–17). Subsequently, a modified strategy was developed for mouse in which the reporter gene mRNA becomes fused to an endogenous transcript. Such 'gene trap' vectors were initially used primarily as a tool to discover genes involved in development (1, 2, 18).

In the last five years there has been a significant shift of GT approaches in mouse to much broader, large scale applications in the context of the analysis of mammalian genomes and 'functional genomics'. Sequencing and physical mapping of both the human and mouse genomes is expected to be completed within the next five years. Already, a large number of mouse and human genes have been identified as expressed sequence tags (ESTs), and very likely the majority of genes will be discovered as ESTs shortly. This vast sequence information contrasts with a rather limited understanding of the *in vivo* functions of these genes. Whereas DNA sequence can provide some indication of the potential functions of these genes and their products, their physiological roles in the organism have to be determined by mutational analysis. Thus, the sequencing effort of the human genome project has to be complemented by efficient functional analyses of the identified genes. One potentially powerful complementation to the efforts of the human genome project would be a strategy whereby large scale random mutagenesis in mouse is combined with the rapid identification of the mutated genes (6, 7, 19, and German gene trap consortium, W. W. unpublished data). Random mutagenesis in ES cells using GT approaches appears to be well suited for this purpose, because tens of thousands of ES cell clones carrying random GT vector integrations can readily be generated *in vitro*, and clonal cell lines can be stored in liquid nitrogen. The integration sites (target genes) can be efficiently cloned using PCR-based approaches and then sequenced. Finally, selected integrations can be introduced into mice by the generation of germline chimeras with the ES cell clones.

In addition to large scale random mutagenesis, GT vectors that allow for the identification of particular classes of genes have been devised, such as those encoding membrane spanning or secreted proteins (5). Strategies have also been developed for identifying genes which are specifically induced or repressed by secreted factors ('induction traps') (8–10) as an approach to unravel genetic networks in mammals. Recently, an induction-type strategy to identify targets of homeobox transcription factors also has been developed (20).

In this chapter, we give an introduction to the principle structure, components, design, and modifications of GT vectors, and provide protocols and background for various types of GT screens in ES cells, and for the molecular analysis of GT integrations. Detailed discussions of other less commonly used entrapment vector types in mouse, such as promoter trap (PT) and enhancer trap (ET) vectors, can be found elsewhere (12, 21, 22), and a recent collection of GT publications can be found in the Developmental Dynamics 1998 special gene trap issue.

2. General properties and design of gene trap vectors

2.1 Basic gene trap vectors

A GT vector in its most basic version contains a splice acceptor sequence linked to the 5′ end of a reporter gene carrying a polyadenylation signal, and a selectable marker gene placed 3′ to the reporter gene (*Figure 1*). A functional reporter gene product can only be produced when integrations occur in introns that allow upstream endogenous gene exons to be spliced to the reporter gene such that the reading frame of the reporter gene is maintained. In most experiments, selector gene-resistant colonies are first selected and then screened for the rare colonies in which the reporter gene is expressed.

The only sequences needed for splicing seem to be short consensus sequences at the exon/intron and intron/exon boundaries (splice-donor and splice-acceptor, respectively) as well as a branch site located about 30 bp upstream of the splice acceptor sequence (for review see ref. 23). Splice acceptor sequences of various genes and of different lengths (e.g. mouse *En-2*, *c-fos*, adenovirus major late gene) (2, 3, 24–33) have been used in GT constructs which produce fusion transcripts and β-galactosidase reporter gene expression, suggesting that any functional splice acceptor sequence can be

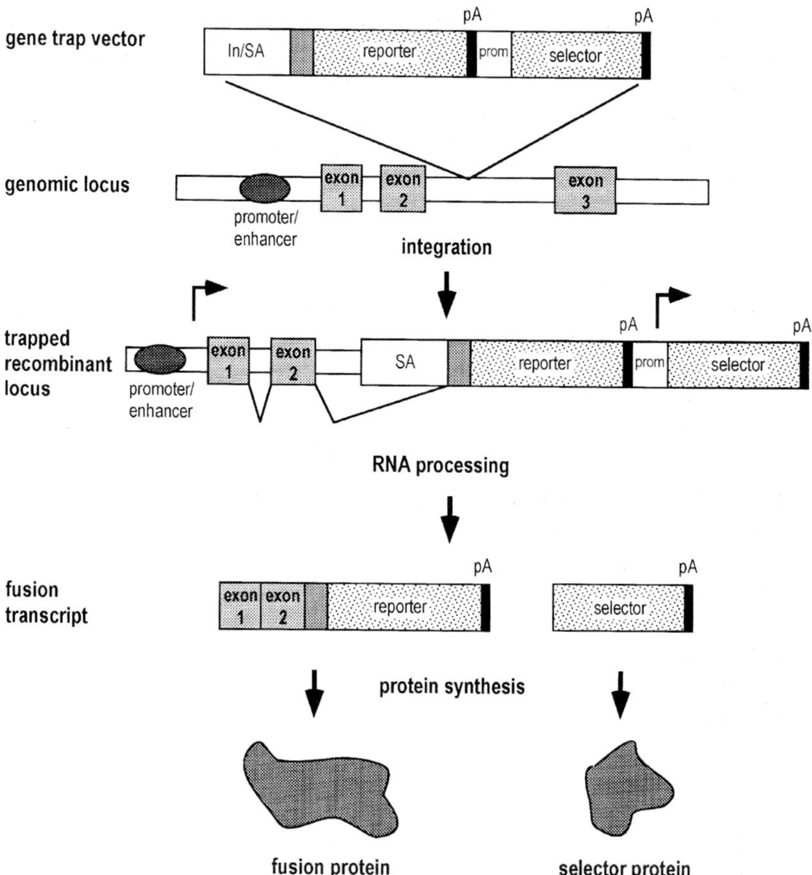

Figure 1. Basic gene trap activation scheme. A splice-acceptor GT vector (top line) represents the most basic form of a GT vector and contains a splice-acceptor site in front of a reporter gene. 3′ to the reporter gene, a polyadenylation sequence is included. The selector gene is driven by a strong promoter followed by a polyadenylation signal. Shown is a schematic representation of integration and activation of a basic splice-acceptor trap into the genome. For simplicity, only one of several possible integrations that could give rise to functional β-galactosidase is shown. Integration of the GT vector into an intron in the correct orientation results in a reporter (*lacZ*) fusion transcript. When an IRES sequence is included upstream of *lacZ*, there will be initiation of translation of the reporter gene independent of the reading frame of the fusion transcript. Moreover, the IRES sequence has the advantage that GT integrations into exons can also be identified. If no IRES is present, the fusion transcript must maintain the appropriate reading frame for a β-galactosidase fusion protein. The reporter gene product can be visualized using an appropriate reporter gene assay. In/SA, intron plus splice-acceptor site; pA, polyadenylation signal.

used. However, the efficiency of splicing to *lacZ* versus the endogenous downstream exon likely varies between splice acceptor sequences. This of course has implications for the number of *lacZ*-expressing clones, the fidelity of the *lacZ* expression pattern, and kind of mutation generated. Unfortunately, a systematic comparison of the efficiency of different splice acceptor sites has not been performed.

2.2 Reporter genes

The reporter gene most commonly used to date in GT vectors is the *E. coli* β-galactosidase gene (*lacZ*). β-Galactosidase (βGal) protein is quite stable and, due to its enzymatic activity, easy to detect by X-Gal staining in cultured cells, mouse embryos, and adult tissues. A limitation of the use of βGal is that the cells/tissues have to be fixed prior to the enzymatic assay with X-Gal. For the analysis of single cells, this limitation can be overcome by an *in vivo* stain (see *Protocol 16*). Similar to βGal, the enzymatic activity of the product of the human placental alkaline phosphatase gene (*HPAP*) has been used as a readily detectable reporter in cells and embryos (34). As with the X-Gal stain, detection of HPAP activity requires fixation of cells or tissues prior to the enzymatic assay (see Chapter 5, *Protocol 14*).

Recently, green fluorescence protein (GFP) has been introduced into mammalian cells and expressed in transgenic animals (see also Chapter 5). The expression of GFP can be visualized in living cells *in vitro* as well as in early mouse embryos (35). This has major advantages for GT screens in ES cells since it allows for analysis of large numbers of cells by fluorescence activated cell sorting (FACS), obviates the need for replica plating (see *Protocol 4*), and also should allow gene expression profiles to be followed in living embryos grown in culture during gastrulation and early organogenesis. However, detection of GFP late in development and in adult tissues seems to be limited to the surface of the tissue. Throughout this chapter we describe protocols for vectors containing *lacZ*.

2.2.1 Selectable marker genes

Different selectable marker genes have been used to identify GT vector integrations in ES cells, including *neomycin phosphotransferase* (*neo*), *histidinol dehydrogenase* (*hisD*), *hygromycin* (*hygro*), and *puromycin* (*puro*). In addition, *hypoxanthine-guanine phosphoribosyltransferase* (*Hprt*) can be used in combination with *Hprt*-defective ES cells (36). Hprt can also be used as a negative selection marker (36) similar to the Herpes simplex thymidine kinase (tk) gene (37). The minimal requirements for the promoters that drive expression of the selectable genes are that they are active in the target cells and exert neither regulatory effects (e.g. enhancer properties) on expression of the reporter gene nor on the endogenous promoter sequences (for examples see *Table 1*). However, even constitutive promoters can be subjected to position effects depending on the site of GT vector integration. For example,

Table 1. Summary and description of gene trap constructs

No.	Construct	Type	Reporter gene elements			Selectable marker elements			Cloning aids	Percentage of target cells expr. reporter gene	Restricted expr. patterns in embryos [%] (abs. Numbers)		Remarks	Ref.
			Promoter/splice acc.	Reporter	Source of pA signal	Promoter	Marker	Source of pA signal		[%] (abs. Numbers)	White ES lines	Blue ES lines		
1	pGT 4.5	GT	En-2 (+ 1.6 kb en-2 intron)	lacZ	SV40	β-actin (human)	neo	SV40	—	1–2 (10/600)	Not analysed	14 (1/7)	3 cell lines showed widespread, 3 ubiquitous expression in embryos; 3 integrations were introduced into the germline, and 2 resulted in a recessive embryonic lethal mutation.	1
2	pSAβGal	GT*	Adeno major late SA	lacZ ATG (Kozak)	bgh	PGK	neo	bgh	—	4.6 (16/350)	[100] (1/1)	(0/2)	The 2 blue lines showed widespread expression in embryos; the white line showedstaining following differentiatioin in embroid bodies; 3 integrations were introduced into the germline, and 1 resulted in arecessive embryonic lethal mutation.	2
3	pSAgeo	GT* lacZ/neo fusion	Adeno major late SA	lacZ	—	—	neo	bgh	—	95 (19/20)	[100] (1/1)	100 (1/1)	13 amino acids between lacZ and neo coding sequences; the white line was neo-resistant, but no staining detectable in ES cells; 2 integrations were introduced into the germline, and 1 resulted in a recessive embryonic lethal mutation.	2
4	AcLac	GT RV	MMLV env SA	lacZ reverse orientation	•	No selectable marker			—	0.1–0.2	Not applicable		Fetal liver cells were infected and reporter gene expressing cells identified by fluorescence activated cell sorting. Δ of regulatory elements of LTR.	82
5	pSAS	GT RV	MMLV env SA	lacZ ATG	SV40	No selectable marker			—	4.1	Not applicable		Tested in cos[7]/NIH 3T3 cells, Δ of enhancer promoter, suicide vector.	40
6	ROSAgal	GT* RV	Adeno major late SA	lacZ reverse orientation	bgh	PGK	neo	bgh	supF in LTR	11.6 (31/268)	Not analysed	Not analysed	Δ of regulatory elements from LTRs of MMLV.	2

Table 1. Summary and description of gene trap constructs

No.	Construct	Type	Reporter gene elements			Selectable marker elements			Cloning aids	Percentage of target cells expr. reporter gene	Restricted expr. patterns in embryos [%] (abs. Numbers)		Remarks	Ref.
			Promoter/splice acc.	Reporter	Source of pA signal	Promoter	Marker	Source of pA signal		[%] (abs. Numbers)	White ES lines	Blue ES lines		
7	ROSAgeo	GT*, RV lacZ/neo fusion	Adeno major late SA	lacZ reverse orientation	—	—	neo	bgh	supF in LTR	95.6 (196/205)	—	27 (6/22)	Δ of regulatory elements from LTRs of MMLV; 16 cell lines showed widespread expression; 32 integrations were introduced into the germline, and 14 resulted in a recessive embryonic lethal mutation.	2
8	PT1	GT	En-2	lacZ	SV40	PGK	neo	PGK	—	2	Not analysed	17 (53/300)	50% showed widespread expressioin in chimereic embryos, 15% no staining.	4
9	PT1 ATG	GT	En-2	lacZ	SV40	PGK	neo	PGK	—	5	Not analysed		As PT1, however, ATG is in front of *lacZ*.	4
10	PT_{10}	GT	c-fos SA 0.5 kb (exon 2)	lacZ	•	HSV-TK	neo	•	—	6–8	Not analysed			28
11	pGTi	GT	MMLV env SA (+ 1.1 kb en-2 intron)	lacZ ATG (Kozak)	SV40	β-actin (human)	neo	SV40	ori, amp	5	(0/14)	22 (4/18)	14 'blue' cell lines showed no expression in embryos.	43
12	pTfu	GT lacZ/neo fusion	c-fos SA 0.5 kb (exon 2)	lacZ	—	—	neo	SV40	—	100 (27/27)	—	46 (6/13)	2 cell lines showed no expression, the remaining lines ubiquitours, widespread, or extraembryonic expression.	43
13	U3LacZ	PT RV	—	lacZ ATG	MMLV U3	HSV TK	neo	MMLV U3	—	0.4	Not applicable		Δ of regulatory elements from LTRs of MMLV; lacZ coding sequences placed into U3 LTR; tested on NIH 3T3 cells	27
14	U3HisD	PT RV	—	His D	MMLV U3	HSV TK	neo	MMLV U3	—	—	Not applicable		Δ of regulatory elements from LTRs of MMLV; hisD coding sequences placed into U3 LTR; *neo* present as an additional selectable marker; 1 integration was introduced into the germline resulted in a recessive embryonic lethal mutation.	18, 25, 26

15	U3 neo	PT RV	—	neo	MMLV U3	No additional selection marker			—	—	Not applicable		Δ of regulatory elements from LTRs of MMLV; neo coding sequences placed into U3 LTR; 3 integrations were introduced into the germline, and 1 resulted in a recessive embryonic lethal mutation.	26
16	U3 tkneo	PT RV	—	neo	MMLV U3	—	tk	— –	MMLV U3	—	Not applicable		Positive/negative selection to identify genes active in specific lineages after EB differentiation.	83
17	pGT 1.8 geo	GT lacZ/neo fusion	En2	lacZ	—	—	neo	SV40	—	46	—	67	33% widespread, 34% restricted expression.	50; Skarnes, personal communication
18	PKC199 geo	GT lac/neo fusion	Hoxc9	lacZ	—	—	neo	SV40	—	42	—	67	33% widespread, 34% restricted expression.	50
19	GFAP	GT RV	EnV	Ap	—	PGK	neo	SV40	—	—	—	3	*In vitro* pre-screen.	34
20	PT-IRES Ap	PT	—	Ap	—	PGK	neo	SV40	—	—	—	5	*In vitro* pre-screen.	34
21	pGT 1.8TM	GT lacZ/neo fusion	En2	lacZ/neo	—	—	neo	bgh	—	20	—	—	Secretary trap vector to identify genes containing a signal peptide.	5
22	pWH14	GT neo/lacZ fusion	pmt	neo/*lacZ*	PGK	—	neo		—	100	—	—	*Neo/lacZ* fusion.	44
23	pPAT	GT	fyn	lacZ	SV40	PGK	neo	Splice donor fyn	—	2.8	—	212	Splice donor trap vector.	47
24	vicTR20	GT RV	Adeno major late SA	lacZ	—	—	neo	bgh	—	—	—	—	*in addition: PGKpuro SD large sequencing effort of integrations.	7
25	pEIT2	PT lacZ/neo	—	lacZ	—	—	neo	SV40	—	9	—	—	Exon trap vector, 259 bp buffer sequence.	50
26	pEIT3	PT lacZ/neo	—	lacZ	—	—	neo	SV40	—	15	—	—	Exon trap vector, 634 bp buffer sequence.	50
27	IRES (gal Neo (-pA)	GT	En2	lacZ	SV40		neo	SD Pax2	—	11 to 77	Not analysed	90	11% of the clones were blue in undifferentiated ES cells, after *in vitro* differentiation 77%.	42

RV: retroviral construct; hsp: heat shock promoter; En-2; engraled-2, sequences and apliced acceptor; bgh: bovine growth hormone gene; His D: histidinol dehygrogenase; PGK phosphoglucerate kinase; neo: neomycine phosphotransferase; HSV TK: herpes simplex virus thymidine kinase gene promoter; MMLV: Moloney Murine Leukaemia Virus; enV: Moloney leukaemia virus-derived splice-acceptor; pmt: polyoma large T-antigen; *: called promoter trap by authors; Δ: deletion of information; •: information not available.

a number of different *neo* cassettes have been tested in ES cells and as much as a tenfold difference in the number of neo^R colonies obtained with different promoters (38). Thus, a weak promoter and a high level of selection pressure can act as a pre-selection for integrations into chromosomal regions that are permissive for selector gene expression. GT constructs containing fusions between the reporter and selectable marker gene (see below) in which the *neo*-fusion gene serves as the selectable marker gene have also been used and such vectors pre-select for genes expressed in ES cells (*Figure 2A*).

2.2.2 'Buffer' sequences

It has been observed in a number of cases that linearized DNA introduced into ES cells by electroporation can lose sequences from either end before integration into the genome. Of importance, GT constructs could lose 5′ sequences required for splicing, or part of the selectable gene from the 3′ end. 'Buffer' sequences can be added to the 5′ end of the vector and plasmid sequences left to protect the 3′ end. Buffer sequences can be obtained from intron sequences in front of the splice acceptor used that are devoid of any regulatory activity which could influence expression of the reporter gene. However, the longer the buffer sequence is the more likely it is to have a cryptic splice site resulting in internal splicing within the vector (39). The length of the sequences 5′ to the splice acceptor may influence the efficiency at which GT integrations can give rise to functional reporter gene expressions. For example, with two GT constructs carrying the *lacZ* reporter and mouse *En-2* splice acceptor with either about 0.2 kb or 1.8 kb of adjacent intron sequences, it was found, that only the construct carrying the long intron sequence gave rise to functional β-galactosidase (W. Skarnes and A. Joyner, unpublished data). These results could be explained either by the 0.2 kb vector missing sequences required for splicing or by our observation that in a considerable number of cases DNA integrated into ES cells following electroporation lose sequences from either end.

2.3 Modifications of basic gene trap vectors

A number of modifications and alterations of the basic gene trap vector have been designed. These are summarized below:

(a) To increase the frequency of gene trap integrations giving rise to a functional reporter gene product, multiple splice acceptor sites have been inserted adjacent to the reporter gene or an ATG or IRES sequence.

(b) To permit direct selection of integrations into genes which are expressed in ES cells reporter/selector fusion cassettes have been used (*Figure 2A*).

(c) To allow selection for integrations into all genes (either active or inactive in ES cells), a selector gene is used without a pA signal and it is followed by a splice donor site (*Figure 2B* and *3*).

A. Reporter / selector fusion

B. Reporter / selector fusion and selector with splice donor

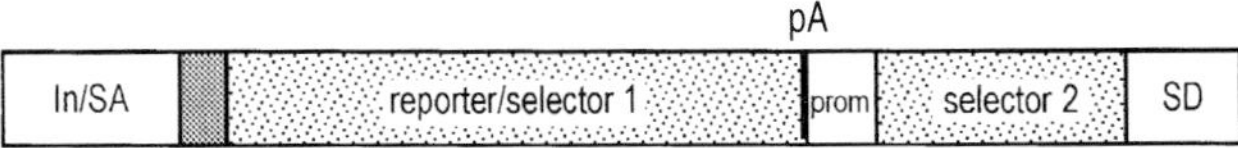

Figure 2. Modified gene trap vectors. (A) Reporter/selector fusion. The reporter gene is fused in-frame to a selector gene. Resistance to the selectable drug will only be obtained if a functional fusion transcript and protein is made. This approach only allows for the selection of integrations into genes which are expressed in the cell type analysed. (B) Reporter/selector fusion with a second selector gene operating as a splice-donor trap. This GT construct is a combination of those shown in (A) and *Figure 3* with a second selector gene which is driven by a strong promoter followed by a splice donor site. In, intron; pA, polyadenylation signal; SD, splice donor signal. Stippled box represents either ATG or IRES sequences.

(d) To obtain a high efficiency of single vector integrations without gross rearrangement at the site of integration, GT vectors have been incorporated into retroviral vectors.
(e) To capture genes encoding proteins with a signal sequence, a transmembrane spanning sequence has been inserted in front of *lacZ* (5).

Components of various GT vectors used and some of the results obtained with them are summarized and referenced in *Table 1*.

2.3.1 Multiple splice acceptor sites, ATG codon, and IRES sequence in front of the reporter gene

To obtain functional β-galactosidase activity with the basic GT vector, the reading frame between *lacZ* and endogenous trapped upstream exon sequences must be maintained. This requirement results in a low rate of integrations with β-galactosidase activity. Several modifications of the basic vector type have been used to increase the frequency of 'productive' integrations. These modifications are:

(a) Introduction of multiple splice acceptor sites in the three reading frames with respect to the reporter gene
(b) Addition of an ATG codon in the context of a Kozak translational start consensus sequence to the reporter gene.
(c) A combination of the two above alterations.
(d) Addition of an IRES sequence in front of the reporter gene.

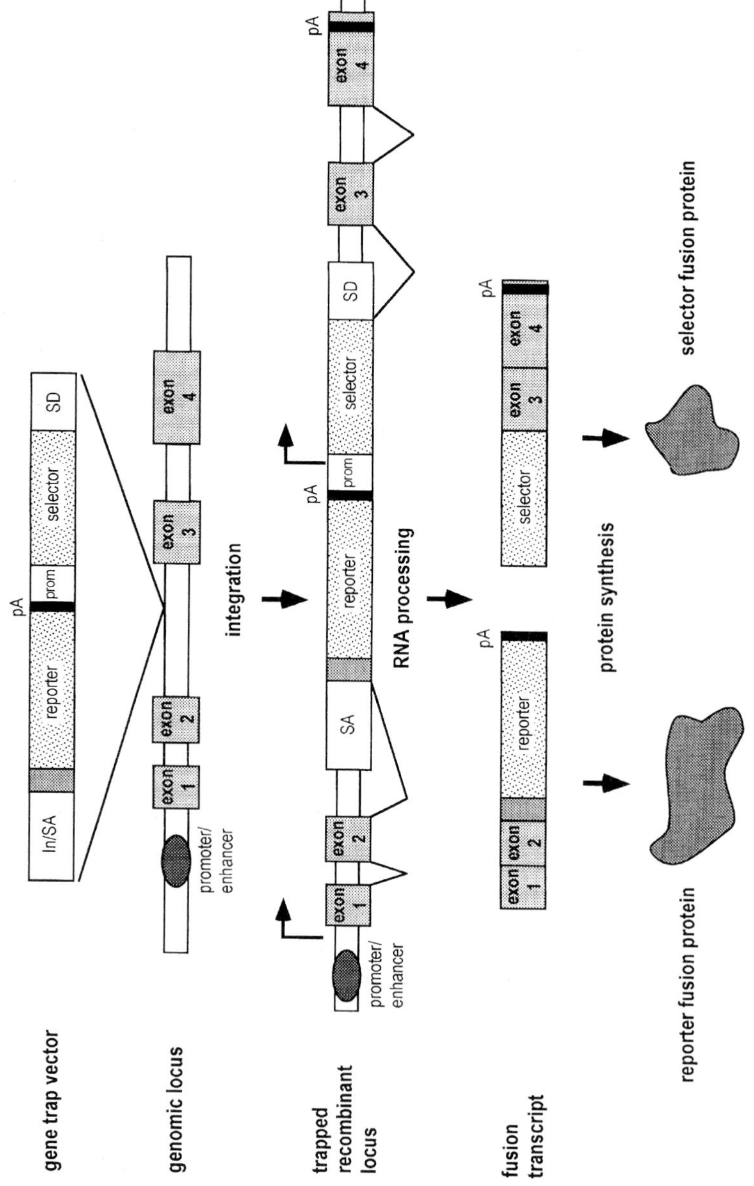

gene trap vector
In/SA
reporter
pA
prom
selector
SD
genomic locus
promoter/
enhancer
exon
1
exon
2
exon
3
exon
4
integration
trapped
recombinant
locus
promoter/
enhancer
exon
1
exon
2
SA
reporter
pA
prom
selector
SD
exon
3
exon
4
pA
RNA processing
fusion
transcript
exon
1
exon
2
reporter
pA
selector
exon
3
exon
4
pA
protein synthesis
reporter fusion protein
selector fusion protein

Figure 3. Gene trap activation using a splice acceptor/reporter and a selector/splice donor vector. The GT vector presented contains a splice acceptor in front of *lacZ* and a splice donor behind the selector gene driven by its own promoter. In this type of vector, the selector gene has to acquire endogenous polyA signals for the cells to become drug resistant. In addition, if IRES sequences are introduced in front of the reporter gene, the expression of the trapped gene can be monitored using the reporter gene. Thus, this vector allows silent genes to be trapped and their expression pattern followed in chimeras or transgenic mice obtained after germline transmission. In, intron; SA, splice acceptor; SD, splice donor; pA, polyadenylation signal.

The reporter gene can be provided with its own efficient start of translation (Kozak consensus sequence). This should extend the range of integration events that lead to functional β-galactosidase activity to integrations into 5′ untranslated exon sequences or into introns that follow untranslated exons. The addition of an ATG in the context of a SA can increase the frequency of reporter gene expression five to ten times with both retroviral (40) and electroporated vectors (W. Wurst and A. Joyner, unpublished results).

Eukaryotic RNA translation is usually dependent on 5′ cap-mediated ribosome binding and subsequent scanning by the ribosomal complex for the site of initiation of translation, in general the first ATG. In contrast, some viruses can produce non-capped RNAs which are efficiently translated by taking advantage of internal ribosomal entry sites (IRES) located within the 5′ UTR (reviewed in ref. 41). Addition of viral IRES sequences between the splice-acceptor site and coding sequence of the reporter gene in a GT vector should allow initiation of translation of the reporter gene independent of the upstream reading frame. About tenfold more G418-resistant clones were obtained using such an IRES-containing gene trap vector compared to a vector without an ATG suggesting that these vectors enhance the success of large scale gene trap screens (7, 42). Triple splice acceptor sequences containing the three reading frames also have been used in an attempt to generate in-frame splicing with *lacZ* for all appropriate integrations into introns. However, that this occurs efficiently has not been demonstrated conclusively.

2.3.2 Reporter/selector fusions

Gene trap vectors containing a protein fusion (β*geo*) between the reporter and selector proteins *lacZ* and *neo* (*Figure 2A*) have been designed to allow for direct selection of gene trap vector integrations into expressed genes (2, 43). With such vectors, G418-resistant colonies will only be obtained when the integration leads to the generation of a functional *lacZ/neo* or *neo/lacZ* (44) fusion protein. The success rate for identifying bona fide GT events is 100% of the G418-resistant colonies, because of the simultaneous expression of the reporter and selector. The main limitation of this approach is that only genes which are expressed in the cell type used for the screen can be identified.

A variety of protein fusions have been found not to interfere with enzymatic activity of β-galactosidase. Thus, various fusions between the *lacZ* gene and selector genes other than *neo* can be expected to produce bifunctional fusion proteins. However, *neo* protein can be affected by fusions, since it has been shown that long amino-terminal fusions can result in profound differences in the enzymatic activity of *neo* (45). If the *neo* gene product is rendered less active in a particular fusion protein, this could result in a preferential recovery of integrations into genes that are expressed at high levels. However, the β*geo* vector which has been used extensively by different groups seems to work efficiently as some integrations that produce enough RNA to render the cells G418^R do not have detectable βGal activity (see Section 3.2.1) (2, 42).

2.3.3 Gene trap vectors containing a polyA trap

One of the limitations of the gene trap vectors described thus far is that for practical reasons predominantly genes which are active in ES cells can be readily identified. GT vectors that carry a selection marker driven by a constitutive promoter however can integrate into genes not expressed in ES cells and provide drug resistance. To identify GT insertions into any gene, polyA trap vectors have been developed (*Figures 2B* and *3*). In these vectors, a selector gene is used that is driven by a constitutive promoter but lacks a polyA signal, and in some cases has a splice donor sequence positioned downstream of the selector gene (7, 42, 46, 47). Thus, the selectable marker gene can be transcribed independently of the transcriptional activity of the tagged gene, but requires capturing an endogenous polyA signal to produce a stable mRNA. Therefore, drug-resistant colonies most likely represent GT insertions into transcription units.

In principle, GT vectors with both a splice acceptor in front of *lacZ* and a splice donor after *neo* without a polyA should allow one to identify and mutate all genes in the genome (7). However, depending on the type of selection cassette, way of delivery, target gene size, and accessibility (chromatin structure) of a gene, not all loci are likely to be tagged with any given GT vector and screen. In addition, there seem to be integration 'hotspots' such as the *jumonji* locus which has been independently trapped five times with different GT vectors and delivery systems (48–50), and twice independently (Wiles, Vauti, and Wurst, unpublished data). Such preferential integrations into a subset of loci are likely to result in the over-representation of some genes at the expense of others, and argue against truly random insertions of GT vectors into the genome.

2.3.4 Retroviral vectors

Retroviral infection as a means of introducing foreign DNA into the genome has a number of advantages compared to electroporation of plasmid vectors. First, the gene trap vector integrates at each site with a defined vector structure and little rearrangement at the site of insertion occurs. Secondly, the infection efficiency can be 100%, which reduces the number of cells that need to be transfected. For some experimental systems (e.g. haematopoetic stem cells) it is a prerequisite to reach this efficiency. Although, there has been some bias in target site selection reported for *MoMuLV*-based vectors (51) no sequence motif that predisposes host DNA for integration has been identified. In addition, in a large scale GT screen using a retrovirus vector there was no bias reported concerning GT vector integration sites (7). The design and application of retroviral GT vectors are described in detail by Hicks *et al.* (52).

2.3.5 Secretion trap vectors

The GT vectors described above are likely to be inefficient at identifying genes that encode secreted proteins, because the N-terminal signal peptide most likely causes the secretion of the reporter fusion protein, resulting in very low or undetectable intracellular reporter gene activity. To allow for identification of this class of genes, a secretory trap vector has been designed which contains the transmembrane region of the CD4 gene fused in-frame to the 5′ end of the *lacZ* gene (5). An appropriate integration of this vector into a gene containing a signal peptide should direct the reporter gene protein product to insert into the cell membrane where it can be visualized by X-Gal staining. However, in a screen using this particular vector, mainly type I membrane molecules containing large introns were identified (5). Furthermore, from six genes identified, two integrations occurred in the receptor-linked protein tyrosine phosphatase (LAR), suggesting that only a limited number of different genes can be readily identified using this particular vector. To detect small transcription units composed of few introns, promoter trap vectors which do not contain a splice acceptor site, have to be utilized (5).

3. Establishment of cell lines carrying reporter gene integrations

3.1 Introduction of reporter gene constructs into ES cells

ES cell clones carrying GT vectors integrated in their genome can be easily produced by various DNA transfer techniques. We use electroporation as a standard procedure to introduce plasmid GT constructs into ES cells. Retroviral vectors are introduced by infection (see *Protocol 3*). The *neo* gene as part of the vector allows in both cases for direct selection of cells which have integrated vector DNA. It is important:

(a) To obtain cell lines which carry a single integration.

(b) To ensure, that the GT vector sequences are intact, i.e. that they contain the elements required for proper splicing and expression of *lacZ*.

Both requirements are intrinsically met when retroviral vectors are used with a multiplicity of infection (m.o.i.) below 1, because of the mechanism of integration of the proviral DNA. When electroporation or other transfection procedures are used, precautions must be taken to meet the conditions mentioned above (see below).

Multiple GT integrations are not necessarily problematic, because the detected 'active' copy is linked to the endogenous gene by the fusion transcript, and it is unlikely that several independent integrations in one cell will give rise to active fusion transcripts. Intact reporter gene integration is necessary to obtain βGal activity, and this is often determined by staining of the $G418^R$ colonies. In cases in which clones which do not express *lacZ* in undifferentiated ES cells are to be further analysed, integrations can be pre-screened for an intact *lacZ* gene using PCR and appropriate sets of primers.

3.1.1 Electroporation

The conditions we use (see *Protocol 1*) usually result in the integration of one, sometimes two to four tandem copies of the GT vector into one site in the genome. Most integrations (> 80%) carry single copies at one site. The following electroporation protocol could be used for all types of GT vectors.

Protocol 1. Electroporation of ES cells with GT constructs

1. The electroporation procedure has been described in Chapter 3 (*Protocol 8*). To ensure single GT vector integrations, however we use different electroporation conditions: 150 μg linearized GT vector DNA with 10^8 cells in 1 ml, and using 3 μF and 0.8 kV electroporation conditions.
2. Seed electroporated cells at a density of $2.5–5 \times 10^6$ cells/90 mm dish[a] on a gelatinized plate in medium containing LIF (Chapter 3, *Protocol 8*).
3. Add selection medium after 36–48 h. Colonies of resistant cells should appear within seven to ten days.

[a] The cell concentration per plate must be adjusted depending on the vector such that no more than 200–500 neo^R colonies are obtained on each plate.

To analyse reporter gene integrations for the presence of sequences essential for reporter gene function by PCR, see *Protocol 2*.

Protocol 2. PCR analysis to identify intact 5′ ends of integrated GT vectors

Equipment and reagents

- Thermocycler
- Mouth-controlled pipette (see Chapter 4)
- Proteinase K (20 mg/ml)
- Appropriate primers for the GT vectors
- *Taq* polymerase
- Taq reaction buffer
- dNTPs

Method

1. Mark the plate on the bottom into four quarters.
2. Pick half of each colony (approx. 50 cells) from each quarter with a mouth-controlled pipette or yellow tip (see Chapter 3, *Protocol 9*) and transfer it into an Eppendorf tube containing 50 μl distilled water.
3. Mix by shaking and freeze at –70 °C for 10 min.
4. Thaw, add 100 μg proteinase K (5 μl of a 20 mg/ml stock), and incubate for 1.5 h at 55 °C.
5. Heat for 10 min at 97 °C, then chill on ice for 5 min.
6. Use 25 μl for PCR and continue as described in *Protocol 17*, steps 4 to 6.
7. Pick the colonies of the positive pools into 96-well plates and expand as in Chapter 3, *Protocol 9*. Repeat the PCR analysis on single clones after expansion.

3.1.2 Retroviral infection

Retroviral infection is easy to perform and does not require special equipment. It is brought about by contact between the viral particles and the cells. A high titre of between 10^5 to 10^7 colony forming units (cfu)/ml (in the absence of helper virus) of the retroviral GT vector is preferred for most experiments. Stable transfection of the retroviral vector sequences into packaging cell lines can produce titres of 10^5 cfu/ml or higher if the transcription and packaging of the vector sequences occurs efficiently (52). Transient transfections lead to titres between 10^1 to 10^3 cfu/ml which may be sufficient for some experiments.

Protocol 3. Infection of ES cells with retroviral vectors

Equipment and reagents

- Gelatinized tissue culture plates (Chapter 3)
- Polybrene (Sigma, Cat. No. H 9268)
- ES cell medium containing LIF (Chapter 3)
- ES cell medium with LIF containing retrovirus

Method

1. Plate ES cells on gelatinized tissue culture dishes at a density of 3×10^6 cells per 90 mm dish in ES cell medium supplemented with LIF.
2. After 24 h aspirate medium, add 5 ml fresh medium containing the retroviral particles at a m.o.i. < 1, and 5 μg/ml Polybrene to obtain single integrations.

Protocol 3. *Continued*

3. After overnight culture (14 h) remove virus-containing medium, add 10 ml fresh medium, and culture for an additional 24 h.
4. Change the medium to selection medium and change medium every other day. Drug-resistant colonies should become visible between seven to ten days.

3.2 Identification of *lacZ*-expressing ES cell clones

In some experiments, only ES cell clones which show β-galactosidase activity in undifferentiated ES cells are to be analysed. Depending on the vector, these ES cell clones usually make up only a small fraction of the clones that carry the vector and are neo^R. Thus, it is desirable to identify *lacZ*-expressing clones as soon as possible in the experiment to minimize the cell culture work, which is costly and labour-intensive. It is preferable to analyse ES cell clones before or during picking of neo^R colonies and then to further expand only the desired cell lines. Several strategies can be used to identify *lacZ*-expressing ES cell clones rapidly.

3.2.1 Direct selection

Direct selection is only applicable for reporter-selector gene fusion GT vectors. With GT *lacZ/neo* fusion vectors, resistance to the drug G418 requires, and indicates, the presence of a β*Gal/neo* fusion protein. However, the sensitivity of detecting both activities can vary and cells which express functional fusions (as indicated by resistance to G418) do not necessarily show detectable X-Gal staining. Thus, if the purpose of the experiment is to analyse clones that express clearly detectable levels of βGal, then X-Gal staining must be performed subsequent to G418 selection. However, even in such cases, the use of fusion vectors with G418 pre-selection provides an easy means to strongly enrich for ES cells which show detectable levels of βGal activity. Alternatively, fusion vectors can be used to isolate genes that are expressed at low levels in the target cells by identifying cells that show low or undetectable levels of *lacZ* expression.

3.2.2 Replica plating

Replica plating is generally applicable for a variety of cultured cell types and involves the generation of a filter bound copy of each neo^R colony that can be assayed for *lacZ* expression. We have found best results using polyester filters as described in *Protocol 4* (21, 53, 54). Depending on the pore size of the polyester cloth, more than one replica can be made from a single plate. The replica plating procedure does not appear to affect the ability of ES cells to colonize the germline of chimeric mice (3, 10).

Protocol 4. Replica plating of ES cells

Equipment and reagents

- Polyester fabric (mesh size 1 μm) (Tetko Inc., Elmsford, NY)
- Glass beads (about 3 mm in diameter) (Fisher Scientific, USA)
- Sterile forceps
- Deionized water
- 9 cm gelatin-coated dishes of G418^R ES cell colonies (from *Protocol 1*)
- PBS (Chapter 3)
- Concentrated HCl
- Absolute ethanol
- ES cell media with G418 and LIF (Chapter 3)

Method

1. Cut the fabric to the appropriate size circles for the plates used and cut small notches in the edges to orient the replicas after their removal.
2. Soak the filters in concentrated HCl in glass dishes overnight.
3. Rinse filters several times in PBS, several times in deionized sterile water, and then a final rinse in absolute ethanol.
4. Air dry the filters and autoclave them.
5. Treat the glass beads as in steps 2–4.
6. Filters should be put on the plates of ES cell colonies approx. seven days after G418 selection is started, at a time when the *neo*^{R} colonies are becoming visible to the eye.
7. Aspirate off the media and gently place the polyester filter on the plates using sterile alcohol-flamed forceps and mark the position of the notches on the bottom of the plate with a water-resistant pen.
8. Overlay the filter completely with a single layer of glass beads and add 10 ml ES cell medium with LIF and G418.
9. Culture cells for two days.
10. To remove the filter, aspirate medium, invert the plate quickly, and tap off the glass beads.
11. With a fine pair of sterile forceps, it is easiest to grasp the filter at a notch and then carefully peel off the filter. Any lateral movement of the filter will dislodge colonies from the plate. Often, a portion of a colony will detach but rarely the entire colony dislodges from the plate.
12. Gently add media to the plates.
13. Fix and stain each replica filter with X-Gal as described for attached cells in *Protocol 11*.

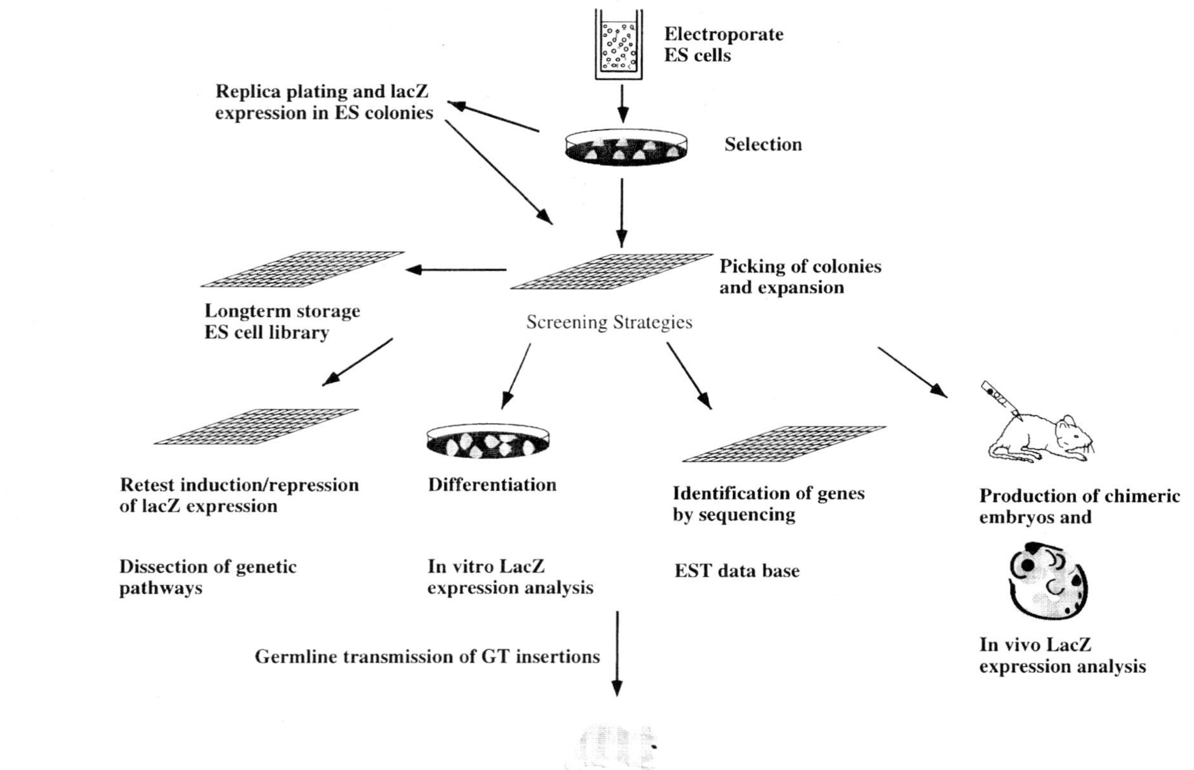

Figure 4. Experimental strategies for the large scale screening of gene trap integrations in ES cells.

4. Screening strategies

Various strategies have been devised to screen GT integrations in ES cells (*Figure 4*) for insertions into particular types of genes. These include the analysis of GT insertions for *lacZ* expression patterns in chimeric embryos, or ES cells differentiating into embryoid bodies, or following various types of induction protocols. In addition, the gene that the GT vector has inserted into can be directly cloned and sequenced or the insertion transmitted into mice for a mutant phenotype analysis.

4.1 Analysis of GT activation patterns in chimeric embryos

4.1.1 Production of chimeras

Chimeric embryos can be produced either by blastocyst or morula injection, or aggregation between ES cell aggregates and morula stage embryos as described in Chapters 4 and 5, respectively. Since X-Gal staining in early embryos, up to about day 12, does not produce endogenous background under the staining conditions given below, any X-Gal staining patterns observed in embryos obtained after chimera formation are the result of expression of the GT reporter gene. In cases in which no staining is observed in presumed ES cell–chimeric embryos, a control is required to distinguish whether the lack of staining is due to no activation of the reporter gene, or due to a lack of contribution of ES cells in the embryos. To monitor the successful generation of chimeras, and to estimate the degree of ES cell contribution in chimeric embryos, a combination of ES cells derived from a pigmented mouse strain with host blastocysts derived from albino mice should be used which allows detection of ES cell contribution to the embryo from about day 12 onwards by analysing eye pigmentation. Any pigmented cells in the eyes are ES cell-derived and are, therefore, indicative of the ES cell contribution to the embryo. For analysis of earlier stages, we proceed according to *Protocol 5*.

Protocol 5. Generation of ES cell–chimeric embryos for *lacZ* expression analysis

1. Inject 10–15 albino-derived blastocysts with ES cells derived from a pigmented mouse for each early embryonic stage to be analysed (Chapter 4).
2. Inject an additional 10–15 blastocysts for a day 12.5 control.
3. Transfer the embryos to foster females in groups of 10–15 (one foster per stage to be analysed and one additional for the control).
4. Recover embryos at desired stages (see *Protocol 6*) and stain for βGal activity (see *Protocol 11*).

Protocol 5. *Continued*

5. Recover control embryos at day 12.5, monitor the embryos for pigmented cells in the eye (which can be easily seen by naked eye or under a dissecting microscope), and stain for βGal activity.

When a given percentage of embryos are chimeric at 12.5 d, it can be assumed that a similar percentage of chimeric embryos were obtained from the same pool of injected blastocysts at earlier stages. When no X-Gal staining is observed at earlier stages, only experiments which had at least 30–50% chimeras with significant ES cell contribution in day 12.5 embryos, and for which at least 15 embryos at each earlier stage were analysed, can be taken as indication of a lack of *lacZ* expression.

4.1.2 Recovery of post-implantation embryos

ES cell–chimeric embryos must be dissected out of the uterus and away from the maternal decidual tissue before βGal staining, to allow for penetration of the substrate. Since embryos are not kept in culture after recovery, no precautions for sterility have to be taken, nor has special care to be taken that chemicals used for the buffer required during dissection are of extreme purity to avoid toxic effects on the embryos. The procedure described here does not take special care to maintain the integrity of the extraembryonic membranes, although those usually stay intact during dissection. Described in some detail is the recovery of embryos between day 6.5–9.5, stages at which the embryo is small compared to the decidual tissue surrounding the embryo. Later stages can be recovered more easily, the first steps being essentially identical to the procedure described.

Protocol 6. Recovery of early post-implantation embryos

Equipment and reagents

- One pair of coarse forceps
- One pair of fine scissors
- Two pairs of fine forceps
- Two pairs of watchmaker's forceps (Dumont No. 5)
- Pasteur pipettes
- Pasteur pipettes with 'wide opening'
- Petri dishes
- Dissecting microscope
- Pregnant female mice
- PBS (Chapter 3)

Method

1. Kill the pregnant female using any acceptable method.
2. Lay mouse on its back, make a large V-shaped incision into the skin with the tip of the V just anterior to the vagina.
3. Fold the skin back, cut the abdominal wall similarly to the skin, and fold back.

4. Find the uterus, hold the uterine horns with coarse forceps at the cervical end, and cut at the uterine-cervical junction (see also Chapter 4).

5. Pull uterine horns slightly, trim away any fat and the mesometrium (part of the broad ligament that is attached to one side of each uterine horn) along the uterine wall, and cut off the uterine horns at their anterior ends.

6. Put uterine horns into a Petri dish containing PBS.

7. Cut the uterine horns between the decidual swellings.

8. Under the dissecting microscope tear the uterine muscle on the opposite side to the attached mesometrium with fine watchmaker's forceps.

9. Free the decidual swelling from the uterine wall by holding the muscle with one fine watchmaker's forceps and sliding along between the torn muscle and deciduum with the second fine watchmaker's forceps (see *Figures 5a, 5b*).

10. Transfer deciduas into fresh dish containing PBS.

11. Hold the deciduum on the mesometrial (broad) end with one fine watchmaker's forceps. Insert the point of the closed second forceps in the midline above the reddish streak (which is the embryo) through the deciduum and open fine watchmaker's forceps splitting the deciduum (see *Figures 5c, 5d*).

12. Grasp the split parts and pull apart. The embryo, surrounded in membranes, usually remains attached to one decidual half.

13. Gently push the embryo with the tip of the closed watchmaker's forceps until the embryo and its membranes are entirely free.

14. Transfer the embryo with a Pasteur pipette (6.5 and 7.5 dpc) or the wide opening of a Pasteur pipette (8.5 and 9.5 dpc). Older specimens can be transferred by 'scooping' them with a curved forceps.

15. Removal of Reichert's membrane (6.5–8.5 dpc). Grasp Reichert's membrane at the extraembryonic portion of the egg cylinder (away from the embryo) with both watchmaker's forceps and gently tear it open (see *Figures 5e, 5f*). Most of the membrane can be torn off leaving behind only some remnants at the ectoplacental cone which do not impair staining and analysis of the embryo. It is advisable to leave the membranes attached to the embryo to discover extra-embryonic staining as well.

16. Removal of extraembryonic membranes ($>$ d 8.5). Grasp visceral yolk sac with both forceps. Tear until the embryo is freed from the yolk sac but still connected with it by the umbilical cord. Hold the cord with one forceps and tear off the yolk sac distally with second forceps. If the amnion, a very thin cellular membrane, is still surrounding the embryo remove that analogously to the yolk sac.

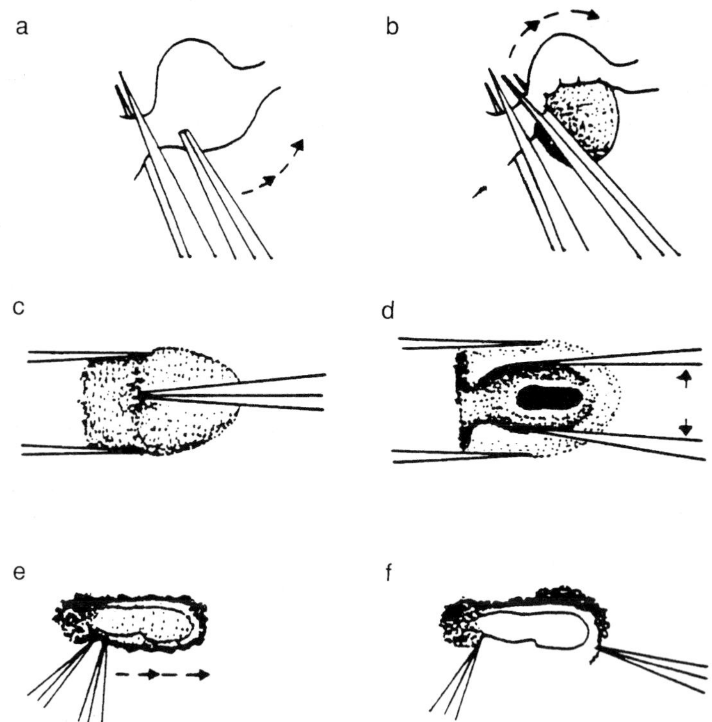

Figure 5. Diagram depicting the dissection of early mouse post-implantation embryos (taken from R. Beddington: Isolation, culture, and manipulation of post-implantation mouse embryos. In *Mammalian development: a practical approach*. IRL Press, Oxford). For details see *Protocol 6*.

4.2 Identification of GT integrations activated or repressed during embryoid body formation

Provided by L. Forrester, Centre for Genome Research, University of Edinburgh, `L.Forrester@ed.ac.uk`).

As described in previous chapters, the pluripotent stem cell character of ES cells can be maintained under appropriate culture conditions (55–58). However, when grown under different conditions, ES cells spontaneously differentiate into embryoid bodies and give rise to various cell lineages derived from all three germ layers (59), thereby recapitulating early embryonic differentiation events *in vitro*. During spontaneous differentiation, numerous cell types can be obtained such as neuronal, endodermal, and mesodermal including cardiac and skeletal muscle, haematopoetic cell types, and endothelial cells (*Table 2*) (60–63) (for review see refs 64–68). In addition, under appropriate culture conditions and by adding different growth and differentiation factors the differentiation of particular cell types can be preferentially stimulated (10, 67).

Table 2. Differentiation of ES cells into different lineages

Cell lineage	Cell type	References
Haematopoiesis	Multiple haematopoietic cells	59, 84, 85
	Erythroblasts	85
	Macrophages, mast cells, neurotrophiles	86
Angiogenesis/vasculogenesis	Endothelial cells, capillary cells	60, 84
Cardiogenesis	Cardiomyocytes	59, 69, 87, 88
Myogenesis	Skeletal muscle	59, 61, 87
Neurogenesis	Neuronal cells	61, 89–92
	Glia cells	91, 92

Embryoid bodies derived from ES cells carrying GT integrations have been used to identify genes whose expression is activated or repressed during differentiation *in vitro* (49, 34). Significantly, the *lacZ* expression pattern in embryoid bodies usually recapitulates the reporter gene expression pattern found *in vivo* in embryos carrying the same GT insertion (34, 49). For procedures that have been devised to specifically induce particular cell types the reader is referred to *Table 2* and *Figure 6*.

4.2.1 *In vitro* differentiation of GT libraries of clones

We have found that the most consistent cardiac differentiation of embryoid bodies is achieved by generating hanging-drops using precise cell numbers as described by Wobus (69). However this technique is prohibitive if a large number of gene trap integrations have to be screened as precise cell concentrations must be determined for individual clones. We have attempted to estimate the cell numbers by eye in individual wells of a 24-well plate prior to trypsinization, then resuspend them in a set volume of medium, and set up 10 μl hanging-drops containing the desired number of cells (i.e. 300). Embryoid bodies were formed in the majority of cases but the consistency in size and subsequent cardiomyocyte differentiation was poor in comparison to that seen when cell numbers were calculated accurately. This was due to the variability in the growth of individual ES cell clones and the difficulties in estimating cell numbers since some clones grow as a confluent monolayer and others as more three-dimensional colonies. Xiong *et al.* (34) have successfully screened a large number (2400) of gene trap integrations using a strategy whereby they pooled a number of gene trap clones then subsequently generated embryoid bodies and analysed the *in vitro* expression pattern of the reporter gene in differentiating cell types. We propose that this protocol (summarized below) could be used as an initial pre-screen to identify pools of clones that show expression in an interesting pattern. Subsequent screening of the individual clones from the pool could then be analysed using a more precise differentiation procedure depending on the lineage of interest. Co-

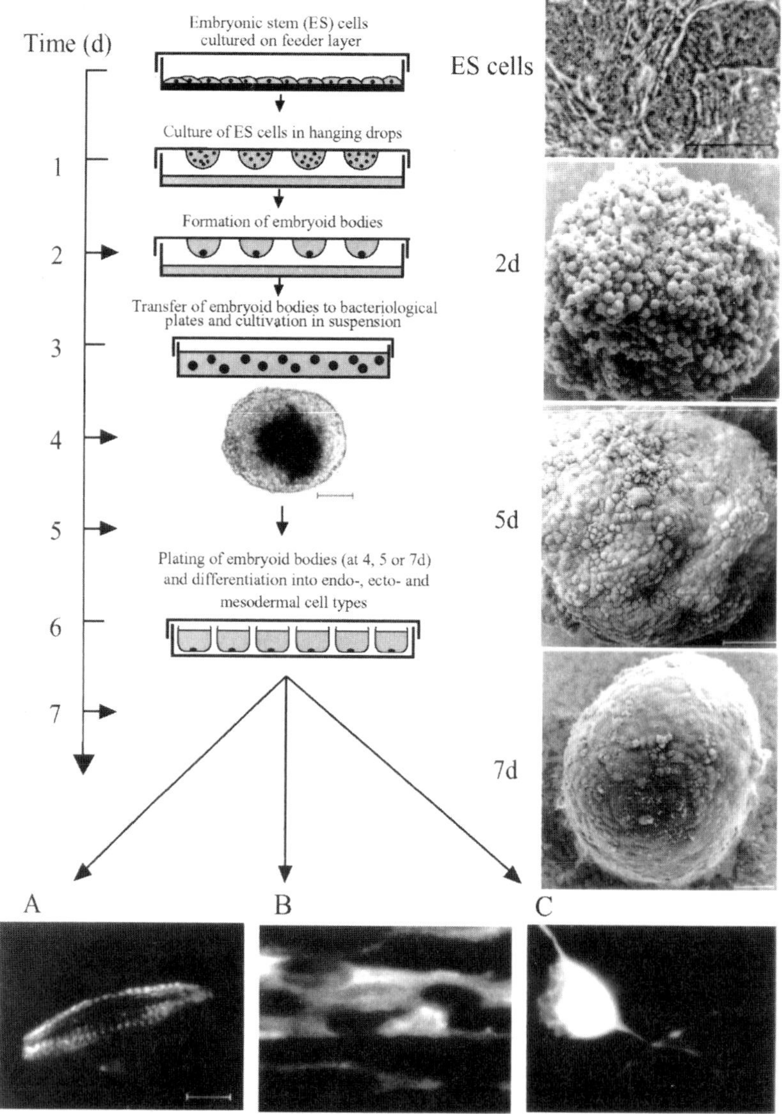

Figure 6. Differentiation protocol for ES cell-derived cardiogenic, myogenic, and neurogenic differentiation. The morphology of ES cells cultivated on a feeder layer (top right) or in suspension for 2d, 5d, and 7d as embryoid bodies (right) are shown visualized by light (ES cell) and scanning electron microscopy (SEM: field emission-SEM; embryoid body 2d, 5d, 7d), respectively. At the bottom are shown ES cells of lines D3 (cardiogenesis) and BLC6 (myogenesis and neurogenesis) that were differentiated in hanging-drops for 2d and cultivated in suspension for 4d (neuronal differentiation) (C), 5d (myogenic differentiation) (B), or 7d (cardiogenic differentiation) (A) before plating. Cardiomyocytes, skeletal muscle, and neuronal cells were characterized using monoclonal antibodies BA-G5, MF-20, and NR-4 respectively. Bars: light microscopy, 100 μm; SEM, 50 μm; IF, 10 μm (bottom). (Kindly provided by A. Wobus, ref. 68.)

expression of known markers could also be used to more accurately analyse the differentiated cell types (60–63).

Protocol 7. Pre-screening of large numbers of gene trap clones (Kindly provided by H. Stuhlman, ref. 34)

1. Grow individual ES cell clones to 30–50% confluency on 24-well plates.
2. Pool six to eight clones and plate on 60 mm gelatin-coated dishes in the absence of feeder cells, in ES cell medium without LIF, for two days.
3. Gently trypsinize and transfer small ES cell clumps into suspension culture (DMEM supplemented with 10% FCS, 20 mM Hepes pH 7.3) in 60 mm Petri dishes.
4. After four days 'simple embryoid bodies' can either be:
   (a) Left in suspension for a further six days.
   (b) Transferred to gelatinized tissue culture plates for a further six days.
5. Stain differentiated cell types for reporter gene expression using standard procedures.

Protocol 8. *In vitro* differentiation into cardiomyocyte (L. Forrester, Centre for Genome Research, University of Edinburgh)

1. Grow ES cells on gelatinized plates in the presence of LIF until confluent (see Chapter 3 or Protocol 7, 1–2).
2. Trypsinize confluent plates of ES cells into a single cell suspension.
3. Resuspend cells to 3×10^4 cells per ml of ES cell medium supplemented with LIF.
4. On a lid of a 100 mm bacterial dish place 10 μl drops of the cell suspension (i.e. 300 cells per drop).
5. Place the lid containing as many drops as possible (approx. 10×10) on the base of a dish containing 10 ml of sterile water, to humidify the plate, and place in incubator for two days. The cells aggregate by the force of gravity to form standard sized embryoid bodies.
6. Transfer the embryoid bodies from one to five lids to a 100 mm bacterial dish containing 15 ml of ES cell medium without LIF and

Protocol 8. *Continued*

leave them for five days. The medium can be changed after two to three days by allowing the embryoid bodies to settle by gravity in a 15 ml Falcon tube and then placing them in a fresh bacterial dish.

7. Plate embryoid bodies on gelatinized 24-well tissue culture dishes (three to four embryoid bodies per well) in ES cell medium without LIF.
8. Check embryoid bodies daily for the presence of contracting cardiomyocytes which should be apparent between two and ten days after plating.

For differentiation into other cell lineages, see *Table 2*, *Figure 6*, and reviews (64, 67).

4.3 Identification of GT integrations into genes active in specific cell lineages and/or downstream of specific factors by 'induction trapping'

Although the differentiation of ES cells into embryoid bodies provides a powerful means to mimic early differentiation events *in vitro*, a screen of large numbers of cell lines by embryoid body formation is laborious and thus limits the number of clones that can be readily analysed. An alternative strategy is 'induction trapping' which circumvents embryoid body formation by triggering differentiation with growth factors leading to alterations of gene expression before morphological changes of the cells are visible (8, 10, 70). With induction gene trap screens, alterations of reporter gene expression can be monitored simultaneously in up to 10000 cell lines with independent GT integrations. This allows for the identification of virtually all genes that are active in a given genetic pathway in ES cells within the time period of induction. The validity of this technique has been proven by identifying genes active in the RA-induced differentiation pathway (8, 10), as well as genes responding to the stimulation of ES cells by brain-derived neurotrophic factor (BDNF) (Karasawa, v. Holst and Wurst, unpublished data), nerve growth factor (NGF) (9), and follistatin (9). Furthermore a cocktail of factors such RA, NGF, and thyroid hormone (T_3) has been used to identify integrations showing inner ear-specific expression *in vivo* (71). In addition, genes responding to genotoxic stress have been identified (72). In principle, all external stimuli which alter gene expression in the cell type under investigation can be used. The obvious limitation with ES cells is that the signalling pathway downstream of a given factor must be intact in the cells. Before starting such a screen with growth factors, it should be determined that the receptors are expressed in ES cells and, if possible, that they can be activated.

Protocol 9. Induction GT screen using replica plating technique

Equipment and reagents

- Polyester fabric (see *Protocol 4*)
- Glass beads (about 3 mm in diameter) (see *Protocol 4*)
- Sterile forceps
- Concentrated HCl
- Absolute ethanol
- Inductive stimuli (e.g. retinoic acid, BDNF)
- ES cell media and plates (see Chapter 3)

Method

1. Process the filters as described (*Protocol 4*, up to step 12).
2. Place the filter with ES cell colonies on it cell side up in a 10 cm Petri dish containing DMEM, 5% FCS, 1000 U/ml LIF, and add the factor to be tested at a biologically active concentration. In cases when your factor is present in FCS, ideally use the artificial serum replacement (knock-out™) developed by Gibco BRL.
3. Incubate the filters at 37 °C, 5% CO_2 for an appropriate time (6 h to 2 days).
4. Assay for βGal activity on the polyester filter (see *Protocol 11*).
5. Pick and expand the βGal-expressing colonies from the original plate and expand them onto 96-well plates to freeze, to re-test, and to keep passaging for further tests (see Chapter 3). Weakly and strongly expressing colonies should be picked.
6. For re-testing, prepare 96-well plates containing βGal-expressing clones in duplicate.
7. Each colony should be cultured with control medium (DMEM, 5% FCS, 1000 U/ml LIF) and a duplicate with medium (DMEM, 5% FCS, 1000 U/ml LIF) containing the induction stimulus.
8. Culture the plates at 37 °C, 5% CO_2 for 6 h to 2 days.
9. Analyse the colonies for βGal activity according to *Protocol 11*.
10. Use a stereomicroscope to screen the stained colonies for modulation (increase or decrease) of βGal activity in the presence of inductive factor.
11. Verify the differential expression of the reporter gene using a quantitative βGal assay (see *Protocol 12*).

4.4 Identification of target genes of homeobox-containing transcription factors

For some homeodomain-containing transcription factors, it has been shown that if the homeodomain is placed outside a cell it is able to translocate through the cell and nuclear membrane and alter target gene expression in the nucleus (73, 74). By combining induction trap screens with the translocation property of homeodomain proteins, target genes can be identified by adding homeoprotein to the culture medium of gene trap vector containing ES cell clones and screening for modulation of reporter gene expression. This type of screen was successfully carried out with the Engrailed homeoprotein as an inducer (20). The amino acid sequence responsible for the translocation has been determined to be located in the third helix of the homeodomain (a 16 aa peptide called penetratin) (75). This oligopeptide when fused to other proteins is sufficient to translocate fusion proteins to the nucleus, and thus should allow for the identification of target genes of non-homeoprotein transcription factors. Various fusion proteins up to 100 aa have successfully been used in translocation experiments (Alain Prochiantz, personal communication).

Alternatively, target genes of transcription factors can be identified by controlling translocation of the transcription factor into the nucleus using an oestrogen receptor fusion protein or using inducible promoter systems such as the tetracycline-dependent induction system (76, 77). Gene trap vectors can be introduced into cell lines carrying such inducible constructs, and reporter gene expression can be studied before and after induction of the transcription factor.

Protocol 10. Identification of target genes of homeobox-containing transcription factors (developed by A. Prochiantz and W. Wurst)

Reagents

- Recombinant homeoprotein
- DNAse (Boehringer Mannheim)

Method

1. To identify target genes of homeoproteins, we basically use *Protocol 9* except that the induction medium contains DMEM, 2% FCS, 1000 U/ml LIF, 1 μg/ml recombinant protein, and 0.5 U/ml DNAse.

5. β-Galactosidase staining

LacZ gene expression is detected by the enzymatic activity of the gene product β-galactosidase. This enzyme is quite stable and remains active even after fixation in 4% paraformaldehyde or after freezing and thawing cells, which allows easy analysis of cells attached to a dish, whole embryos, and tissues, as well as analysis of frozen sections. When early embryos are to be analysed, then staining of whole embryos is the method of choice to analyse *lacZ* expression as it:

(a) Is fast and easy to perform.

(b) Allows a rapid assessment of most aspects of the expression patterns since early embryos are almost transparent.

(c) Permits subsequent processing by sectioning for histological examination to determine cell and tissue types expressing *lacZ*.

Staining whole embryos and tissues is limited by the size of the specimen. Embryos up to day 12 can be stained as whole-mounts. Older embryos tend to give staining problems either due to impaired penetration of the substrate or due to endogenous βGal activity. Therefore, for older embryos dissected tissues or frozen sections can be stained. The solutions and reagents for staining are identical in all cases.

Background staining (endogenous βGal) can be encountered in embryos from day 12.5 on. In most cases this is due either to:

(a) Insufficient fixation (too short, too little volume).

(b) A pH that is too acidic for the reaction (buffer capacity not sufficient, too little volume).

Fixation in 4% PFA is more efficient in reducing background than glutaraldehyde, but over-fixation can reduce the activity of the bacterial enzyme. Using Tris-buffered saline as a basis for the staining solution reduces background but in our hands also decreases the sensitivity of the staining reaction. In cases where endogenous background causes serious problems, antibodies against the bacterial βGal protein (5 Prime→3 Prime Inc., Boulder, CO, Cat. No. 7–063100) and an indirect immunohistochemical assay can be used on sections to detect and distinguish a 'signal' (the *E. coli* enzyme) from background (endogenous enzyme).

5.1 Staining of attached cells, embryos, and tissues

Cells attached to the dish, embryos, and tissue samples can be stained without any special treatment prior to the procedure given below. Depending on the level of expression of the *lacZ* gene, staining becomes visible by eye after incubation for as little as 15 min to as long as overnight. Staining times of longer than 24 h do not appear to increase the sensitivity.

Protocol 11. β-Galactosidase staining of attached cells, embryos, and tissues

Reagents

- PBS (Chapter 3)
- Solution A (KPP): 100 mM potassium phosphate buffer pH 7.4 (store at room temperature; rt)
- Solution B[a] ('fix'): 0.2% glutaraldehyde (GDA) in solution A, containing 5 mM EGTA and 2 mM $MgCl_2$ (store at –20 °C)
- Solution C ('wash'): 0.01% Na deoxycholate and 0.02% Nonidet P-40 in solution A, containing 5 mM EGTA and 2 mM $MgCl_2$ (store at rt)
- Solution D[b,c] ('stain'): 0.5 mg/ml X-Gal,[d] 10 mM $K_3[Fe(CN)_6]$,[e] and 10 mM $K_4[Fe(CN)_6]$[e] in solution C (store at –20 °C in dark)

Protocol 11. *Continued*

Method

1. Wash in PBS.
   (a) For cells: aspirate the medium and replace with same volume PBS. Repeat.
   (b) For embryos: transfer into large volume PBS, gently swirl around.
   (c) For tissues: transfer into large volume PBS, gently swirl around.
2. Fix in buffer B.
   (a) For cells: aspirate PBS, add sufficient buffer B to the plate such that the cells are well covered, and leave for 5 min at rt.
   (b) For embryos: up to day 9.5[f] put 10–20 embryos in 1 ml buffer B and leave for 5 min at rt. For day 10.5–12.5 embryos[g] use 5–10 ml buffer B for ten embryos and leave for 15 min at rt.
   (c) For tissues: use about ten times the volume of the tissue of buffer B, and leave 15–60 min (depending on size) at rt.
   (d) For all fixation steps: aspirate well wash buffer before adding buffer B to prevent dilution.
3. Wash three times with 10 ml buffer C at room temperature.
   (a) For cells: 5 min each.
   (b) For embryos: up to day 9.5, 5 min each.
   (c) For embryos: day 10.5–12.5, 15 min each.
   (d) For tissues: 15–60 min (depending on size).
4. Replace buffer C with buffer D and incubate at 37 °C.
   (a) Before adding buffer D, aspirate well buffer C.
   (b) For cells: add sufficient buffer to the plate such that the cells are well covered and that solution will not evaporate.
   (c) For embryos: up to day 9.5 add 1 ml buffer for 10–20 embryos. For day 10.5–12.5 embryos add 5–10 ml for ten embryos.
   (d) For tissues: add about ten times the volume of the tissue.
5. After staining wash samples three times in 10 ml buffer C.
6. Samples can be stored for short-term (a few days) in solution C at 4 °C. For prolonged storage the specimens should be fixed again in 4% paraformaldehyde for 2 h at rt and kept in 70% ethanol at 4 °C.

[a] Always prepare fresh prior to use or store at −20 °C. Other fixations can also be used, e.g. 2% glutaraldehyde or 4% paraformaldehyde (PFA) are possible.
[b] When prepared freshly, chill on ice for 10 min and spin down precipitate, aliquot supernatant.
[c] Solution D can be reused several times. Filter after each use and keep in dark at −20 °C in between.
[d] X-Gal stock solution can be made at 50 mg/ml X-Gal in dimethyl sulfoxide or dimethyl formamide (DMF). Keep at −20 °C. DMF stays liquid at −20 °C.
[e] $K_3[Fe(CN)_6]$ and $K_4[Fe(CN)_6]$ are kept as 0.5 M stock solutions in dark bottles (−20 °C).
[f] Convenient for staining are 30 mm (or smaller) Petri dishes.
[g] Convenient for staining are 60 mm Petri dishes.

For induction trap screens, the βGal response (induction or repression) of GT vector integrations can be determined using a quantitative βGal assay.

Protocol 12. Quantitative β-galactosidase assay

Reagents

- Buffer Z: 25 mM Tris–HCl pH 8.0, 5 mM DTT
- Buffer P: 60 mM Na_2HPO_4, 40 mM NaH_2PO_4
- *o*-nitrophenyl β-D-galactopyranoside (ONPG)

Method

1. Harvest cells from a confluent 6 cm dish by trypsinization (see Chapter 3).
2. Resuspend cell pellet (approx. 5×10^6 cells) in 170 μl buffer Z.
3. Freeze–thaw the sample three times and centrifuge at 13000 *g* for 5 min at 4°C.
4. Transfer the supernatant to a fresh assay tube and add 10 μl ONPG (10 mg/ml).
5. Add 90 μl of buffer P.
6. Incubate at 37°C overnight.
7. Add 500 μl of 2 M Na_2CO_3 and determine optical density at 420 nm (OD_{420}).

5.2 Staining of cryostat sections

The staining of frozen sections is required when older embryos or large tissues are to be analysed. In addition, frozen sections are desirable when tissues are to be analysed for β-galactosidase activity and for a second marker (e.g., with antibodies, to identify particular cell types in the sample). Cryostat sections can be made from freshly frozen material or from a tissue which was frozen and stored at –70°C. When material is stored at –70°C, care should be taken that samples do not dry out (e.g. store in screw cap tubes sealed with Parafilm). We have not observed a loss of β-galactosidase activity in samples which were stored for up to 6 months under these conditions. We routinely cut sections at 10 μm thickness. For most tissues, sectioning at temperatures between −15 to −20°C is fine. Slow freezing of embryos gives better tissue morphology compared to rapid freezing in liquid nitrogen or on dry ice. Stained sections can either be mounted directly or can be counter stained first.

5.3 Histological analysis of stained whole-mount embryos and tissues

Stained embryos and tissues can be processed further for histological analysis by sectioning after refixation and dehydration. Fixation and dehydration times are given in *Table 3*. Both methacrylate and paraffin embedding give good results. We provide a method (*Protocol 14*) for paraffin sectioning since it is used most widely.

Table 3. Fixation and dehydration times for embryos and tissues

Specimen	Fixation times in 4% PFA[a]	Dehydration times in EtOH[b] EtOH concentration 50%	70%	96%	100%
Embryos up to day 10.5	30 min	30 min	30 min	30 min	30 min
Embryos between day 11.5–12.5	1 h	1 h	2 h	2 h	1 h
Embryos between day 13.5–15.5	2 h	2 h	4 h	4 h	2 h to overnight
Embryos older than day 15.5 and whole adult tissues	Overnight	2 h	2 × 4 h	Overnight	Overnight

[a] 4% paraformaldehyde in phosphate-buffered saline.
[b] Wash specimen after fixation and prior to dehydration for 2 h with water.

Protocol 13. Production and staining of frozen sections

You will need the following:

- OTC embedding medium (e.g. Tissue Tek, Miles)
- scalpel blades, corse forceps, small spatula
- gelatinized slides
- cryostat
- histological staining trays
- embedding medium (e.g. Eukitt, Riedel-DeHaan, FRG)

Method

1. Place (about 0.5 ml) OTC embedding medium onto the pre-cooled (–20 °C) mounting block of the cryostat. When this layer is almost frozen, add more embedding medium and the specimen which should be free of excess liquid. Excess liquid can be removed from specimens by transferring them through several drops of embedding medium at room temperature and then to the block (early embryos can be frozen in the deciduum).
2. Orient the specimen with small spatula while the embedding medium is still liquid. Often it is easier to first freeze the specimen (without or in a small amount of embedding medium) and then mount the frozen sample onto the block as is done with pre-frozen and stored material.
3. When the specimen is properly oriented and completely frozen, add sufficient embedding medium to surround and slightly cover the sample. Avoid air bubbles.
4. Trim away excess parts of frozen embedding medium with scalpel blade leaving behind a few mm around the sample.

5. Fix freezing block onto the block holder of microtome (usually the block is put on the holder and fixed by a screw) and fine adjust the angle between the specimen and knife (this option depends on the make of the cryostate).
6. Move the knife to the specimen and trim away embedding medium covering the outer edge of the sample (this can be done by setting the section thickness to 50–100 μm). When you reach the sample, or desired area of the sample go down to 10 μm and start to collect sections.
7. Collect sections on gelatinised slides by approaching the section (which still lays on the surface of the knife) with the slide (slide at room temperature). When slide just touches the section the latter will melt onto the glass. Do not press the slide onto the section and knife! Several sections can be collected on one slide.
8. Place the slide at room temperature or on a heating plate at 37 °C and allow it to dry. This will only take a few seconds. Store dry sections at room temperature while the rest of sections are cut.
9. Fix slides in soln B (see *Protocol 11*) for 5 min at rt (convenient are histological staining trays).
10. Wash 3 times in soln C (see *Protocol 11*), 5 min each.
11. Stain in soln D (see *Protocol 11*) overnight[a].
12. Wash slides first 5 min in soln C (see *Protocol 11*), then 5 min in distilled water. When sections are to be counterstained proceed according to *Protocol 15*. Otherwise:
13. Dehydrate sections in ethanol (70%, 96%, 100% EtOH 5 min each).
14. Pass through xylene: EtOH 1:1 (1 min) and xylene (1 min). Do not dry, only allow excess xylene to rinse off the slide and wipe the back.
15. Place slide horizontally, pour embedding medium (for a cover slip of 24 × 50 mm about 4 drops) on sections and place the cover slip first with one end on the slide. Keep the other end up with a needle and slowly go down allowing air bubbles to escape.

[a]Histochemical staining trays are very convenient and allow easy handling of 20 slides at a time when two slides are put back to back.

5.3.1 Paraffin embedding and sectioning

Any paraffin embedding protocol that avoids solvents which affect the blue staining is fine. Short times of exposure to xylene are tolerable, prolonged incubation washes out the staining. We give a method (*Protocol 14*) that uses isopropanol for pre-infiltration, since even incubation for three to four days in isopropanol does not affect the blue staining.

Protocol 14. Paraffin embedding of β-galactosidase stained tissues

Equipment and reagents

- Screw-cap tubes (e.g. 50 ml Falcon tubes)
- Casting mould (e.g. Reichert-Jung, FRG; several sizes are available)
- Small spatula
- Isopropanol
- Microtome and holder for fixing the paraffin block to the microtome
- Paraffin for histology (e.g. Histowax, Reichert and Jung, FRG)
- Xylene

Method

1. Place the dehydrated sample (ten embryos) in 10 ml 100% isopropanol for 2 h with one change of the isopropanol at rt. Alternatively, samples can be dehydrated directly in isopropanol similar to procedure given in *Table 3* for ethanol. (Using isopropanol the steps are 50%, 75%, 90%, and two times 100% with incubation times as in *Table 3*.)
2. Pre-infiltrate with paraffin:isopropanol (1:1) at 60 °C. We use the following incubation times:
   (a) Embryos up to day 10.5: 2 h.
   (b) Embryos day 11.5–16.5: 12 h.
   (c) Embryos $>$ than day 16.5: 24 h.
   (d) Adult tissue, e.g. brain: 24 h.
3. Infiltrate with paraffin at 60 °C. The times given above apply.
4. Place into pre-warmed (60 °C) mould and orientate the specimen with a needle or spatula and fill mould with paraffin.
5. Depending on the mould used, directly cast the holder for fixing the paraffin block to the microtome. After hardening, remove the paraffin block from the mould and prepare 10 μm sections.
6. Dewax sections 1–2 min in xylene. Sections can now be embedded or processed for counter-staining.

5.3.2 Counter-stains

All staining procedures which do not remove or mask the X-Gal staining or affect the embedding, can be used. We routinely counter-stain with haematoxylin/eosin under conditions which give only a weak colour. When *lacZ* staining is in the nucleus eosin stain alone may be preferable. If the X-Gal staining is very faint, we counter-stain only every other slide. This allows for comparison of *lacZ* expression in the otherwise uncoloured sections with adjacent counter-stained sections and facilitates identification of *lacZ*-expressing cells. Since methacrylate sections are affected by ethanol, we give a procedure with ethanol-free reagents which we use for all sections including frozen sections. Paraffin sections have to be rehydrated after dewaxing in descending ethanol concentrations (100%, 96%, 80%, 60% ethanol, distilled water, 2 min each).

Protocol 15. Counter-staining sections of X-Gal stained material

Equipment and reagents

- Histological staining trays
- Haematoxylin and eosin solution (for recipes see *Table 4*)
- 60%, 80%, 96%, and 100% ethanol
- Xylene
- Embedding medium (e.g. Eukitt)

Method

1. Submerge sections for 1 min in haematoxylin solution in staining tray.
2. Rinse in staining tray for 5 min with running tap-water.
3. Submerge sections for 2 min in eosin solution.
4. Rinse 5 min with distilled water.
5. Dehydrate frozen sections and paraffin sections in ascending ethanol (60%, 80%, 96%, twice in 100%, 2 min each).
6. Remove the ethanol by two incubations in xylene (1 min each), add Eukitt, and put coverslip on. Any other commercially available embedding solution can be used. Methacrylate sections can be dried and embedded directly after step 4.

5.4 *In vivo* staining of cells to detect *lacZ* expression

Staining for βGal activity with X-Gal requires fixation of cells prior to the enzymatic assay. Living cells can be analysed for βGal activity (34) using fluorescein di-β-D-galactopyranoside (FDG) or 'Imagene', an FDG derivative (Molecular Probes). FDG is introduced into the cells by hypotonic loading while Imagene passes freely across cell membranes, but becomes highly hydrophobic after galactosidase cleavage and does not diffuse outside of the cell. Positive cells can be identified by the fluorescence of the metabolized substrate.

Table 4. Recipes for counter-stains

Haematoxylin solution		**Eosin solution**	
Haematoxylin	1.5 g	Eosin (water soluble)	0.5 g
Sodium iodate	0.15 g	Glacial acetic acid	2–3 drops
Aluminium sulfate	13.2 g	Distilled water	100 ml
Distilled water	172.5 ml		
Ethylene glycol	162.5 ml		
Glacial acetic acid	15.0 ml		
	350.0 ml		100 ml

Protocol 16. *In vivo* staining of *lacZ*-expressing ES cell colonies (provided by Greg Barsh, adapted from ref. 34)

Equipment and reagents

- Loading medium for FDG: dilute FDG stock solution (20 mM FDG in 10% DMSO) 1:10 with sterile water. A 1:1 mixture of tissue culture medium and the 1:10 diluted FDG solution is used as loading medium. FDG stocks may vary depending on batch and supplier.
- Fluorescence microscope
- FDG or Imagene
- ES cells grown on tissue culture dishes that fit between the stage and the objective of the microscope

Method

1. (a) FDG: aspirate tissue culture medium and add sufficient loading medium[a] to cover the cells (i.e. 1 ml/30 mm dish, 2 ml/60 mm dish, 3 ml/90 mm dish). Incubate for 1 min. Change media back to regular tissue culture medium.

   (b) Imagene: add dye directly to the medium at a final concentration of 33 μM.
2. Incubate at 37 °C for 1 h (FDG) or 2 h (Imagene). For Imagene, change medium to fresh medium after incubation.
3. Identify fluorescing colonies with a fluorescence microscope using × 10 or × 20 objectives and filters for fluorescein. Mark the positions of positive clones[a] with a dot on the bottom of the dish.

[a] Verify positive clones afterwards by X-Gal staining a group of cells. The fluorescence signal of a true positive clone can vary tremendously and most of the signal often localizes only to a portion of the colony. Use a cell line known to express *lacZ* as a positive control.

6. Identification of trapped genes

A primary aim of any GT strategy is to clone and sequence the gene that directs *lacZ* expression. With GT vectors, since a *lacZ*–fusion transcript is produced, most cloning strategies initially involve cloning the gene as a *lacZ*–fusion cDNA. However, in cases where the vector has inserted into 5′ untranslated sequences (promoter or enhancer trap vectors) it is desirable to clone the genomic sequences flanking the inserted vector.

6.1 Cloning genomic DNA flanking insertion sites by inverse PCR

A cloning strategy using inverse PCR is schematically outlined in *Figure 7*. Different batches of circularized DNA can result in different efficiencies

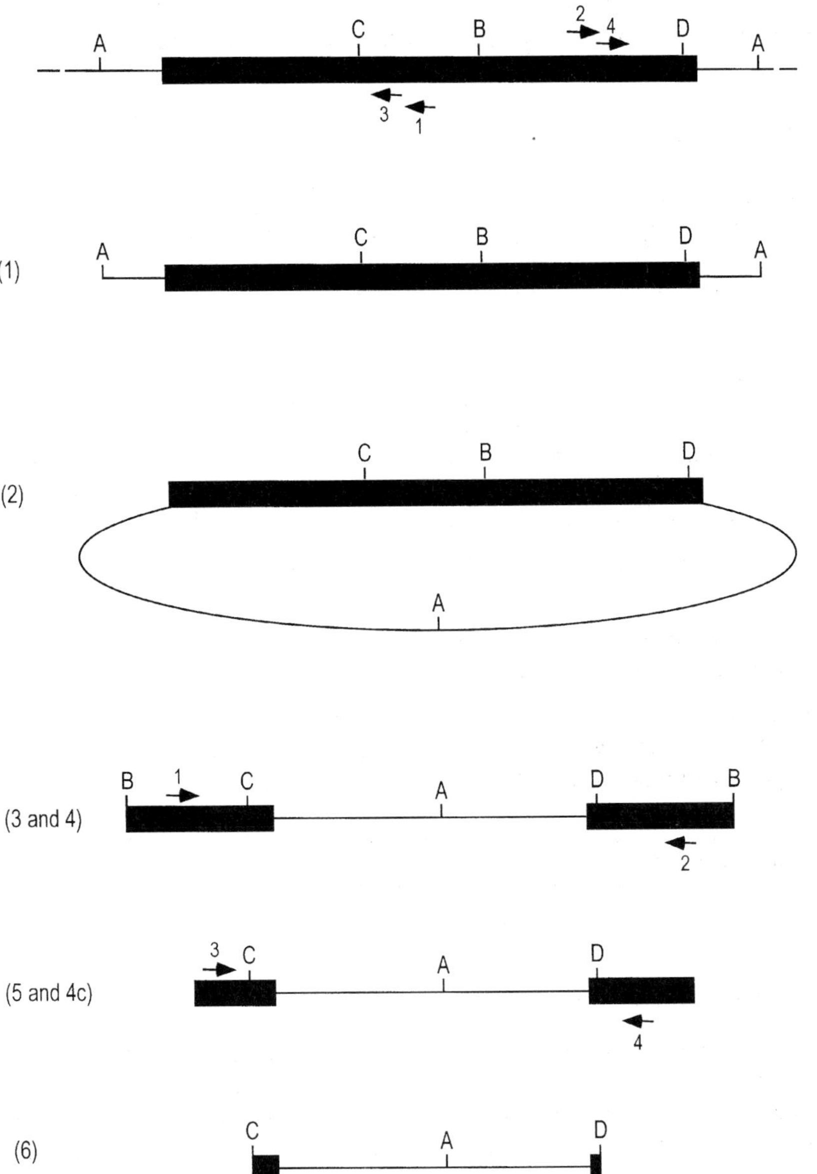

Figure 7. Principle of cloning flanking genomic sequences by inverse PCR. Numbers in brackets correspond to steps in *Protocol 17*. A–D: sites for restriction enzymes. 1–4: oligonucleotide primers. Bar: known GT vector sequences. Line: genomic sequences flanking the site of integration.

during the PCR amplification. If the DNA from one ligation works in a test PCR then there should be sufficient material to amplify enough of the desired fragment for cloning. In our hands fragments between 1.5–2 kb have been amplified under the conditions given below. These conditions may have to be altered if larger fragments are to be amplified.

The efficiency of the first steps of the protocol can be estimated on agarose gels. If multiple bands are visible after the PCR step, try less input DNA or raise the annealing temperature above T_m. Primer–dimer artefacts and other factors (buffer, dNTP concentration, etc.) influencing the specificity of amplification are discussed in detail in ref. 78.

Protocol 17. Cloning flanking sequences using inverse PCR

Equipment and reagents

- *Taq* polymerase, appropriate reagents, and equipment for PCR
- Genomic DNA carrying the GT vector insertion
- Oligonucleotide primers (1 and 2 in *Figure 7*)
- Restriction enzymes (A–D in *Figure 7*)
- T4 ligase (Boehringer Mannheim)

Method

1. Digest genomic DNA to completion with a restriction enzyme (A in *Figure 7*) that does not cut within vector DNA sequences between oligonucleotides 1 and 2, and that cuts at least 0.5–1 kb outside the vector in the genomic flanking DNA.[a] Heat inactivate the restriction enzyme (15 min, 65 °C) when digestion is complete.
2. Ligate the digested DNA with 1–2 Weiss units of T4 DNA ligase at 14 °C, for 16 h, at a concentration of 1 μg DNA/ml. This DNA concentration favours intramolecular as opposed to intermolecular ligation. Heat inactivate the ligation mix (15 min, 65 °C).
3. Linearize DNA with appropriate restriction enzyme[b] (B in *Figure 7*). Add the enzyme and its 10 × incubation buffer (see manufacturer's data sheet) directly to the ligation reaction. After appropriate incubation at 37 °C heat inactivate (15 min, 65 °C) the restriction enzyme.
4. Precipitate the DNA in high salt and isopropanol in the presence of 10 μg/ml tRNA.
5. Set up the PCR amplification reaction by combining:
   - 1–100 ng genomic DNA[c] (cut, ligated, and re-cut)
   - 5 mM each dNTP
   - 1–2 U *Taq* polymerase
   - 0.1–0.5 μM each primer (1 and 2 in *Figure 7*)

- 5 μl 10 × *Taq* incubation buffer (manufacturer's recommended buffer)
- H_2O to a total volume of 50 μl

The following conditions apply:

No. of cycles	94°C (denaturation)	T_m (annealing)[d]	72°C (extension)
1	10 min[e]	2 min	2 min
30	1.5 min	2 min	2 min
1	1.5 min	2 min	10 min

6. Analyse 20 μl of each test reaction on a 1% agarose gel. A band corresponding in size to the fragment deduced from primer placement and genomic Southern blot data should be visible on an ethidium bromide stained gel.[f] Confirm that the correct piece of DNA has been amplified by Southern blot analysis of the gel by hybridization with vector sequences contained in the amplified fragment.

7. Digest the rest of the test reaction, or additional reactions, with restriction enzymes which lie outside the first primer pair but within vector sequences (C and D in *Figure 7*). Alternatively PCR primers (step 5) that contain internal restriction sites can be designed. Analyse an aliquot of this digestion on an agarose gel. The amplified fragment should have shifted downward by the anticipated number of base pairs.

8. Purify the DNA fragment from an agarose gel (by electroelution, low melting point agarose, etc.).

9. Ligate the digested and purified PCR DNA fragment to a plasmid vector cut with the same enzymes. If no convenient restriction enzymes (C, D; which do not cut in the genomic flank) exist, clone the purified PCR product by blunt-end ligation into a plasmid vector[g] or use appropriate plasmids for the direct cloning of PCR products.[h]

10. Transform competent bacteria with an aliquot of the ligation reaction and screen transformants for the presence of the desired clone.

[a] Shorter flanking sequences may not contain enough unique sequences to be used as a probe.
[b] Linearization of the circular DNA in the vector sequences between the primers 1 and 2 may improve the following amplification step.
[c] 50–100 ng of genomic input DNA works best in our hands.
[d] As a rough estimate the primer melting temperature (T_m) = 2(X) + 4(Y); where X = • of dATPs + dTTPs, and Y = • of dGTPs + dCTPs; an applicable annealing temperature is 5°C below the true T_m.
[e] Optional without *Taq* polymerase; add enzyme before annealing step of first cycle.
[f] A second round of amplification with nested primers (3 and 4, *Figure 7*) is optional if the material obtained by the first round of PCR is not sufficient for cloning. Use 0.1–1 μl of the first PCR reaction and add new primer, dNTPs, 10 × reaction buffer, and *Taq* polymerase.
[g] Pre-treatment with T4 Pol (36) enhances the efficiency of this step.
[h] e.g. TA cloning® kit (Invitrogen), AdvanTAge™ PCR cloning kit (Clontech).

6.2 Cloning cDNAs containing endogenous sequences fused to *lacZ* or selector gene transcripts by RACE–PCR

For GT approaches in which there is a physical link between *lacZ* or *neo* RNA and endogenous gene transcript sequences the ES cells provide a convenient source of RNA for cloning the genes marked by *lacZ* expression. The endogenous portion of the fusion transcript can be cloned by means of the rapid amplification of cDNA ends (RACE) (see *Protocol 18*) (3, 79, 80). In cases where GT vectors are used that produce *lacZ*–fusions but the GT ES cell lines do not express *lacZ*, cloning by the same method can be attempted from RNA isolated from embryoid bodies, chimeric embryos or, after germline transmission, from transgenic embryos expressing *lacZ*.

There are a number of different PCR-based protocols available to clone cDNAs representing the trapped endogenous gene and to determine the partial nucleotide sequence of the trapped gene. All protocols take advantage of the capability of the GT vectors to capture endogenous exons by forming fusion transcripts between the endogenous gene and the reporter or selector gene. Depending on the vector type used, either the reporter gene can be used as a molecular tag using 5′ RACE (reporters containing splice acceptor sites), or the selector gene using 3′ RACE (selector genes containing splice donor sites) to clone parts of the tagged gene. We routinely use 5′ RACE based on the protocol described in ref. 3. In principle, the first strand cDNA synthesis is made using primers complementary to the 5′ end of *lacZ*. The second strand cDNA is then prepared with an anchor primer and subsequently amplified with a second nested primer (*Figure 8*). The use of nested primers for the PCR reaction is essential to obtain the desired cDNA products since the conditions used in synthesizing the first strand can be relatively non-specific. Routinely, we obtain 200 bp to 500 bp of endogenous gene sequence. Subsequently, these fragments are cloned in pAMP1 vectors (Gibco BRL) and the nucleotide sequence is determined. Alternatively, direct sequencing can be performed using a method for solid phase sequencing of 5′ RACE products (80).

Protocol 18. 5′ RACE to clone the endogenous sequences fused to *lacZ* transcripts[a]

Equipment and reagents

- Glass DNA Isolation Spin Cartridge System (Gibco, Cat. No. 15590–037)
- NACS column (Gibco BRL)
- Oligonucleotide primers (1–4 in *Figure 8*)
- Superscript II RT reverse transcriptase (Gibco)
- DEPC-treated deionized water (ddH_2O)
- RNASEmix (Gibco)
- Terminal transferase TdT (Gibco)
- Terminal transferase reaction buffer (Gibco)
- 2 mM dCTP (Gibco)
- *Taq* polymerase (Boehringer Mannheim), appropriate reagents, and equipment for PCR
- 25 mM $MgCl_2$ (Gibco)
- 10 mM dNTP mix
- 70% ethanol
- RNase inhibitor: 40 U/μl (Boehringer Mannheim, Cat. No. 799017)
- Amp PaMP1 System (Gibco BRL, 18381–012)

A. *First strand synthesis*

1. Purify total RNA from each ES cell line using guanidinium thiocyanate followed by centrifugation through a cesium chloride cushion as described (81).
2. Resuspend 0.5 μg total RNA and 5 ng of lacZRT in a total volume of 14.5 μl DEPC-ddH_2O. Incubate at 70 °C for 10 min.
3. Cool on ice and add to a total volume of 24 μl:
   - 2.5 μl 10 × reaction mix (Gibco BRL)
   - 2.5 μl of 0.1 M dithiothreitol (DTT)
   - 1 μl mixture of deoxynucleotides (each at 10 mM)
   - 0.5 μl RNasin (40 U/μl)
   - 3 μl of 25 mM $MgCl_2$
4. Incubate at 42 °C for 2 min.
5. Add 1 μl Superscript II RT and incubate at 42 °C for 30 min, and then 70 °C for 15 min.
6. Incubate reaction mix at 37 °C for 2 min, add 1 μl RNase mix, incubate at 37 °C for 30 min.

B. *Purification of the cDNA*

1. Using the reagents provided in the Glass DNA Isolation Spin Cartridge System, add 75 μl of ddH_2O and 120 μl NaI solution.
2. Transfer the cDNA/NaI solution to a glass max spin cartridge and centrifuge at 13000 *g* for 20 sec.
3. Wash three times with 400 μl 1 × wash buffer (4 °C) and centrifuge at 13000 *g* for 20 sec.
4. Wash two times with 400 μl 70% EtOH (4 °C) and centrifuge at 13000 *g* for 20 sec the first time, and 1 min the second time.
5. Transfer the spin cartridge into a fresh sample tube, add 50 μl of ddH_2O at 65 °C, and centrifuge 13000 *g* for 20 sec.

C. *TdC tailing of cDNA*

1. Mix together in an Eppendorf tube (total volume 24 μl):
   - 9.25 μl ddH_2O
   - 1.25 μl of 10 × reaction buffer
   - 1 μl of 25 mM $MgCl_2$
   - 2.5 μl of 2 mM dCTP
   - 10 μl cDNA sample

   Incubate reaction at 94 °C for 3 min and transfer it onto ice.

Protocol 18. *Continued*

2. Add 1 µl terminal transferase TdT (10 U) and incubate at 37 °C for 10 min. Then incubate it at 70 °C for 10 min and transfer it onto ice. At this step the material can be stored at –80 °C.

D. *First PCR amplification of cDNA*

1. Mix together in an Eppendorf tube (total volume 50 µl):
   - 35 µl ddH_2O
   - 4.5 µl of 10 × PCR buffer (1.5 mM $MgCl_2$) (Boehringer Mannheim)
   - 2 µl AAP primer (20 pmole)
   - 2 µl L232 primer (20 pmole)
   - 1 µl of 10 mM dNTP mix
   - 5 µl tailed cDNA
   - 0.5 µl *Taq* polymerase (5 U/µl) (Boehringer Mannheim)
2. Incubate at 94 °C for 4 min. Then perform 35 cycles of:
   (a) 94 °C (denaturation) for 1 min.
   (b) Td (annealing, 58 °C) for 30 sec.
   (c) 72 °C (extension) for 2 min.
3. The final fill-in reaction should be at 72 °C for 10 min.

E. *Second PCR amplification*

1. Mix together in an Eppendorf tube (total volume 50 µl):
   - 38.5 µl ddH_2O
   - 5 µl of 10 × PCR buffer (1.25 mM $MgCl_2$)
   - 2 µl UAP primer (20 pmole)
   - 2 µl Nested2AU primer (20 pmole)
   - 1 µl of 10 mM dNTP
   - 1 µl solution from the first amplification
   - 0.5 µl *Taq* polymerase (5 U/µl)
2. Incubate the reaction mix at 94 °C for 4 min. Then perform 35 cycles of:
   (a) 94 °C (denaturation) for 30 sec.
   (b) Td (annealing, 62 °C) for 30 sec.
   (c) 72 °C (extension) for 3 min.
3. The final fill-in reaction should be at 72 °C for 10 min.
4. To determine whether the desired cDNAs were amplified, analyse 5 µl of the PCR reaction by agarose gel electrophoresis followed by Southern blot analysis with a probe from the *lacZ* or splice-acceptor portion of the amplified fusion transcript. In general, a distinct band or a

smear can be seen on the blot. It is useful to know the size of the fusion transcript and therefore the size of the expected cDNA products.[b]

5. Gel purify the largest PCR products that hybridize with the probe following digestion with the appropriate restriction enzymes and clone the fragments into a suitable DNA plasmid vector.

F. *Cloning of the PCR products*

To clone the PCR products we take advantage of uracil DNA glycosylase (UDG) to facilitate directional cloning of the PCR products using the clone Amp PaMP1 System.

1. Mix together in an Eppendorf tube (total volume 20 μl):
   - 15 μl of 1 × annealing buffer (20 mM Tris–HCl pH 8.4, 50 mM KCl)
   - 2 μl of pAmp1 vector (Gibco BRL)
   - 2 μl of the PCR product
   - 1 μl UDG

   Incubate at 37 °C for 30 min.
2. Ligated plasmid can be transformed into bacteria using standard protocols (81). Cloning the PCR products has a great advantage because potential PCR artefacts can be easily eliminated by Southern blot hybridization before subjecting the samples to sequence analysis. We usually hybridize with a splice-acceptor specific probe as well as with intron probes located 5′ to the splice-acceptor to determine proper splicing occurred. In addition, we use probes for *neo* and *lacZ* to eliminate splice products which might occur after tandem integration of two or more GT vectors. For large scale screens, direct sequencing of the PCR amplified product is recommended (80).

[a] The protocol is based on a 5′ RACE manual supplied by Gibco BRL (Cat. No. 18374–041).

[b] For example, if the fusion transcript, as determined by Northern blot analysis with a *lacZ* probe, contains a large fragment from the endogenous gene (i.e. greater than 1–2 kb), then a heterogeneous smear of amplified products is likely to be seen on the gel.

In order to clone the 3′ end of selector gene fusion transcripts the 3′ RACE protocol supplied in the Clontech Marathon cDNA amplification kit (User Manual, Catalogue K1802–1) can be used. This kit provides for cloning 5′ or 3′ RACE products by simply adding adapters to the initial double-stranded cDNAs that are produced. The adapters are blocked such that the adapter primer can not prime from cDNAs. However, after the first strand of the PCR reaction is synthesized using a gene trap vector-specific primer, then the adapter primer can be used to further amplify the single PCR product. Thus by using gene-specific primers oriented in the appropriate direction, endogenous sequences can be obtained from gene trap fusion transcripts (see *Figure 9*). Furthermore, the adapters contain specific restriction sites and if

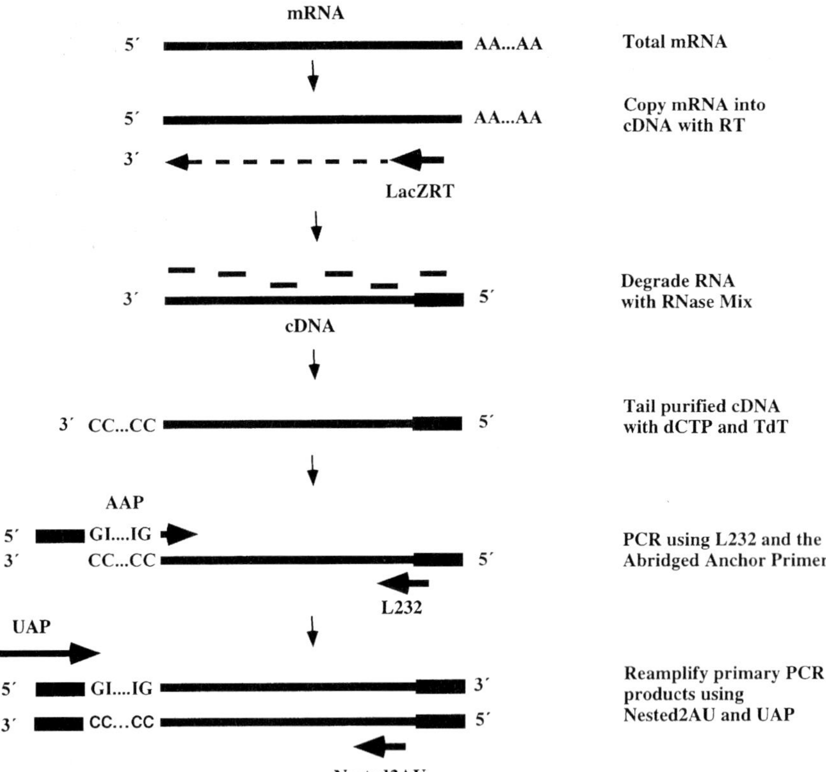

Figure 8. Principle of 5′ RACE–PCR. Total mRNA is transcribed into a first strand cDNA using a reporter gene-specific primer (LacZRT; 5′-TGGCGAAAGGGGGATGTG-3′) and, subsequently, a polyC tail is added using terminal transferase. The first PCR amplification is performed using a nested *lacZ* primer (L232; 5′-GATGTGCGTCAAGGCGATTA-3′) and the APP primer (5′-GGCCACGCGTCGACTAGTACGGGIIGGGIIGGGIIG-3′) (Gibco) complementary to the C tail. For the second PCR amplification, a third GT vector nested primer is used (nested2AU; 5′-CAUCAUCAUCAUTTGTCGACCTGTTGGTCTGAAACTC-AGC CT-3′) and a primer complementary to the minigene portion of the AAP primer (UAP; 5′-CUACUACUACUAAGGCCACGCGTCGACTAGTAC-3′) (Gibco). The PCR products can be directly ligated into pAMP1 (Gibco BRL) following RE digestion specific for sequences in primers ZAU and UAP.

the gene trap vector-specific primers also do, the RACE products can be easily cloned.

Acknowledgements

We wish to thank Drs Greg Barsh, Véronique Blanquet, Lesley Forrester, Mika Karasawa, Alain Prochiantz, Franz Vauti, and Anna Wobus for providing some of the protocols. We thank all colleagues who contributed unpublished data on vectors and methods and helped to improve this manuscript; thanks

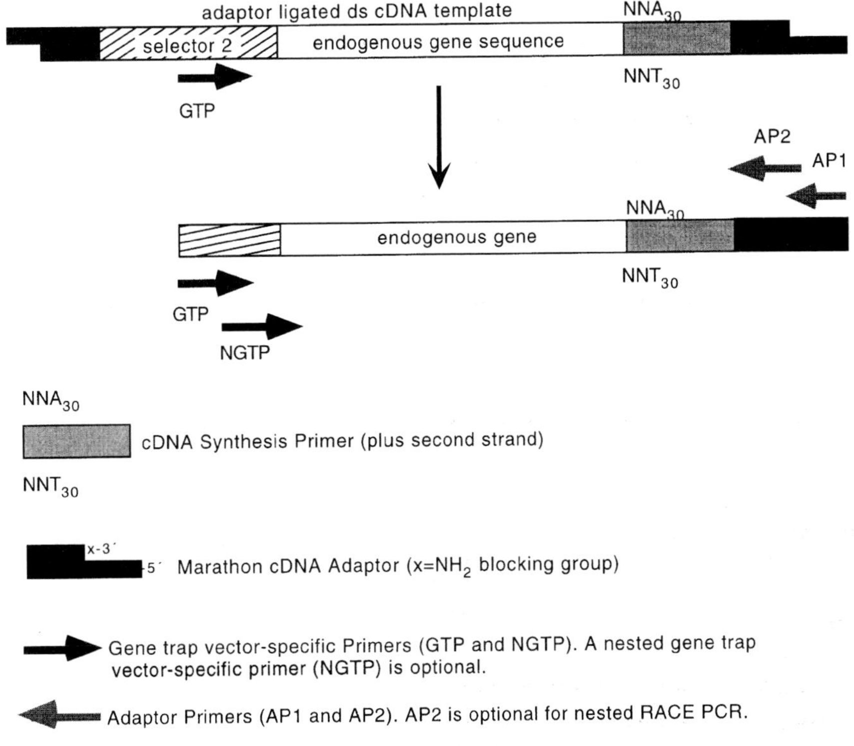

Figure 9. Principle of 3′ RACE–PCR. 3′ RACE and scheme shown is based on the Marathon cDNA amplification kit (Clontech, K1802–1). Adaptors are ligated to the ends of the cDNAs and contain primer binding sites AP2 and AP1. Specific PCR fragments are then produced using adaptor-specific primers AP1 or AP2 and gene-specific primers GTP or NGTP. To clone endogenous 3′ sequences in an experiment using a GT vector with a splice donor after the selector gene, the GSP2 primer and the nester NGTP primer would be specific to 3′ sequences in the selector gene.

also to Kostantinos Zarbalis for artwork, Sandra Rengsberger for secretarial assistance, and Jochen Zachgo for his contribution to the previous version of this chapter.

References

1. Gossler, A., Joyner, A. L., Rossant, J., and Skarnes, W. C. (1989). *Science*, **244**, 463.
2. Friedrich, G. and Soriano, P. (1991). *Genes Dev.*, **5**, 1513.
3. Skarnes, W. C., Auerbach, A., and Joyner, A. L. (1992). *Genes Dev.*, **6**, 903.
4. Wurst, W., Rossant, J., Prideaux, V., Kownacka, M., Joyner, A., Hill, D. P., *et al.* (1995). *Genetics*, **139**, 889.
5. Skarnes, W. C., Moss, J. E., Hurtley, S. M., and Beddington, R. S. (1995). *Proc. Natl. Acad. Sci. USA*, **92**, 6592.

6. Hicks, G. G., Shi, E.-G., Li, X.-M., Li, C.-H., Pawlak, M., and Ruley, H. E. (1997). *Nature Genet.*, **16**, 338.
7. Zambrowicz, B. P., Glenn, A. F., Buxton, E. C., Lilleberg, S. L., Person, C., and Sands, A. T. (1998). *Nature*, **392**, 608.
8. Hill, P. D. and Wurst, W. (1993). In *Methods in enzymology* (ed. P. M. Wassarman and M. L. De Pamphilis), Vol. 225, pp. 664–81. Academic Press Ltd., UK.
9. Stroykova, A., Chowdhury, K., Bonaldo, P., Torres, M., and Gruss, P. (1998). *Dev. Dyn.*, **212**, 198.
10. Forrester, L. M., Nagy, A., Sam, M., Watt, A., Stevenson, L., Bernstein, A., *et al.* (1996). *Proc. Natl. Acad. Sci. USA*, **93**, 1677.
11. Skarnes, W. C. (1990). *Biotechnology*, **8**, 827.
12. Gossler, A. and Zachgo, J. (1993). In *Gene targeting: a practical approach* (ed. A. L. Joyner), pp. 181–227. Oxford University Press.
13. Casadaban, M. J., Chou, J., and Cohen, S. N. (1980). *J. Bacteriol.*, **143**, 971.
14. O'Kane, C. J. and Gehring, W. J. (1986). *Proc. Natl. Acad. Sci. USA*, **84**, 9123.
15. Bier, E., Vaessin, H., Shepherd, S., Lee, K., McCall, K., Barbel, S., *et al.* (1989). *Genes Dev.*, **3**, 1273.
16. Bellen, H. J., O'Kane, C. J., Wilson, C., Grossniklaus, U., Kurth Pearson, R., and Gehring, W. J. (1989). *Genes Dev.*, **3**, 1288.
17. Wilson, C., Kurth Pearson, R., Bellen, H. J., O'Kane, C. J., Grossniklaus, U., and Gehring, W. J. (1989). *Genes Dev.*, **3**, 1301.
18. von Melchner, H., Reddy, S., and Ruley, H. E. (1990). *Proc. Natl. Acad. Sci. USA*, **87**, 3733.
19. Chowdhury, K., Bonaldo, P., Torres, M., Stoykova, A., and Gruss, P. (1997). *Nucleic Acids Res.*, **25**, 1531.
20. Mainguy, G., Erno, H., Montesinos, M. S., Lesaffre, B., Wurst, W., Volovitch, M., and Prochiantz, A. (1999). *J. Invest. Derm.*, in press.
21. Hill, D. H. P. and Wurst, W. (1993). *Curr. Top. Dev. Biol.*, **28**, 181.
22. Evans, M. J., Carlton, M. B. L., and Russ, A. P. (1997). *Trends Genet.*, **13**, 370.
23. Sharp, P. A. (1994). *Cell*, **77**, 805.
24. Hope, I. A. (1991). *Development*, **113**, 399.
25. von Melchner, H. and Ruley, H. E. (1989). *J. Virol.*, **63**, 3227.
26. von Melchner, H., DeGregori, J. V., Rayburn, H., Reddy, S., Friedel, C., and Ruley, H. E. (1992). *Genes Dev.*, **6**, 919.
27. Reddy, S., DeGregory, J. V., von Melchner, H., and Ruley, H. E. (1991). *J. Virol.*, **65**, 1507.
28. Rijkers, T., and Rüther, U. (1996). *Biochimica et Biophysica Acta*, 1307(3), 294.
29. Korn, R., Schoor, M., Neuhaus, H., Hännseling, U., Soininen, R., Zachgo J., *et al.* (1992). *Mechanisms Dev.*, **39**, 95.
30. Soininen, R., Schoor, M., Henseling, U., Tepe, C., Kisters-Woike, B., Rossant, J., *et al.* (1992). *Mech. Dev.*, **39**, 111.
31. Perry, M. D. and Moran, L. A. (1987). *Proc. Natl. Acad. Sci. USA*, **84**, 156.
32. Wagner, E. F., Stewart, T. A., and Mince, B. (1981). *Proc. Natl. Acad. Sci. USA*, **78**, 5016.
33. Bhat, K., McBurney, M. W., and Hamada, H. (1988). *Mol. Cell. Biol.*, **8**, 3251.
34. Xiong, J.-W., Battaglino, R., Leahy, A., and Stuhlmann, H. (1998). *Dev. Dyn.*, **212**, 181.
35. Zernicka-Goetz, M., Pines, J., McLean Hunter, S. M., Dixon, J. P. C., Siemering, K. R., Haseloff, J., *et al.* (1997). *Development*, **124**, 1133.

36. Hooper, M., Hardy, K., Handyside, A., Hunter, S., and Monk, M. (1987). *Nature*, **326**, 292.
37. Mansour, L. S., Thomas, K. R., and Capecchi, M. R. (1988). *Nature*, **336**, 348.
38. Wurst, W., Auerbach, A. B., and Joyner, A. L. (1994). *Development*, **120**, 2065.
39. McClive, P., Pall, G., Newton, K., Lee, M., Mullins, J., and Forrester, L. (1998). *Dev. Dyn.*, **212**, 267.
40. Brenner, D. G., Lin-Chao, S., and Cohen, S. N. (1989). *Proc. Natl. Acad. Sci. USA*, **86**, 5517.
41. Mountford, P. S. and Smith, A. G. (1995). *Trends Genet.*, **11**, 179.
42. Salminen, M., Meyer, B. I., and Gruss, P. (1998). *Dev. Dyn.*, **212**, 326.
43. Schuster-Gossler, K., Zachgo, J., Soininen, R., Schoor, M., Korn, R., and Gossler, A. (1994). *Transgene*, **1**, 281.
44. Kang, H. M., Kang, N.-G., Kim, D.-G., and Shin, H.-S. (1997). *Mol. Cells*, **7**, 502.
45. Reiss, B., Sprengel, R., and Schaller, H. (1984). *EMBO J.*, **3**, 3317.
46. Niwa, H., Araki, K., Kimura, S., Taniguchi, S., Wakasugi, S., and Yamamura, K. (1993). *J. Biochem.*, **113**, 343.
47. Yoshida, M., Yagi, T., Furuta, Y., Takayanagi, K., Kominami, R., Takeda, N., *et al.* (1995). *Transgenic Res.*, **4**, 277.
48. Takeuchi, T., Yamazaki, Y., Katoh-Fukui, Y., Tsuchiya, R., Kondo, S., Motoyama, J., *et al.* (1995). *Genes Dev.*, **9**, 1211.
49. Baker, R. K., Haendel, M. A., Swanson, B. J., Shambaugh, J. C., Micales, B. K., and Lyons, G. E. (1997). *Dev. Biol.*, **185**, 201.
50. Voss, A. K., Thomas, T., and Gruss, P. (1998). *Dev. Dyn.*, **212**, 171.
51. King, W., Patel, M. D., Lobel, L. I., Goff, S. P., and Nguyen Huu, M. C. (1985). *Science*, **288**, 554.
52. Hicks, G. G., Shi, E.-G., Chen, J., Roshon, M., Williamson, D., Scherer, C., *et al.* (1995). In *Methods in enzymology*, Vol. 254, p. 263.
53. Raetz, C. R. H., Wermuth, M. M., McIntyre, T. M., Esko, J. D., and Wing, D. C. (1982). *Proc. Natl. Acad. Sci. USA*, **79**, 3223.
54. Gollesmann, S. (1987). In *Methods in enzymology*, Vol. 151, p. 104.
55. Martin, G. (1981). *Proc. Natl. Acad. Sci. USA*, **78**, 7634.
56. Evans, M. J. and Kaufman, M. H. (1981). *Nature*, **292**, 154.
57. Smith, A. G. and Hooper, M. L. (1987). *Dev. Biol.*, **121**, 1.
58. Williams, R. L., Hilton, D. J., Pease, S., Wilson, T. A., Stewart, C. L., Gearing, D. P., *et al.* (1988). *Nature*, **336**, 684.
59. Doetschman, T. C., Eistetter, H., Kutz, M., Schmidt, W., and Kemler, R. (1985). *J. Embryol. Exp. Morphol.*, **87**, 27.
60. Risau, W., Sariola, H., Zerwes, H.-G., Sasse, J., Ekblom, P., Kemler, R., *et al.* (1988). *Development*, **102**, 471.
61. Wobus, A. M., Grosse, R., and Schöneich, J. (1988). *Biomed. Biochim. Acta*, **47**, 965.
62. Rohwedel, J., Maltsev, V., Bober, E., Arnold, H.-H., Hescheler, J., and Wobus, A. M. (1994). *Dev. Biol.*, **164**, 87.
63. Wiles, M. V. and Keller, G. (1991). *Development*, **111**, 259.
64. Pedersen, R. A. (1994). *Reprod. Fertil. Dev.*, **6**, 543.
65. Wiles, M. V. (1993). In *Methods in enzymology* (ed. P. M. Wassarman and M. L. De Pamphilis), Vol. 225, pp. 900–18. Academic Press Ltd., UK.
66. Baker, R. K. and Lyons, G. E. (1996). *Curr. Top. Dev. Biol.*, **33**, 263.
67. Wobus, A. M., Rohwedel, J., Strübing, C., Shan, J., Adler, K., Maltsev, V., *et al.*

(1997). In *Methods in developmental toxicology and biology* (ed. S. Klug and R. Thiel), pp. 1–17. Blackwell Science, Berlin, Vienna.

68. Wobus, A. M. and Kaomei, G. (1998). *Trends Cardiovasc. Med.*, **8**, 64.
69. Wobus, A. M., Rohwedel, J., Maltsev, V., and Hescheler, J. (1994). *Roux's Arch. Dev. Biol.*, **204**, 36.
70. Yamada, G., Kioussi, C., Schubert, F. R., Eto, Y., Chowdhury, K., Pituello, F., *et al.* (1994). *Biochem. Biophys. Res. Commun.*, **199**, 552.
71. Yang, W., Musci, T. S., and Mansour, S. L. (1997). *Hear. Res.*, **114**, 53.
72. Menichini, P., Viaggi, S., Gallerani, E., Fronza, G., Ottaggio, S., Comes, A., *et al.* (1997). *Nucleic Acids Res.*, **25**, 4803.
73. Joliot, A., Pernelle, C., Deagostini-Bazin, H., and Prochiantz, A. (1991). *Proc. Natl. Acad. Sci. USA*, **88**, 1864.
74. Joliot, A., Trombleau, A., Raposo, G., Calvet, S., Volovitch, M., and Prochiantz, A. (1997). *Development*, **124**, 1865.
75. Derossi, D., Joliot, A. H., Chassaing, G., and Prochiantz, A. (1994). *J. Biol. Chem.*, **269**, 10444.
76. Metzger, D., Clifford, J., Chiba, H., and Chambon, P. (1995). *Proc. Natl. Acad. Sci. USA*, **92**, 6991.
77. Gossen, M., Freundlieb, S., Bender, G., Müller, G., Hillen, W., and Bujard, H. (1995). *Science*, **268**, 1766.
78. Innis, M. A., Gelfand, D. H., Sninsky, J. J., and White, T. J. (1989). In *PCR protocols* (ed. Innis, Gelfand, and Sninsky), p. 3–12. Academic Press. (1990).
79. Frohman, M. A., Dush, M. K., and Martin, G. R. (1988). *Proc. Natl. Acad. Sci. USA*, **85**, 8998.
80. Townley, D. J., Avery, B. J., Rosen, B., and Skarnes, W. C. (1997). *Genome Res.*, **7**, 293.
81. Sambrook, J., Fritsch, E. F., and Maniatis, T. (ed.) (1989). *Molecular cloning: a laboratory manual*, 2nd edn. Cold Spring Harbor Laboratory Press, Cold Spring Harbor, NY.
82. Kerr, W. G., Nolan G. P., Serafini, A. T., and Herzenberg, L. A. (1989). *Cold Spring Harbor Symp. Quant. Biol.*, **LIV**, 767.
83. Muth, K., Bruyns, R., Thorey, I. S., and von Melchner, H. (1998). *Dev. Dyn.*, **212**, 277.
84. Wang, R., Clark, R., and Batch, V. L. (1992). *Development*, **114**, 303.
85. Schmitt, R. M., Bruyns, E., and Snodgrass, H. R. (1991). *Dev. Dyn.*, **5**, 728.
86. Keller, G., Kennedy, M., Papayannopoulou, T., and Wiles, M. V. (1993). *Mol. Cell. Biol.*, **13**, 473.
87. Miller-Hance, W. C., LaCorbiere, M., Fuller, S. J., Evans, S. M., Lyons, G., Schmidt, C., *et al.* (1993). *J. Biol. Chem.*, **268**, 25244.
88. Wobus, A. M., Wallukat, G., and Hescheler, J. (1991). *Differentiation*, **48**, 173.
89. Bain, G., Kitchens, D., Yao, M., Huettner, J. E., and Gottlieb, D. I. (1995). *Dev. Biol.*, **168**, 342.
90. Finley, M. F. A., Kulkarni, N., and Huettner, J. E. (1996). *J. Neurosci.*, **16**, 1056.
91. Strübing, C., Ahnert-Hilger, G., Shan, J., Wiedenmann, B., Hescheler, J., and Wobus, A. M. (1995). *Mech. Dev.*, **53**, 275.
92. Fraichard, A., Chassande, O., Bilbaut, G., Dehay, C., Savatier, P., and Samarut, J. (1995). *J. Cell Sci.*, **108**, 3181.

7

Classical genetics and gene targeting

SCOTT BULTMAN and TERRY MAGNUSON

1. Introduction

Gene targeting has provided considerable insight into the functions of numerous genes since it was developed a decade ago (1–4). A listing of the diversity of targeted genes and breadth of mutant phenotypes characterized to date can be obtained through mouse mutation databases (`http://www.bis.med.jhmi.edu/Dan/tbase.html` and `http://biomednet.com/cgi-bin/mko/mkohome.pl`) (5, 6). However, in order to take full advantage of this technology, classical genetic methods should be utilized to extend our knowledge of individual genes to genetic pathways. In this chapter, the significance of genetics in gene targeting and phenotype interpretation are discussed. We describe how Mendelian and quantitative genetics can be exploited to map modifier loci or generate animals carrying mutations in two or more genes. We also discuss the development and application of classical genetic approaches towards elucidating gene function such as generation of allelic series and creation of deletion complexes throughout the genome in ES cells and mice.

2. Genetic considerations in gene targeting

2.1 Implications of genetic heterogeneity among 129 mouse strains

Several genetic considerations should be taken into account during the initial stages of a gene targeting experiment. In order to maximize homologous recombination efficiency, both arms of a targeting vector should be isolated from the same strain of mice as that of the ES cells (see Chapter 1 for details). Although most ES cell lines have been isolated from 129 strains of mice, significant genetic variation exists among the different substrains, which is sometimes evident by pronounced differences in coat colour (see Chapter 4, *Table 1*) (7, 8). For instance, 129/Sv mice (A^w/A^w, $+^{c\text{-}Tyr}$ $+^p/+^{c\text{-}Tyr}$ $+^p$) have a white-bellied agouti (A^w) phenotype, whereas 129/SvJ mice (A^w/A^w, Tyr^c $p/Tyr^{c\text{-}ch}$ p) have a cream colour owing to the effect of mutant tyrosinase (*Tyr*) and pink-eyed dilution (*p*) alleles which are epistatic to A^w (9).

Molecular analysis of different 129 substrains using microsatellite markers has provided insight into their genetic differences and revealed that 129/SvJ is particularly divergent and actually contaminated with genomic regions of non-129 origin (*Figure 1*) (7, 8). Before the significance of this heterogeneity was appreciated, many targeting vectors were constructed from 129/SvJ DNA for use in 129/Sv ES cell lines. DNA sequence differences between 129/SvJ and 129/Sv, and of non-isogenic vectors in general, reduce targeting efficiencies and even make some loci appear 'untargetable'. For example, attempts to target a region tightly linked to *Tyr* on chromosome 7 were unsuccessful (0

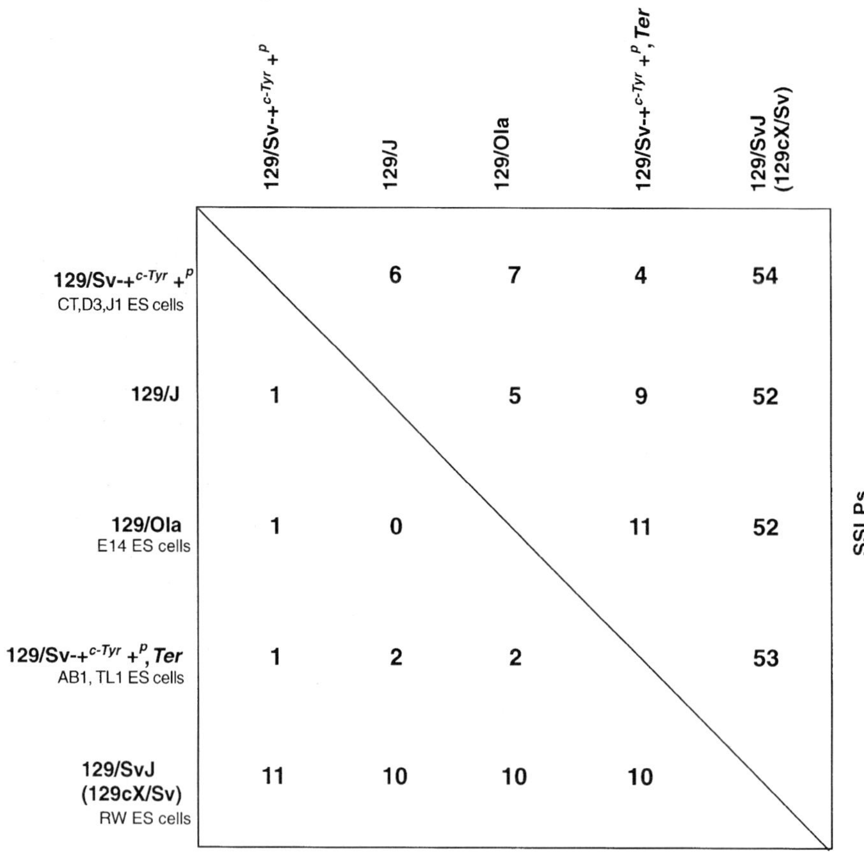

Figure 1. Simple sequence length polymorphisms (SSLPs) between 129 substrains. Number of SSLPs out of 212 above the diagonal and chromosome regions below the diagonal that differ between selected 129 substrains. ES cell lines derived from the various substrains are indicated on the left below the substrain designation. The R1 ES line (not included in the diagram) is derived from a (129/Sv- $+^{c\text{-}Tyr}$, $+^{p}$ $\times$ 129/SvJ)F_1 blastocyst. Figure is adapted with permission from ref. 8.

targeted colonies/418 colonies screened) when a 129/SvJ-derived construct was electroporated into a 129/Sv ES cell line (CT cells); however, the same construct underwent homologous recombination very efficiently (16 targeted colonies/179 colonies screened) in a (129/Sv- $+^{c\text{-}Tyr}$ $+^{p}$ × 129/SvJ)F_1 hybrid ES cell line (R1 cells) (S. Kendall and T. M., unpublished data). This discrepancy is not surprising considering that *Tyr* and sequences flanking it are highly divergent between 129/SvJ and 129/Sv (7, 8). These data suggest that it is best to prepare a genomic library using DNA isolated directly from ES cells used in the laboratory. Alternatively, given the fact that many ES cell lines are derived from the inner cell masses of 129/Sv blastocysts, it is worth noting that P1 artificial chromosome (PAC) and lambda phage libraries prepared from 129/Sv (129/SvEvTac) DNA are commercially available from Research Genetics and Stratagene, respectively.

2.2 Molecular and genetic nature of targeted mutations

2.2.1 Caveats regarding putative null mutations

The vast majority of targeted mutations generated to date have been null mutations consisting of small deletions or insertions which ablate part of the open reading frame (ORF) predicted to be crucial for function of the encoded protein. An often overlooked caveat to this approach, however, is that deletions can perturb neighbouring genes (10–12). The likelihood of unwittingly creating a polygenic mutation is relatively high if the targeted gene is embedded in the intron of another gene or lies in a multigene complex, like the homeo box (*Hox*) or globin genes.

Disruption of a closely linked gene is exemplified by a case in which three groups independently targeted myogenic factor 6 (*Myf6*) (also known as *MRF4* or herculin) (12). Each mutation deleted *Myf6* coding sequence and was produced on a similar genetic background, but the phenotypes were dramatically different due to varying degrees of down-regulation of myogenic factor 5 (*Myf5*), which lies only 8 kb downstream of *Myf6*. These data underscore the importance of knowing the number and position of transcription units and their cognate regulatory elements in a given region of the genome. Although few regions have been characterized this thoroughly to date, sequencing of the human and mouse genomes will likely increase the resolution of gene maps.

The *Myf6* mutations do not delete *Myf5* coding sequence but instead exert position effects (12). Position-effect mutations are known to affect regulatory elements or chromatin configuration in spontaneous or agent-induced mutants (13), and insertion of *neo* has been shown to contribute to, and even cause, phenotypes in targeted mutants (12, 14–18). The position effects associated with *Myf6* mutations act *in cis* (19), and there is evidence that a *Myf5* regulatory element(s) is located within an exon or intron of *Myf6* that was deleted by the two targeted mutations causing the most severe phenotypes (20, 21). To prevent the disruption of closely linked genes by targeted lesions, hit-and-run or Cre-*loxP* targeting strategies can be utilized to create point mutations in which *neo* is not retained by the mutant allele (see Chapters 1 and 2 for details).

A second caveat regarding the production of putative null mutations by gene targeting is that in some cases, particularly if the transcription unit undergoes alternative splicing, the transcriptional machinery can splice around the targeted lesion resulting in a dominant-negative (antimorph allele) or partial loss-of-function (hypomorph allele) mutation. Both normal and aberrant mRNAs have been observed in targeted mutants (22–26), and some of these transcripts apparently encode proteins with biological activity (23, 26). In order to reduce the likelihood of having the targeted mutation spliced out, targeting vectors should be constructed with the *neo* cassette inserted into an exon without perturbing the preceding splice acceptor site. Alternatively, introduction of a premature stop codon into the 5′ end of an ORF often will prevent mRNAs with coding potential from being transcribed. Verification of whether alternative transcripts are being produced can be accomplished using RT-PCR (22, 24–27).

2.2.2 Generating an allelic series via gene targeting

Although null alleles are important for elucidating gene function, much more can be gleaned by analysing additional alleles. A limitation of null alleles is that they sometimes confer embryonic-lethal phenotypes which preclude analysis of gene function at later stages of development or in adults. This problem can be circumvented by analysing hypomorphic or null alleles induced in a tissue- or stage-specific manner because they are generally less severe. These alleles can also reveal the significance of individual amino acid residues or specific protein motifs and provide insight into the regulation of gene expression. Other types of alleles, such as dominant-negative or gain-of-function mutations (dominant and neomorph alleles), can provide information about potential protein–protein interactions (28) and the importance of gene dosage or protein levels (29). Finally, another advantage of allelic series is that compound heterozygotes can be generated in which two distinct mutant alleles (e.g. null/hypomorph) produce novel, often intermediate, phenotypes, revealing even more about gene function (17, 18).

Allelic series in mice have been of great importance for a number of classical loci but have not been used extensively in gene targeting. Furthermore, even though few hypomorphic alleles have been generated by gene targeting, several have been serendipitous due to aberrant splicing and exon skipping as described above. Some 'leaky' mutations have resulted from duplications caused by introduction of targeting vectors into endogenous loci using insertion rather than replacement mechanisms (14, 23). In some of these cases as well as in certain other hypomorphs, cryptic splice sites within *neo* are thought to perturb normal splicing and reduce the quantity of full-length transcripts (17, 18, 25). Although conventional and hit-and-run targeting strategies have been used to create hypomorphic alleles, they have been underrepresented thus far (15, 30–35).

Allelic series are becoming more common in gene targeting. Tissue-specific mutations have been produced by deleting enhancer elements (36, 37) as well as conditionally by exploiting the Cre-*loxP* or FLP-*frt* site-specific recombinase systems (see Chapter 2 for details). A drawback to conditional targeting is that threshold levels of Cre or Flp recombinase are required to mediate recombination, and, in cases where insufficient levels are produced, only a subset of the expected cells will carry the targeted mutation. Furthermore, the lesion in each mutant animal is usually acquired *de novo* instead of being inherited. As a result, individuals can display phenotypic variability, particularly if the disrupted gene product functions in a cell-autonomous manner (38). This shortcoming notwithstanding, there are situations where cellular heterogeneity can be desirable since it resembles the use of genetic mosaics in *Drosophila* (39). By marking those cells which have undergone Cre- or Flp-mediated recombination with lacZ or green fluorescent protein (GFP) (40–43), it should be possible to examine cell fates during development or determine whether a gene product functions cell autonomously.

An efficient approach for generating allelic series by gene targeting has been demonstrated for fibroblast growth factor 8 (*Fgf8*) and neuroblastoma myc-related oncogene 1 (*Nmyc1*) (*Figure 2*) (17, 18). In each case, a single construct was targeted which utilized *neo* both as a mutagen and a positive-selectable marker and incorporated Cre and/or Flp recombination sites (*loxP* and *frt* sites, respectively). For *Fgf8*, a mouse line was produced in which *neo* was inserted into the first intron resulting in a hypomorphic allele (17). Moreover, *neo* was flanked by *frt* sites, and *loxP* sites were placed upstream and downstream of coding exons two and three, respectively. As a result, the hypomorphic allele could be either reverted to wild-type using Flp recombinase to excise *neo* or converted to a null allele using Cre recombinase to delete exons two and three. The *Nmyc1* targeting strategy was similar to *Fgf8* except *neo* was flanked by *loxP* sites to mediate reversion to wild-type with Cre recombinase (18). The other notable difference is that the null allele resulted from a point mutation introduced into an exon in the 3′ homology arm that was incorporated via homologous recombination in some ES cell clones but not others. Crossing the *Fgf8* or *Nmyc1* hypomorphic mice to transgenic lines expressing *flp* or *cre* respectively, in a tissue-specific manner could produce mosaic progeny exhibiting tissue-specific rescue. If it can be demonstrated that insertion of *neo* into introns is a generally applicable mutagenic strategy, these approaches will represent an important advance in targeting technology and bring gene targeting a step closer to classical mutagenesis.

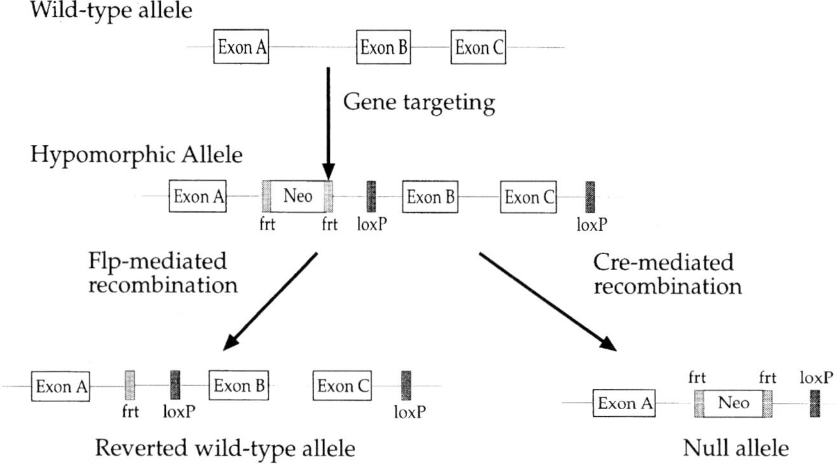

Figure 2. Production of an allelic series. A mouse line is produced by first generating a potential hypomorphic allele with one gene targeting event, whereby a *neo* gene flanked by *frt* sites is introduced into an intron, and exons essential for protein function (depicted here as exons B and C) are flanked by *loxP* sites. The presence of *neo* in an intron has the potential of affecting splicing, thereby down-regulating wild-type mRNA levels. Flp-mediated recombination can then convert the hypomorphic allele to a putative reverted wild-type allele by removing the *neo* gene and leaving behind one *frt* site. This reversion could also be generated in a tissue-specific manner if the Flp recombinase is expressed tissue specifically. Cre-mediated recombination could then convert the hypomorphic allele to a null allele by removing the exons critical for protein function. This conversion could also be done tissue specifically.

3. The significance of genetic background on phenotypes created by gene targeting

A targeted mutation can be established on an inbred genetic background by breeding chimeras which transmit the mutation at a relatively high frequency to the appropriate 129 substrain (see Chapter 4 for details). Mutant phenotypes are often more pronounced on inbred genetic backgrounds, and phenotypic analyses can be more straightforward due to less variability. Analysing the mutation on an inbred 129 background also allows the 129 parental strain to serve as a co-isogenic, wild-type control. However, it is also wise to introduce targeted mutations onto different genetic backgrounds.

The mutant phenotypes conferred by targeted mutations are often influenced by genetic background (*Table 1*). For example, a targeted mutation of the epidermal growth factor receptor (*Egfr*) results in three non-overlapping phenotypes depending on genetic background: perimplantation lethality on a 129/Sv-CF1 background, midgestation lethality on an inbred 129/Sv background, or perinatal lethality on a 129/Sv-CD1 or -C57BL/6J

Table 1. Effects of genetic background on the phenotypes of targeted mutations

Targeted gene[a]	**Genetic background**	**Phenotype**	**Reference**
Cftr	(129/Sv × DBA/2J)	Neonatal lethality	45–48
	129/Sv	Weaning-stage lethality	
	(129/Sv × CD1, C57BL/6J, or BALB/cJ)	Adult-stage lethality	
Egfr	(129/Sv × CF-1)	Perimplantation lethality	24, 44
	129/Sv	Midgestation lethality	
	(129/Sv × CD1 or C57BL/6J)	Neonatal lethality	
Igf1	129/Sv	10% viability	136
	(129/Sv × C57BL/6J)	16% viability	
	(129/Sv × MF1)	68% viability	
mK8	129/Sv and (129/Sv × C57BL/6J)	1.5% viability	137
	FVB/N	55% viability	
Tgfb1	C57BL/6J/Ola	Pre-organogenesis lethality	50
	(C57BL/6J/Ola × NIH/Ola)	Midgestation lethality	
	NIH/Ola	Postnatal lethality	
Tgfb3	129/Sv or (129/Ola × CF-1)	Mild cleft palate	138
	C57BL/6J	Severe cleft palate	
Trp53	129/Sv	65% lymphomas, 35% testicular tumours	139
	(129/Sv × C57BL/6J)	75% lymphomas, 9% testicular tumours	

[a] *Cftr*, cystic fibrosis transmembrane regulator homologue; *Egfr*, epidermal growth factor receptor; *Igf1*, insulin-like growth factor 1; *mK8*, keratin 8; *Tgfb1*, transforming growth factor β1; *Tgfb3*, transforming growth factor β3; *Trp53*, transformation related protein 53.

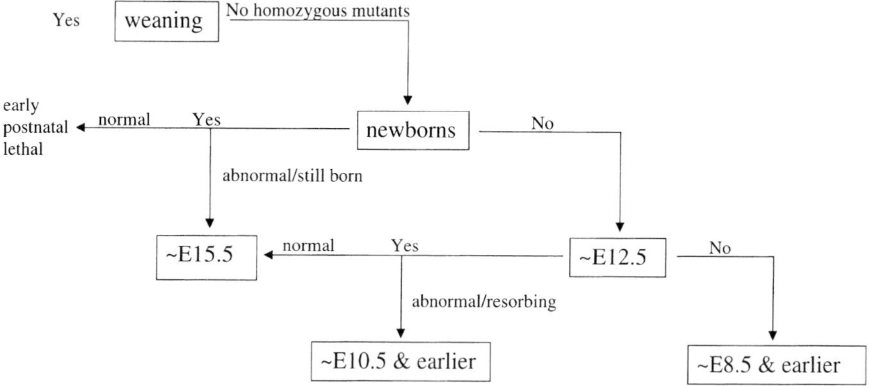

Figure 3. Flowchart for determining the onset of a mutant phenotype and the time of embryonic or early postnatal lethality. Initially, an intercross is performed and progeny are analysed and genotyped at weaning. If homozygotes are not recovered, then the procedure is repeated on progeny collected at the early postnatal stage (newborns) and, if necessary, progressively earlier embryonic (E) stages until homozygotes are recovered and represented at the expected frequency (25%).

background (24, 44). The time of onset of a mutant phenotype and embryonic lethality can be determined by simply analysing and genotyping embryos at various stages of development (*Figure 3*). Moreover, determining the stage of embryonic lethality often provides insight into the cause of death (*Table 2*).

Changing genetic backgrounds can also have a profound effect on the penetrance or expressivity of postnatal phenotypes. For instance, cystic fibrosis transmembrane conductance regulator homologue (*Cftr*)-deficient mice fall into three distinct phenotypic categories where the distribution is dependent on genetic background: lethality within ten days of birth (129/Sv-DBA/2J), lethality at around weaning (129/Sv), or prolonged survival beyond six weeks of age (129/Sv-CD1, -C57BL/6J, or -BALB/cJ) (*Table 1*) (45–49). Although such variability may confound initial phenotypic analyses, the significance of this can be outweighed by the potential benefits. First, in developing mouse models of human disease, a mutation can be crossed onto different genetic backgrounds in an attempt to 'fine tune' the phenotype such that the time of onset and pathology more closely resemble a given human disease or syndrome (48). Secondly, phenotypic variation provides an opportunity to map and clone modifier genes (see Section 3.2) which are likely to encode proteins that function in either the same or a parallel biochemical pathway as that of the disrupted gene product. In fact, modifier genes have already been mapped which exacerbate or ameliorate the severity of *Cftr* and transforming growth factor β1 (*Tgfb1*) mutant phenotypes (49, 50). This approach is conceptually analogous to enhancer and suppressor screens in genetically more tractable organisms such as *C. elegans* and *Drosophila*.

Table 2. Potential causes of embryonic lethality at various stages of development[a]

Stage[b]	Potential cause of death
E3.5	Formation of blastocyst (compaction, inner cell mass versus trophectoderm development)
E4.5–5.5	Implantation
E6.5–7.5	Growth and gastrulation
E8.5	Establishment of yolk sac vasculature
E9.0	Establishment of embryonic circulation
E9.5	Formation of chorio-allantoic placenta
E10.0 to term	Fetal circulation and placenta organogenesis

[a] Adapted from ref. 135.
[b] E, embryonic stage, days post-coitum

3.1 Developing mouse models of human diseases

3.1.1 Altering genetic background

In developing mouse models of human disease, genetic background can affect how closely a mutant phenotype recapitulates a human condition (48, 51–53). For example, the non-obese diabetic mouse strain (NOD), a model of insulin-dependent diabetes mellitus (IDDM) or type I diabetes in humans, is highly dependent on genetic background. Crosses between NOD and the C57BL/10SnJ diabetes-resistant strain indicated that the insulin-dependent diabetes 1 locus (*Idd1*) in the major histocompatibility complex (MHC) is necessary, but not sufficient, to confer type I diabetes (54, 55). Similar crosses identified 14 additional, unlinked loci (*Idd2–Idd15*) which contribute to susceptibility of NOD mice to type I diabetes, although apparently none are necessary on their own (54, 55). Furthermore, based on the many regions of the mammalian genome exhibiting conserved synteny (56), the human homologue of *Idd5* may map to chromosome 2q31-q33 which has been implicated in susceptibility to human IDDM (54, 55, 57).

Most mouse models of atherosclerosis using transgenic or gene targeting technology have been created on a pure or mixed C57BL/6J background. C57BL/6J is advantageous because it is prone to atherosclerotic plaques when placed on a high fat diet (58) and a number of susceptibility genes have been identified genetically (59).

In general, however, mutations generated by gene targeting have not exploited the genetic diversity of different mouse strains to create better models of disease states. Nevertheless, this approach is likely to become more common in the near future. It would be interesting to determine, for example, whether mice carrying a ΔF508 mutation in *Cftr* on a CD1, C57BL/6J, or BALB/cJ background develop lung pathology similar to that of human cystic fibrosis patients (31–33, 52). The ΔF508 mutation, which is currently on mixed genetic backgrounds, is a better model than previously generated null mutants because ΔF508 is responsible for about 70% of human cystic fibrosis cases. Moreover, both the mouse and human ΔF508 alleles produce a Cl^- channel protein defective in intracellular trafficking. Most importantly, gastrointestinal obstruction and early postnatal lethality is less prevalent in the mouse ΔF508 mutation than null mutants. This lethality may be decreased even further if the ΔF508 allele is crossed onto a C57BL/6J background because it produces relatively high levels of an alternative Cl^- channel which mitigates the gastrointestinal defect and reduces the early postnatal lethality in *Cftr* null mutants (49, 52, 60). CD1 and BALB/cJ also have modifiers that reduce the early postnatal lethality of *Cftr* null mutants which may include Cl^- channel genes.

3.1.2 Congenic strains

In order to obtain mice which consistently display an appropriate phenotype, a targeted mutation can be transferred, or introgressed, from a 129/Sv or mixed genetic background to a given inbred strain of mice by repeated backcrossing (61). In so doing, the genetic contribution of the donor strain is serially diluted such that the percentage remaining after n generations is $(1/2)^n$. After ten generations, the strain is 99.9% pure and considered congenic. At this point, it is nearly identical to the host strain, and most or all modifier genes will be of host strain origin. In fact, only the targeted mutation and approximately 20 cM of flanking host strain DNA should persist. In some instances, it might not be necessary to create a congenic strain. Because most of the donor strain DNA is lost in the first few backcross generations, a strain is 93% pure after four generations and this may suffice.

Although the creation of congenic strains has been an arduous, long-term procedure since being conceived by Snell in the 1940s, the process can be expedited by marker-assisted selection (62–64) or superovulation and embryo transfer (65). In the case of marker-assisted selection (i.e. genome scans, see Section 3.2.1), those mice with the least amount of donor sequence are identified and bred. This is a useful strategy because the required number of backcross generations can be reduced from ten to five, thereby saving nearly two years of time. In the case of superovulation and embryo transfer, congenic strains are produced without performing labour-intensive genome scans. Instead, the backcross process is sped up by superovulating and mating prepubescent females at three weeks of age and transferring the embryos to foster mothers. Because it takes three weeks to produce a pup from a fertilized egg and three additional weeks to produce a female that can be superovulated, each generation only takes six weeks. Thus, a congenic strain can be established in only 60 weeks, which is comparable to marker-assisted selection. Moreover, marker-assisted selection and superovulation/embryo transfer are not mutually exclusive and can be used in concert to create congenic strains in just over six months.

Additional rounds of backcrossing beyond ten generations can be useful for high-resolution genetic mapping, which has been a prerequisite for most successful positional cloning efforts. It is important to note that mapping and cloning genes which suppress or enhance disease phenotypes are of potential clinical significance because they may be more amenable to gene therapy or therapeutic intervention than the primary disease gene *per se*. Manipulation of modifier genes or the proteins they encode may therefore provide an effective means of diagnosing, treating, or even curing some diseases.

3.2 Genetic mapping of modifier loci

3.2.1 Genome scans

Over 60 years ago, Reed and Dunn described the effects of genetic background on the tail length and coat colour phenotypes of Fused ($Axin^{Fu}/+$) and piebald ($Endrb^{s}/Endrb^{s}$) mice, respectively (66–68). Grüneberg subsequently investigated modifier loci in great detail during the 1950s and 1960s (69), but it was not until the advent of molecular genotyping methods and a relatively dense genetic map that mapping modifier genes became feasible in the mouse (with a couple exceptions) (70, 71).

To detect the existence of modifier gene(s) which either enhance or suppress the severity of a homozygous mutant phenotype, crosses are performed to produce homozygotes of mixed genetic background (61). Heterozygous F_1 hybrids between 129 and another strain (Strain B) are either backcrossed to 129 heterozygotes (*Figure 4A*) or intercrossed (*Figure 4B*). One advantage of the intercross strategy is that both parental meioses are informative so fewer mutant animals need to be genotyped. Another advantage is that both dominant and recessive modifiers can be uncovered. The backcross scheme, in contrast, can only reveal Strain B alleles which are dominant or co-dominant to 129. Recessive Strain B alleles will not be detected since the genotype at any given locus will be either 129/129 or 129/Strain B.

On the other hand, the advantage of the backcross strategy is that genetic mapping and statistical analyses are more straightforward since only the F_1 meioses are informative. There are two possible genotypes at each locus (129/129 or 129/Strain B) instead of three for an intercross (129/129, 129/Strain B, or Strain B/Strain B), and this disparity is striking as the number of modifier genes increases. For three unlinked modifiers, there are 8 possible genotypes in a backcross (2^3) compared to 27 possible genotypes (3^3) for an intercross. The simplicity of the backcross strategy enables the number of segregating modifiers to be estimated (see below). In addition, a higher proportion of animals will display extreme phenotypes which are useful for mapping mulitiple loci using selective genotyping (see below).

To map segregating modifier genes, genome scans are employed such that homozygotes of a particular phenotypic subclass (e.g. mild phenotype for mapping suppressors or severe phenotype for mapping enhancers) are genotyped by PCR for 75–100 microsatellite markers distributed throughout

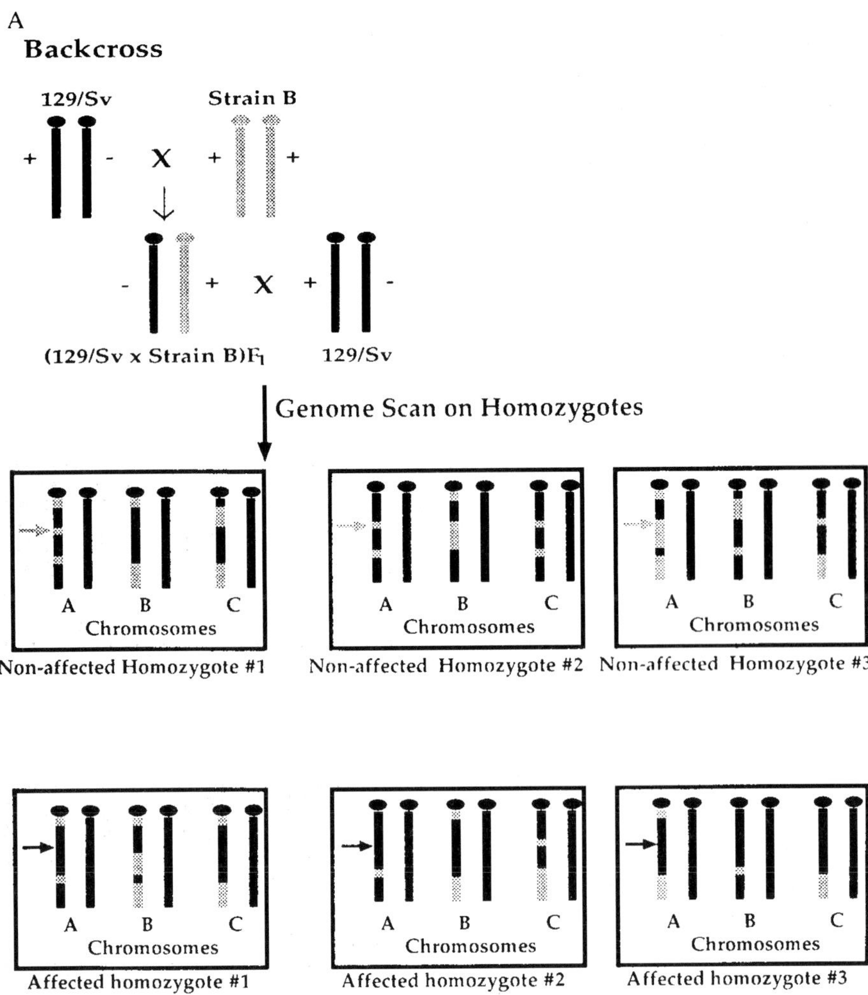

Figure 4. Mapping modifier loci. (A) Backcross: 129/Sv (black chromosomes) heterozygous mice are outcrossed to Strain B (light chromosomes) mice to produce (129/Sv × Strain B)F_1 mice which contain one set of 129/Sv chromosomes and one set of Strain B chromosomes. F_1 mice which are heterozygous for the targeted mutation are backcrossed to 129/Sv heterozygotes to produce wild-type (+/+), heterozygous (+/−), and homozygous (–/–) progeny. These animals will have one set of 129/Sv chromosomes (black) and one set of chromosomes that are a mix of 129/Sv (black) and Strain B (light) due to meiotic recombination in the F_1 germline. If homozygous animals showing a 'non-affected' versus 'affected' phenotype are detected in this generation, a Strain B allele at one or more modifier loci acting in a dominant manner is responsible for the 'non-affected' condition. A genome scan is performed on the 'non-affected' homozygotes using microsatellite markers polymorphic between 129/Sv and Strain B. The goal is to identify regions in common of Strain B origin in the 'non-affected' homozygotes (light arrows for chromosome A). These regions likely contain the Strain B allele necessary for rescue of the phenotype. To confirm these results, a genome scan on the 'affected' homozygotes is done if it is possible to retrieve tissue. At least a portion of the regions of Strain B required

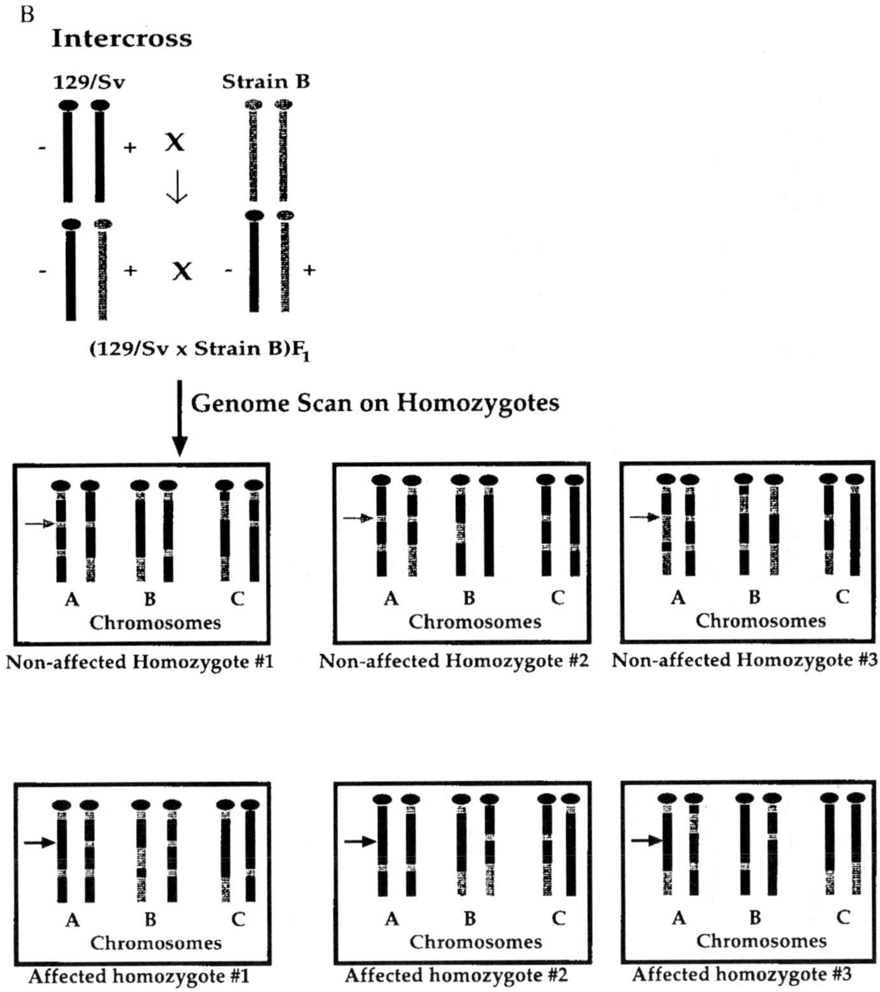

B

Intercross

129/Sv **Strain B**

\- + X

\- + X - +

(129/Sv x Strain B)F_1

Genome Scan on Homozygotes

A B C
Chromosomes
Non-affected Homozygote #1

A B C
Chromosomes
Non-affected Homozygote #2

A B C
Chromosomes
Non-affected Homozygote #3

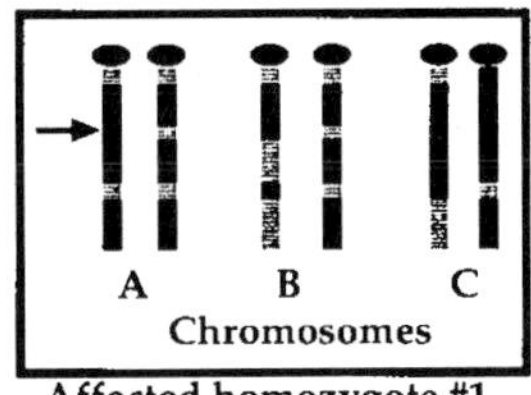

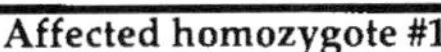

Affected homozygote #1

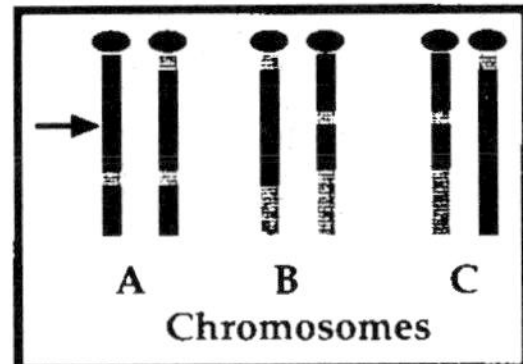

Affected homozygote #2

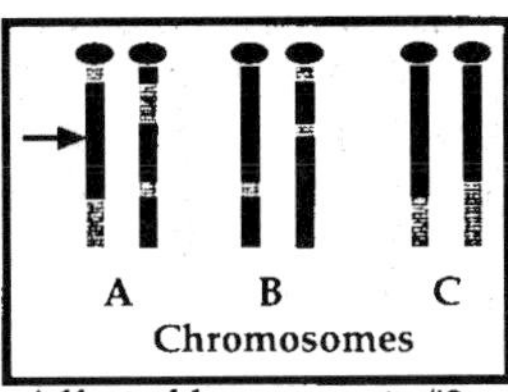

Affected homozygote #3

for rescue should be 129/Sv (dark arrows for chromosome A) in these animals. (B) Intercross: 129/Sv (black chromosomes) heterozygous mice are outcrossed to Strain B (light chromosomes) mice to produce (129/Sv × Strain B)F_1 mice which contain one set of 129/Sv chromosomes and one set of Strain B chromosomes. F_1 heterozygotes are crossed *inter se* to produce wild-type (+/+), heterozygous (+/−), and homozygous (–/–) animals in the next generation. These animals will have two sets of chromosomes that are a mix of 129/Sv (black) and Strain B (light) due to meiotic recombination in the F_1 germline. If homozygous animals showing a 'non-affected' versus 'affected' phenotype are detected in this generation, a Strain B allele at one or more modifier loci is acting either in a dominant or recessive manner. A genome scan is performed on 'non-affected' versus 'affected' homozygotes using microsatellite markers polymorphic between 129/Sv and Strain B. The goal is to identify regions in common of Strain B origin in 'non-affected' homozygotes (light arrows) that are not present in 'affected' homozygotes (dark arrows) as described in A. These regions likely contain Strain B alleles responsible for rescuing the phenotype. In this example, a recessive Strain B modifier is depicted and the region highlighted by the light arrow is of Strain B origin for both chromosomes.

the genome at approximately 15–20 cM intervals. Microsatellite markers are $(CA)_n$ and other simple sequence repeats of strain-dependent lengths which are polymorphic between the two parental strains (i.e. simple sequence length polymorphisms or SSLPs) (72–74). A website maintained by the Whitehead Institute/MIT Center for Genomic Research (`http://www.genome.wi.mit.edu`) is a useful resource in this regard because it lists the genetic and/or physical map positions for over 6000 markers plus the expected PCR product sizes for many of these markers in various strains of mice. PCR amplified markers are resolved on acrylamide gels, scored to compile a haplotype for each chromosome of every specimen, and subjected to computer analyses—such as *Mapmaker* and *Mapmaker/QTL* (75), *Gene-link* (76), and *Map Manager* (77)—to determine linkage relationships. The efficiency of performing genome scans can be improved by pooling DNA samples or analysing multiple microsatellite markers simultaneously using multiplex PCR. Furthermore, genome scans are amenable to automation using fluorescently labelled dNTPs or PCR primers and automated sequencing systems. Markers must be typed in every animal included in the dataset, however, to provide accurate mapping information.

Due to independent assortment, each marker should be 129/129 in 50% of the mice and 129/Strain B in the other 50% if a backcross is performed as described above. If an intercross is performed, then 25% of the mice should be 129/129, 50% should be 129/Strain B, and 25% should be Strain B/Strain B. Deviation from the expected 1:1 or 1:2:1 ratios suggests that a given marker is not inherited in a Mendelian fashion and implies it might be linked to a modifier gene. Every chromosomal region showing potential linkage is analysed in more detail by genotyping as many mutants as possible with all of the polymorphic markers in that region of the genome. By increasing the sample size and using markers more tightly linked to a putative modifier gene, statistically significant values are more likely to be obtained.

3.2.2 Statistical analysis

Microsatellite markers must deviate from the expected 1:1 (backcross) or 1:2:1 (intercross) ratios at a rigorous statistical level to avoid reports of spurious linkage and futile positional cloning efforts. Adopting a null hypothesis (H_0) that a sample comes from a population having a 1:1 or 1:2:1 ratio, a chi-squared goodness-of-fit value (χ^2) is calculated and the level of significance (P value) determined using computer software listed above. Although $P < 0.05$ is often sufficient to reject a H_0 and infer linkage, it is not adequate under these conditions because the likelihood of encountering a relatively large degree of deviation by chance is much greater on a genome-wide scale than at a specific locus.

Whereas $P = 0.05$ means that, assuming the H_0 is true, there is only a 5% chance of encountering the observed deviation at a specific locus, deviation of this magnitude is expected to occur by chance alone at up to two dozen sites throughout the genome (78). As a result, it has been proposed that for a backcross, $P < 3.4 \times 10^{-3}$ is necessary for suggestive linkage and $< 1.0 \times 10^{-4}$ for significant linkage (78). Similarly, $P < 1.6 \times 10^{-3}$ and $< 5.2 \times 10^{-5}$ have been adopted for intercrosses. Although these values may seem too stringent, they are equivalent to generally accepted log likelihood of the odds (LOD) score values of 2.8 and 4.3 for suggestive and significant linkage, respectively, used for many human mapping studies (78). Moreover, adopting these P values normalizes significance to that of 0.05 at the single locus level. This level of significance also means there is only a 5% chance of committing a type I error by erroneously rejecting the H_0.

3.2.3 Complexities associated with multiple modifiers

The number of mice required to map modifer loci is determined by several factors including the degree of phenotypic variability, the density and distribution of microsatellite markers, and the desired level of statistical significance. Perhaps the most critical factors are the number and relative contribution of segregating modifier genes, but these are difficult or impossible to determine a priori. In the case of a single modifier gene inherited in Mendelian fashion, 20 progeny can be enough to perform an initial genome scan in progeny from an F_1 intercross since both meioses are informative (79). If the alleles exhibit a simple dominant-recessive relationship and are associated with a discrete trait, then the phenotype of F_2 progeny should conform to a 3:1 ratio. If the alleles are co-dominant, then there should be severe, intermediate, and mild phenotypic classes occurring at approximately a 1:2:1 ratio in the F_2 generation. X-linked or imprinted genes can be detected if the phenotype is skewed by sex of mutant progeny or the orientation of the crosses, respectively.

More often than not, however, matters are complicated by multiple modifier loci. Consider a null mutation of *Egfr* which confers a recessive, midgestation lethal phenotype that is completely penetrant on a 129/Sv background but incompletely penetrant on 129/Sv-CD1 or 129/Sv-C57BL/6J mixed genetic backgrounds (24, 44). The number of segregating modifier genes was estimated by backcrossing (129/Sv × CD1)F_1 heterozygotes to 129/Sv heterozygotes. If 12.5% of the live-born progeny had been –/–, then the simplest interpretation would have been that a single modifier gene (*Mdfr*) was 'rescuing' the embryonic lethality. Half of the –/– progeny would have been $Mdfr^{129}/Mdfr^{129}$ and presumably died at midgestation, while the other half would have been $Mdfr^{129}/Mdfr^{CD1}$ and presumably lived. Similarly, identifying 6.25% of the N_2 progeny as –/– would suggest the presence of two modifiers, 3.13% would suggest three modifiers, and so on. In the *Egfr* backcross, 5.5% of the N_2 progeny were rescued homozygotes, suggesting that two to three modifier genes influence the time of death (P. Green, J. Om, D. Yee, D. Threadgill, and T. M., unpublished data).

Estimating the number of modifier loci as described above is not always accurate for several reasons: only dominant and co-dominant Strain B (CD1 in this case) alleles can be detected (in a backcross), surviving backcross progeny are assumed to be heterozygous for every modifier (e.g. *Mdfr1*^{129}/*Mdfr1*^{CD1}; *Mdfr2*^{129}/*Mdfr2*^{CD1}; etc.), and intergenic interactions such as epistasis are not taken into account. In the case of *Egfr*, two to three genes were estimated but at least six putative modifiers have been mapped. This discrepancy is probably due to the fact that none of the six modifier genes are absolutely required to rescue the embryonic lethal phenotype of –/– embryos. Instead of acting insularly, the modifiers act collectively with different combinations being sufficient on an animal-by-animal basis. Many scenarios could be postulated, but one possibility is that any three modifier genes are dispensable as long as the other three are represented by genotypes which confer rescue.

As described above for *Egfr*, multiple modifier loci can be difficult to map because they do not necessarily co-segregate with a particular trait and, therefore, might not deviate much from the expected 1:1 or 1:2:1 ratios. As a result, genome scans must be performed on relatively large numbers of mice and more sophisticated statistical approaches, such as calculation of LOD scores, need to be used (80). Although more than 20 animals are required, the number can be kept to a minimum by selective genotyping. Many F_2 intercross or N_2 backcross mutant progeny are generated and examined phenotypically but only those with the most extreme phenotypes undergo genome scans. Of course, it will always be more difficult and require more animals to map minor modifiers than major modifiers.

In some instances, exclusion mapping can be useful for confirming the position of modifier genes. This approach simply consists of performing genome scans on F_2 or N_2 animals that were not rescued or which exhibit a phenotype opposite to those individuals chosen for selective genotyping. In a backcross, for example, rescued or non-affected homozygotes might appear to have a 129/Strain B genotype for a particular microsatellite marker more often than would be expected by chance alone. However, the sample size might be too small for the data to be statistically significant. If the marker is, in fact, linked to a modifier locus, then the genotype of non-rescued or affected homozygotes should not be biased towards 129/Strain B. Instead, they should be 129/129 and 129/Strain B with equal prevalence (*Figure 4*). In some cases, however, exclusion mapping might identify microsatellite markers which are not randomly inherited. In these cases, the predominant genotype likely will be different than that of the rescued or non-affected individuals. One explanation for such a situation is that one allele (e.g. Strain B) suppresses the phenotype of the targeted mutation, while the other allele (e.g. 129) enhances it.

3.2.4 Cloning modifier genes

Despite the intrinsic value of mapping modifiers, genes must be cloned and characterized at the molecular level for most experiments to reach fruition. A number of approaches, including direct selection and exon trapping, can be implemented to clone modifier genes once the region of the gene has been isolated in bacterial or yeast vectors. However, high-throughput sequencing coupled with gene identification algorithms (81) have made the candidate gene approach particularly attractive. This approach also has been made possible by expressed sequence tag (EST) mapping efforts in both mouse and human as well as the Jackson Laboratory's Mouse Genome Database (`http://www.informatics.jax.org`) which posts updated genetic maps. In fact, the first mammalian modifier gene to be characterized at the molecular level, *Mom1* (modifier of *Min-1*), was revealed by the candidate gene approach. The *Pla2g2a* secretory phospholipase gene was identified as *Mom1* (82), which influences the number of intestinal tumours conferred by Apc^{Min1} (83). Apc^{Min} is an *N*-ethyl-*N*-nitrosourea (ENU)-induced mutation, but the theoretical and practical aspects of cloning or identifying genes which modify the phenotype of targeted mutations are essentially the same.

Although intriguing candidate genes can be readily selected, it is difficult to demonstrate unequivocally which, if any, modify a mutant phenotype. The success of *Pla2g2a* and *Mom1* stems from the fact that it was a straightforward case. First, genetic factors are responsible for more of the phenotypic variation in $Apc^{Min1}/+$ mice than environmental or stochastic factors (83). Secondly, *Mom1* is a major modifier of Apc^{Min1}, accounting for 50% of the genetic variation (83). Thirdly, the tumour-sensitive C57BL/6J strain carries a frameshift mutation that abolishes expression of *Pla2g2a* (82). However, even though a *Pla2g2a* transgene derived from the tumour-resistant AKR strain decreased tumour number in $Apc^{Min1}/+$ mice on a C57BL/6J background, there is evidence that *Mom1* consists of *Pla2g2a* plus at least one other closely linked gene (84). Interestingly, several other phospholipase genes map in the vicinity of *Pla2g2a* and may be involved (84, 85).

As opposed to *Pla2g2a* and *Mom1*, many modifier genes might not have significant differences in coding sequence or expression levels in sensitive versus resistant strains. Moreover, identification of minor modifier genes will be much more difficult to map than major modifiers like *Mom1* and may not be amenable to functional analyses. Consider a mutant phenotype modified by five to ten genes where any one, two, or three are dispensable if the other genes are represented by the appropriate alleles. Performing a knock-out or knock-in experiment on any given gene will be obfuscated by the presence of the other modifier genes. Validation of modifier genes also will be complicated by the fact that some may have other functions and might cause lethal phenotypes if perturbed. For the reasons listed above, it seems likely that validation of candidate genes will be the crux of mapping and cloning modifier genes.

4. Targeting multiple genes

4.1 Redundancy and analysis of multiple mutations

Since the advent of gene targeting, a recurring theme has been that targeted mutations often result in surprisingly mild phenotypes or no phenotype at all (see for example refs 86 and 87). In many cases, the targeted gene arose through duplication and subsequent divergence of an ancestral gene, suggesting that other, presumably related, gene(s) can compensate for the mutated gene. Gene targeting experiments have provided much evidence in support of this hypothesis, which was proposed by Muller in 1935 to explain a viable two-gene deficiency in *Drosophila* (88). Expression of related genes is increased in some targeted mutants (see for example refs 89 and 90), and inactivating two members of the same gene family often confers a synergistic phenotype. Knock-in experiments have also been informative (91, 92). For example, introducing an additional copy of the myogenin (*Myog*) coding sequence into the *Myf5* locus, while concomitantly disrupting *Myf5*, completely rescues the truncated ribs and perinatal lethality caused by *Myf5* deficiency (92).

Although some models predict that true genetic redundancy can persist for considerable periods of time during evolution, it seems more likely, given the role of natural selection, that related genes share partially overlapping functions (88, 93). This thought has also recently been borne out in gene targeting experiments and is exemplified by the aforementioned *Myog* knock-in allele (*Myf5-Myog-Jae2* or $Myf5^{Myg\text{-}ki}$). When another myogenic regulatory gene, myogenic differentiation 1 (*Myod1*), is also disrupted, the double mutant mice that have a null *Myod1* mutation and *Myf5-Myog-Jae2* knock-in allele die at birth with underdeveloped skeletal muscle (94). In contrast, *Myod1* null mutant mice on a wild type background are viable with fully developed muscle (89). However, the knocked-in *Myog* gene clearly does contribute to myogenesis since *Myf5*;*Myod1* double null mutants, without the knock-in copy of *Myog*, lack skeletal muscle altogether (95). These data indicate that although *Myog* can fully compensate for *Myf5* in rib development, it is only partially able to do so in myogenesis.

It may be necessary to examine mice with no overt phenotype in a more sophisticated manner using molecular markers, *in vitro* cellular assays, physiological tests, or behavioural evaluations. Alternatively, it might be necessary to prime the conditions by altering the genetic background or targeting additional genes (96, 97). For instance, administration of neuropeptide Y (NPY) results in increased food intake and weight gain in rodents, yet no change in feeding behaviour or body weight was observed in *Npy* knock-out mice (98). However, NPY levels are up-regulated in leptin-deficient, classical obese (Lep^{ob}) mutants, and loss of *Npy* subsequently was shown to attenuate the increased food intake and body weight of Lep^{ob}/Lep^{ob} mice (96). Therefore, *Npy* does play a role in feeding behaviour and energy homeostasis and apparently lies downstream of, and is inhibited by, leptin and its receptor. The *Npy*;*Lep*^{ob} double mutants also demonstrate the importance of combining targeted and classical mutations and creating lines of mice with mutations in genes unrelated at the sequence level.

By simply breeding together mice carrying mutations in different genes, many lines of mice have been generated which carry mutations in two related genes. The *Hox*, Rous sarcoma oncogene (*src*), insulin-like growth factor (*Igf*), and retinoid acid receptor gene families have undergone extensive genetic analyses, and different combinations of double mutants have helped establish or reshape conceptual frameworks for how some of the corresponding proteins are thought to function (99–105). Mutations in several *Hox* gene paralogues have resulted in ablation of structures in addition to homeotic transformations. Mice nullizygous for *Hoxa-11* and *Hoxd-11* virtually lack radius and ulna bones in their forelimbs (102), and *Hoxa-3* and *Hoxd-3* double mutants are missing their first cervical vertebrae (101). These findings led to a hypothesis that HOX proteins control cell proliferation and that altered cell fates which manifest as homeotic transformations might be a consequence of aberrant cell proliferation in groups of precursor cells (101, 106, 107).

4.2 Limitations, controls, and caveats

The number of targeted mutations which can be incorporated into a single mutant line is restricted by independent assortment and the limited reproductive capacity of mice compared to *C. elegans*, *Drosophila*, or zebrafish. In a double-heterozygous intercross, only 1 out of 16 progeny will be homozygous for both mutations. In a triple-heterozygous intercross, only 1 out of 64 progeny will be homozygous for each mutation, and, in a quadruple-heterozygous intercross, only 1 out of 256 progeny will be homozygous for all four mutations. Although a triple-gene mutation has been reported (108), it will be prohibitively expensive (due to animal costs) and time-consuming for many laboratories to go beyond a three or four gene knock-out if the single or double mutants are not viable. In cases where double or triple homozygotes are viable and fertile, however, it should be possible to disrupt five or more genes in a single line by crossing mice homozygous for two or three loci and heterozygous for several other loci.

Because different targeted mutants are often maintained on different, sometimes mixed, genetic backgrounds, the more severe phenotype of a 'double mutant' might be due to changes in genetic background rather than the mutations in question. Therefore, each genotypic class of progeny should be analysed to control for differences in genetic background. In a double-heterozygous intercross, for example, 2 out of 16 progeny will be homozygous for one gene while being wild-type for the other gene (+/+; –/– or –/–; +/+). These mice, along with the other predicted classes of progeny (+/–; +/–, +/–; –/–, –/–; +/–, +/+; +/+, –/–; –/–), will provide insight into the relative contribution of each mutation toward the mutant phenotype. Analysis of the different genotypic classes might also uncover dosage-dependent quantitative effects such that the mutant phenotype becomes progressively more severe as wild-type alleles are replaced by mutant alleles (102, 108). Such intergenic interactions can be additive or synergistic in nature, the latter suggesting that the gene products interact directly or indirectly. However, it should be noted that this is not necessarily the case; it is possible that two mutations compromise the general health of a common subset of cells by unrelated mechanisms and that these cells in double mutants exhibit a much more severe phenotype or even die.

5. Application of ENU and radiation mutagenesis in gene targeting experiments

5.1 Generating allelic series

As described in Section 2.2.2, allelic series are of critical importance for elucidating gene function. Although gene targeting can be utilized to generate various types of alleles, the lesions have been confined to well characterized motifs or domains and probably will continue to be with the possible exception of transposon-based strategies (109). This is expected, owing to the nature of gene targeting, but introduces considerable bias since the significance of only a short segment of a gene and corresponding protein is evaluated in each experiment. *neo* has also been targeted into introns, but these mutations often are difficult to interpret and do not delineate structure–function relationships (see Section 2.2.2).

The limitations associated with creating an allelic series by gene targeting do not apply to chemical mutagenesis. For example, the highly efficient mutagen ENU introduces point mutations randomly throughout genes, although only those base pair substitutions which impair gene function are detected in a typical screen (except for non-functional, sequence-based screens such as SSCP assays). Small intragenic deletions are also generated by ENU, albeit infrequently. Although ENU has generated many interesting mutations, e.g. Apc^{Min1} (110), *Clock* (111), Eed^{1989} (112), *Quaking* (113), *Kreisler* (114), and *Wheels* (115), it has not been used in conjunction with gene targeting. Nevertheless, these disparate approaches can be combined to generate mutant alleles not biased towards a particular DNA sequence. For instance, it is easy to envisage a one or two generation screen in which ENU-mutagenized chromosomes are balanced with a targeted chromosome to identify mutations which fail to complement the targeted null mutation (*Figure 5*). It should also be possible to perform second-site enhancer/suppressor screens by mutagenizing viable targeted mutants and identifying progeny with an increase or decrease in phenotype severity.

Screening for ENU-induced mutations in genes which have already been

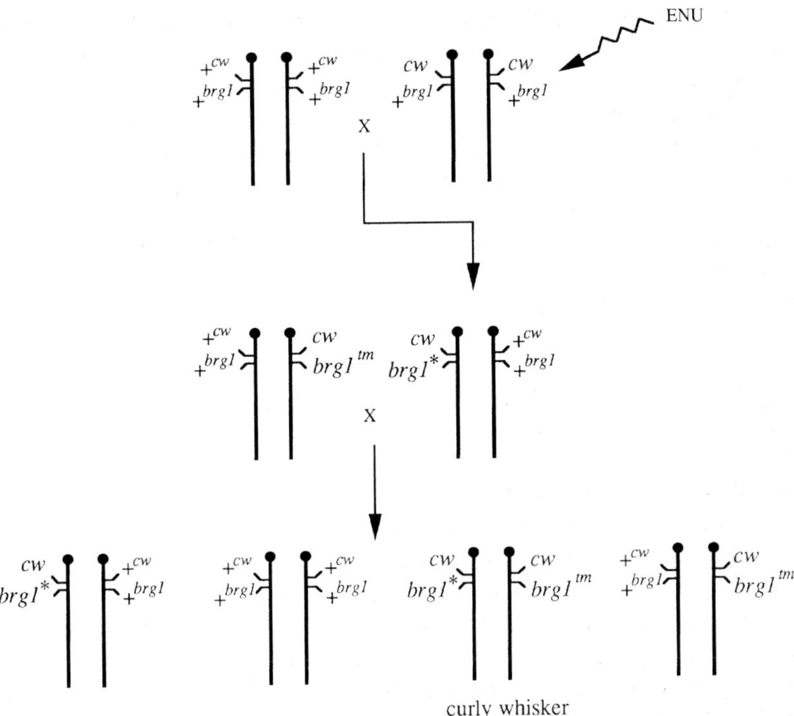

Figure 5. Breeding scheme for identifying and recovering ENU-induced mutations in cloned genes with a targeted mutation, exemplified by the brahma related gene 1 (*brg1*) on chromosome 9 (S. B. and T. M., unpublished data). Male mice homozygous for the recessive marker, curly whisker (*cw*), are mutagenized and bred with wild-type females. Resulting G_1 progeny are heterozygous for *cw* which is potentially linked to an ENU-induced *brg1* mutation (*brg1**). G_1 males are bred with females that have one chromosome 9 that is wild-type and one chromosome 9 carrying the *cw* marker and a targeted mutation of *brg1* ($brg1^{tm}$). 25% of the G_2 progeny are expected to have curly whiskers (*cw/cw*) and may exhibit a visible phenotype if an ENU-induced mutation of *brg1* arose in the treated male progenitor in G_0. It is also possible that curly whisker mice will be absent in G_2 if the ENU-induced mutation is lethal when balanced with the targeted mutation ($brg1^{*}/brg1^{tm}$). If this is the case, the *brg1** allele can be recovered by mating the appropriate G_1 male to *cw/cw* mice.

targeted should be fruitful based on the observation that a number of spontaneous classical mutants are due to hypomorphic mutations in genes that have also been completely inactivated by gene targeting: wingless-related MMTV integration site 1 (*Wnt1*) and *Swaying*, *Wnt3a* and *Vestigial tail*, *Egfr* and *Waved*-2, Fas antigen (*Fas*) and *Lymphoproliferation*. The classical mutations have shed insight into gene function not provided by the knock-out alleles because the phenotypes are largely non-overlapping. Additionally, the molecular lesions identified in several of these mutants (as well as other classical mutants which are either nulls or whose genes have not been targeted) lie outside of conserved motifs and likely would never be generated by gene targeting.

Although ENU-induced mutation frequencies vary considerably among different loci (116), one can realistically expect to recover one new mutation per locus for every 200–1000 gametes screened (117). By determining the concentration of ENU spectrophotometrically and optimizing the dose, even the less conservative 1/200 figure is attainable (118) (M. Justice, unpublished data). Outbred and F_1 hybrid (e.g. C3H/101) mouse strains often have been mutagenized because they can withstand relatively high doses of ENU without becoming permanently sterile. However, the BTBR strain has comparable vigour and the advantage of being inbred (119).

Even 1/200 may seem like a low frequency, but visible genetic markers are usually incorporated into genetic screens such that new mutations can be identified by simply examining G_1 or G_2 progeny without performing time-consuming molecular analyses (*Figure 5*). It might also be possible to increase the mutation frequency by mutagenizing primordial germ cells instead of spermatagonial stem cells (120) or by using strains of mice deficient in DNA repair enzymes such as O^6-ethyltransferase which removes alkyl adducts introduced by ENU.

Genetic screens in the mouse might also be made more efficient and cost-effective by being conducted *in vitro*. ENU can induce mutations at the hypoxanthine guanine phosphoribosyl transferase (*Hprt*) locus in P19 embryonal carcinoma cells (121) and ES cells (122) (D. Yee and T. M., unpublished data). If a sufficiently high mutation frequency can be obtained, then DNA from mutagenized ES cell clones could be PCR amplified and sequenced using new, highly efficient DNA microchip technologies (123) to identify mutations in non-selectable genes. ES cell clones with mutations in a gene of interest would be subsequently passed through the germline so that the phenotypes could be examined *in vivo*.

5.2 Deletion complexes in ES cells

The specific-locus test (SLT) was designed nearly 50 years ago by W. L. Russell to quantitate the effects of radiation exposure and chemical mutagens on mutation rates in the mammalian germline (124) but has also provided a wealth of biological information regarding a number of chromosomal regions in the mouse, totalling approximately 3% of the genome in aggregate (125). In the SLT, wild-type mice are mutagenized and bred to a tester stock homozygous for a number of visible, recessive mutations (non-agouti (*a*), brown ($Tyrp1^b$), chinchilla ($Tyr^{c\text{-}ch}$), dilute ($Myo5a^d$), pink-eyed dilution (*p*), piebald ($Ednrb^s$), and short ear ($Bmp5^{se}$) were employed in the original SLT). Mutations at any of these loci are revealed by F_1 progeny displaying the appropriate mutant phenotype. Over the years, a large number of mutations have been generated in the original seven marker genes as well as in additional morphological and biochemical marker genes that have been incorporated into SLTs (126).

Radiation frequently creates large deletions which ablate housekeeping or developmental genes tightly linked to visible markers in the SLT. Consequently, many deletions conferred lethal or viable mutant phenotypes when homozygous. Complementation analyses of numerous deletions demarcated functional units, and ENU mutagenesis was subsequently utilized to identify mutable genes in several of the deletion intervals (125). These deletions and point mutations facilitated the molecular cloning of several interesting genes in recent years. In the *Tyr*-locus (formerly *c*-locus) deletion complex, genes involved in pre-implantation survival, extraembryonic development, mesoderm induction, axial patterning, and haematopoeisis have been identified genetically (127), and genes involved in anterior-posterior pattern formation and regulation of liver-specific enzymes have been cloned (128–131). Additionally, allelic series exist for many of the genes which lie in the deletion complexes. Over 20 distinct alleles exist for some of the genes.

Although the SLT has been a powerful tool for decades, its utility has been limited by the fact that many regions of the genome lack visible markers which are vital for efficient mutagenesis screens. Furthermore, performing large scale mutagenesis experiments in the mouse is relatively inefficient and feasible only at institutions with large mouse colonies. An alternative strategy that could circumvent these limitations is to perform mutagenesis *in vitro* in ES cells. Relatively large deletions have been generated by gene targeting using both conventional and Cre-*loxP* targeting vectors (see Chapter 2). However, in order to produce deletion complexes consisting of random, partially overlapping deletions, a better approach is to irradiate ES cells that have a negative selectable marker such as HSV-TK targeted into a specific locus or chromosomal region. In fact, by selecting for irradiated clones that have lost HSV-TK, deletion complexes have already been generated in the *t* complex region of chromosome 17, around the neural cell adhesion molecule (*Ncam*) on chromosome 9, and *Hprt* on the X chromosome (132–134).

The proximal-distal boundaries of each deletion have been ascertained by identifying microsatellite markers which are polymorphic in the parental ES cell line but are no longer heterozygous in the irradiated clone. An ES cell line recently derived from a (129/Sv $+^{c\text{-}Tyr}$ $+^{p}$ × CAST/Ei)F_1 hybrid is a useful reagent for deletion mapping since it is highly polymorphic (133). Like in the SLT experiment, many deletions have been shown to span centimorgan intervals in ES cells (132–134). Most importantly, ES cells harbouring radiation-induced deletions have been transmitted through the germline (132, 134). This, of course, enables the phenotype of deletions created *in vitro* to be assessed *in vivo* and sets the foundation for performing complementation analyses and ENU mutagenesis screens. Ultimately, deletion complexes should provide a valuable resource for identifying the functions of numerous genes along megabase segments of DNA throughout the entire mouse genome.

6. Future prospects

Although the ability to exploit different genetic backgrounds in order to manipulate phenotypes of mutant mice has existed for more than half a century, the advent of microsatellite markers and a dense genetic map has made it possible to map modifier loci. Utilizing the complete sequence of the mouse genome should enable these genes to be cloned more efficiently. Analysis of modifier genes coupled with classical and targeted mutagenesis to create allelic series and multiple gene mutations will undoubtedly reveal genetic pathways and molecular mechanisms underlying fundamental biological processes.

Acknowledgements

We would like to thank Dr Nicholas Schork and members of the Magnuson laboratory for critical reading of this manuscript, and Dr Achim Gossler for contributing the design of *Figure 3*. S. B. was supported by the American Cancer Society and the work described was supported by grants from the NIH (HD26722 and HD24462) and March of Dimes to T. M.

References

1. Thomas, K. R. and Capecchi, M. R. (1987). *Cell*, **51**, 503.
2. Doetschman, T., Maeda, N., and Smithies, O. (1988). *Proc. Natl. Acad. Sci. USA*, **85**, 8583.
3. Thompson, S., Clarke, A. R., Pow, A. M., Hooper, M. L., and Melton, D. W. (1989). *Cell*, **56**, 313.
4. Cappecchi, M. R. (1989). *Science*, **244**, 1288.
5. Woychik, R. P., Wassom, J. S., Kingsbury, D., and Jacobson, D. A. (1993). *Nature*, **363**, 375.
6. Brandon, E. P., Idzerda, R. L., and McKnight, G. S. (1995). *Curr. Biol.*, **5**, 569.
7. Simpson, E. M., Linder, C. C., Sargent, E. E., Davisson, M. T., Mobraaten, L. E., and Sharp, J. J. (1997). *Nature Genet.*, **16**, 19.
8. Threadgill, D. W., Yee, D., Matin, A., Nadeau, J. H., and Magnuson, T. (1997). *Mamm. Genome*, **8**, 390.
9. Silvers, W. K. (ed.) (1979). *The coat colors of mice: a model for mammalian gene action and interaction*. Springer–Verlag, New York, NY.
10. Rijli, F. M., Dolle, P., Fraulob, V., LeMeur, M., and Chambon, P. (1994). *Dev. Dyn.*, **201**, 366.
11. Leighton, P. A., Saam, J. R., Ingram, R. S., Stewart, C. L., and Tilghman, S. M. (1995). *Genes Dev.*, **9**, 2079.
12. Olson, E. N., Arnold, H.-H., Rigby, P. W. J., and Wold, B. J. (1996). *Cell*, **85**, 1.
13. Bedell, M. A., Jenkins, N. A., and Copeland, N. G. (1996). *Nature Genet.*, **12**, 229.
14. Moens, C. B., Auerbach, A. B., Conlon, R. A., Joyner, A. L., and Rossant, J. (1992). *Genes Dev.*, **6**, 691.

15. Ramirez-Solis, R., Zheng, H., Whiting, J., Krumlauf, R., and Bradley, A. (1993). *Cell*, **73**, 279.
16. Fiering, S., Epner, E., Robinson, K., Zhuang, Y., Telling, A., Hu, M., *et al.* (1995). *Genes Dev.*, **9**, 2203.
17. Meyers, E. N., Lewandoski, M., and Martin, G. R. (1998). *Nature Genet.*, **18**, 136.
18. Nagy, A., Moens, C., Ivanyi, E., Pawling, J., Gertsenstein, M., Hadjantonakis, A. K., *et al.* (1998). *Curr. Biol.*, **8**, 661.
19. Floss, T., Arnold, H.-H., and Braun, T. (1996). *Dev. Biol.*, **174**, 140.
20. Yoon, J. K., Olson, E. N., Arnold, H.-H., and Wold, B. J. (1997). *Dev. Biol.*, **188**, 349.
21. Zweigerdt, R., Braun, T., and Arnold, H.-H. (1997). *Dev. Biol.*, **192**, 172.
22. Luetteke, N. C., Qiu, T. H., Peiffer, R. L., Oliver, P., Smithies, O., and Lee, D. C. (1993). *Cell*, **73**, 263.
23. Dorin, J. R., Stevenson, B. J., Fleming, S., Alton, E. W., Dickinson, P., and Porteous, D. J. (1994). *Mamm. Genome*, **5**, 465.
24. Threadgill, D. W., Dlugosz, A. A., Hansen, L. A., Tennenbaum, T., Lichti, U., Yee, D., *et al.* (1995). *Science*, **269**, 230.
25. Carmeliet, P., Ferreira, V., Breier, G., Pollefeyt, S., Kieckens, L., Gertsenstein, M., *et al.* (1996). *Nature*, **380**, 435.
26. Blendy, J. A., Kaestner, K. H., Schmid, W., Gass, P., and Schutz, G. (1996). *EMBO J.*, **15**, 1098.
27. Hakem, R., de la Pompa, J. L., Sirard, C., Mo, R., Woo, M., Hakem, A., *et al.* (1996). *Cell*, **85**, 1009.
28. Rabinowitz, J. E., Rutishauser, U., and Magnuson, T. (1996). *Proc. Natl. Acad. Sci. USA*, **93**, 6421.
29. Smithies, O. and Kim, H.-S. (1994). *Proc. Natl. Acad. Sci. USA*, **91**, 3612.
30. Tomasiewicz, H., Ono, K., Yee, D., Thompson, C., Goridis, C., Rutishauser, U., *et al.* (1993). *Neuron*, **11**, 1163.
31. Colledge, W. H., Abella, B. S., Southern, K. W., Ratcliff, R., Jiang, C., Cheng, S. H., *et al.* (1995). *Nature Genet.*, **10**, 445.
32. van Doorninck, J. H., French, P. J., Verbeek, E., Peters, R. H., Morreau, H., Bijman, J., *et al.* (1995). *EMBO J.*, **14**, 4403.
33. Zeiher, B. G., Eichwald, E., Zabner, J., Smith, J. J., Puga, A. P., McCray, P. B., *et al.* (1995). *J. Clin. Invest.*, **96**, 2051.
34. Delaney, S. J., Alton, E. W., Smith, S. N., Lunn, D. P., Farley, R., Lovelock, P. K., *et al.* (1996). *EMBO J.*, **15**, 955.
35. Lakhlani, P. P., MacMillan, L. B., Guo, T. Z., McCool, B. A., Lovinger, D. M., Maze, M., *et al.* (1997). *Proc. Natl. Acad. Sci. USA*, **94**, 9950.
36. Shivdasani, R. A., Fujiwara, Y., McDevitt, M. A., and Orkin, S. H. (1997). *EMBO J.*, **16**, 3965.
37. Zákány, J., Gérard, M., Favier, B., and Duboule, D. (1997). *EMBO J.*, **16**, 4393.
38. Tarutani, M., Itami, S., Okabe, M., Ikawa, M., Tezuka, T., Yoshikawa, K., *et al.* (1997). *Proc. Natl. Acad. Sci. USA*, **94**, 7400.
39. Betz, U. A., Vosshenrich, C. A., Rajewsky, K., and Muller, W. (1996). *Curr. Biol.*, **6**, 1307.
40. Brocard, J., Warot, X., Wendling, O., Messaddeq, N., Vonesch, J.-L., Chambon, P., *et al.* (1997). *Proc. Natl. Acad. Sci. USA*, **94**, 14559.
41. Gagneten, S., Le, Y., Miller, J., and Sauer, B. (1997). *Nucleic Acids Res.*, **25**, 3326.

42. Akagi, K., Sandig, V., Vooijs, M., Van der Valk, M., Giovannini, M., Strauss, M., *et al.* (1997). *Nucleic Acids Res.*, **25**, 1766.
43. Zinyk, D. L., Mercer, E. H., Harris, E., Anderson, D. J., and Joyner, A. L. (1998). *Curr Biol.*, **8**, 665.
44. Sibilia, M. and Wagner, E. F. (1995). *Science*, **269**, 234.
45. Snouwaert, J. N., Brigman, K. K., Latour, A. M., Malouf, N. N., Boucher, R. C., Smithies, O., *et al.* (1992). *Science*, **257**, 1083.
46. O'Neal, W. K., Hasty, P., McCray, P. B., Casey, B., Rivera-Perez, J., Welsh, M. J., *et al.* (1993). *Hum. Mol. Genet.*, **2**, 1561.
47. Ratcliff, R., Evans, M. J., Cuthbert, A. W., MacVinish, L. J., Foster, D., Anderson, J. R., *et al.* (1993). *Nature Genet.*, **4**, 35.
48. Smithies, O. (1993). *Trends Genet.*, **9**, 112.
49. Rozmahel, R., Wilschanski, M., Matin, A., Plyte, S., Oliver, M., Auerbach, W., *et al.* (1996). *Nature Genet.*, **12**, 280.
50. Bonyadi, M., Rusholme, S. A. B., Cousins, F. M., Su, H. C., Biron, C. A., Farrall, M., *et al.* (1997). *Nature Genet.*, **15**, 207.
51. Clarke, A. R. (1994). *Curr. Opin. Genet. Dev.*, **4**, 453.
52. Erickson, R. P. (1996). *BioEssays*, **18**, 993.
53. Bedell, M. A., Jenkins, N. A., and Copeland, N. G. (1997). *Genes Dev.*, **11**, 1.
54. Wicker, L. S., Todd, J. A., and Peterson, L. B. (1995). *Annu. Rev. Immunol.*, **13**, 179.
55. Vyse, T. J. and Todd, J. A. (1996). *Cell*, **85**, 311.
56. DeBry, R. W. and Seldin, M. F. (1996). *Genomics*, **33**, 337.
57. Copeman, J. B., Cucca, F., Hearne, C. M., Cornall, R. J., Reed, P. W., Ronningen, K. S., *et al.* (1995). *Nature Genet.*, **9**, 80.
58. Paigen, B., Ishida, B. Y., Verstuyft, J., Winters, R. B., and Albee, D. (1990). *Atherosclerosis*, **10**, 316.
59. Paigen, B. (1995). *Am. J. Clin. Nutr.*, **62**, 458.
60. Sweet, A., Erickson, R. P., Huntington, C., and Dawson, D. (1992). *Biochem. Med. Metab. Biol.*, **47**, 97.
61. Silver, L. M. (ed.) (1995). *Mouse genetics: concepts and applications*. Oxford University Press, New York, NY.
62. Visscher, P. M., Haley, C. S., and Thompson, R. (1996). *Genetics*, **144**, 1923.
63. Hospital, F. and Charcosset, A. (1997). *Genetics*, **147**, 1469.
64. Markel, P., Shu, P., Ebeling, C., Carlson, G. A., Nagle, D. L., Smutko, J. S., *et al.* (1997). *Nature Genet.*, **17**, 280.
65. Behringer, R. (1998). *Nature Genet.*, **18**, 136.
66. Dunn, L. C. (1937). *Genetics*, **22**, 43.
67. Dunn, L. and Charles, D. (1937). *Genetics*, **22**, 14.
68. Reed, S. C. (1937). *Genetics*, **22**, 1.
69. Grüneberg, H. (1963). *The pathology of development*. Wiley, New York, NY.
70. Wallace, M. E. (1976). *Genetica*, **46**, 529.
71. Sweet, H. O. (1983). *J. Hered.*, **74**, 305.
72. Weber, J. L. and May, P. E. (1989). *Am. J. Hum. Genet.*, **44**, 388.
73. Love, J. M., Knight, A. M., McAleer, M. A., and Todd, J. A. (1990). *Nucleic Acids Res.*, **18**, 4123.
74. Jacob, H. J., Lindpaintner, K., Lincoln, S. E., Kusumi, K., Bunker, R. K., Mao, Y.-P., *et al.* (1991). *Cell*, **67**, 213.

75. Lander, E. S., Green, P., Abrahamson, J., Barlow, A., Daly, M. J., Lincoln, S. S., *et al.* (1987). *Genomics*, **1**, 187.
76. Montagutelli, X. (1990). *J. Hered.*, **81**, 490.
77. Manly, K. F. (1993). *Mamm. Genome*, **4**, 303.
78. Lander, E. S. and Kruglyak, L. (1995). *Nature Genet.*, **11**, 241.
79. Lander, E. S. and Schork, N. J. (1994). *Science*, **265**, 2037.
80. Lander, E. S. and Botstein, D. (1989). *Genetics*, **121**, 185.
81. Borsani, G., Ballabio, A., and Banfi, S. (1998). *Hum. Mol. Genet.*, **7**, 1641.
82. MacPhee, M., Chepenik, K. P., Liddell, R. A., Nelson, K. K., Siracusa, L. D., and Buchberg, A. M. (1995). *Cell*, **81**, 957.
83. Dietrich, W. F., Lander, E. S., Smith, J. S., Moser, A. R., Gould, K. A., Luongo, C., *et al.* (1993). *Cell*, **75**, 631.
84. Cormier, R. T., Hong, K. H., Halberg, R. B., Hawkins, T. L., Richardson, P., Mulherkar, R., *et al.* (1997). *Nature Genet.*, **17**, 88.
85. Tischfield, J. A., Xia, Y.-R., Shih, D. M., Klisak, I., Chen, J., Engle, S. J., *et al.* (1996). *Genomics*, **32**, 328.
86. Joyner, A. L., Herrup, K., Auerbach, B. A., Davis, C. A., and Rossant, J. (1991). *Science*, **251**, 1239.
87. Laird, P. W., van der Lugt, N. M., Clarke, A., Domen, J., Linders, K., McWhir, J., *et al.* (1993). *Nucleic Acids Res.*, **21**, 4750.
88. Muller, H. J. (1935). *J. Hered.*, **26**, 469.
89. Rudnicki, M. A., Braun, T., Hinuma, S., and Jaenisch, R. (1992). *Cell*, **71**, 383.
90. Zhang, W., Behringer, R. R., and Olson, E. N. (1995). *Genes Dev.*, **9**, 1388.
91. Hanks, M., Wurst, W., Anson-Cartwright, L., Auerbach, A. B., and Joyner, A. L. (1995). *Science*, **269**, 679.
92. Wang, Y., Schnegelsberg, P. N. J., Dausman, J., and Jaenisch, R. (1996). *Nature*, **379**, 823.
93. Nowak, M. A., Boerlijst, M. C., Cooke, J., and Smith, J. M. (1997). *Nature*, **388**, 167.
94. Wang, Y. and Jaenisch, R. (1997). *Development*, **124**, 2507.
95. Rudnicki, M. A., Schnegelsberg, P. N., Stead, R. H., Braun, T., Arnold, H.-H., and Jaenisch, R. (1993). *Cell*, **75**, 1351.
96. Erickson, J. C., Hollopeter, G., and Palmiter, R. D. (1996). *Science*, **274**, 1704.
97. Filosa, S., Rivera-Peréz, J. A., Gómez, A. P., Gansmuller, A., Sasaki, H., Behringer, R. R., *et al.* (1997). *Development*, **124**, 2843.
98. Erickson, J. C., Clegg, K. E., and Palmiter, R. D. (1996). *Nature*, **381**, 415.
99. Baker, J., Liu, J. P., Robertson, E. J., and Efstratiadis, A. (1993). *Cell*, **75**, 73.
100. Filson, A. J., Louvi, A., Efstratiadis, A., and Robertson, E. J. (1993). *Development*, **118**, 731.
101. Condie, B. G. and Capecchi, M. R. (1994). *Nature*, **370**, 304.
102. Davis, A. P., Witte, D. P., Hsieh-Li, H. M., Potter, S. S., and Capecchi, M. R. (1995). *Nature*, **375**, 791.
103. Lowell, C. A. and Soriano, P. (1996). *Genes Dev.*, **10**, 1845.
104. Kastner, P., Mark, M., Ghyselinck, N., Krezel, W., Dupe, V., Grondona, J. M., *et al.* (1997). *Development*, **124**, 313.
105. Louvi, A., Accili, D., and Efstratiadis, A. (1997). *Dev. Biol.*, **189**, 33.
106. Duboule, D. (1995). *Curr. Opin. Genet. Dev.*, **5**, 525.
107. Rijli, F. M. and Chambon, P. (1997). *Curr. Opin. Genet. Dev.*, **7**, 481.

108. Horan, G. S., Ramirez-Solis, R., Featherstone, M. S., Wolgemuth, D. J., Bradley, A., and Behringer, R. R. (1995). *Genes Dev.*, **9**, 1667.
109. Westphal, C. H. and Leder, P. (1997). *Curr. Biol.*, **7**, 530.
110. Moser, A. R., Pitot, H. C., and Dove, W. F. (1990). *Science*, **247**, 322.
111. Vitaterna, M. H., King, D. P., Chang, A.-M., Kornhauser, J. M., Lowrey, P. L., McDonald, J. D., *et al.* (1994). *Science*, **264**, 719.
112. Rinchik, E. M. and Carpenter, D. A. (1993). *Mamm. Genome*, **4**, 349.
113. Ebersole, T. A., Chen, Q., Justice, M. J., and Artzt, K. (1996). *Nature Genet.*, **12**, 260.
114. Cordes, S. P. and Barsh, G. S. (1994). *Cell*, **79**, 1025.
115. Nolan, P. M., Sollars, P. J., Bohne, B. A., Ewens, W. J., Pickard, G. E., and Bucan, M. (1995). *Genetics*, **140**, 245.
116. Hitotsumachi, S., Carpenter, D. A., and Russell, W. L. (1985). *Proc. Natl. Acad. Sci. USA*, **82**, 6619.
117. Rinchik, E. M. (1991). *Trends Genet.*, **7**, 15.
118. Shedlovsky, A., McDonald, J. D., Symula, D., and Dove, W. F. (1993). *Genetics*, **134**, 1205.
119. Shedlovsky, A., Guenet, J. L., Johnson, L. L., and Dove, W. F. (1986). *Genet. Res.*, **47**, 135.
120. Shibuya, T., Horiya, N., Matsuda, H., Sakamoto, K., and Hara, T. (1996). *Mutat. Res.*, **357**, 219.
121. Sehlmeyer, U. and Wobus, A. M. (1994). *Mutat. Res.*, **324**, 69.
122. Dobrovolsky, V. N., Casciano, D. A., and Heflich, R. H. (1996). *Environ. Mol. Mutagen.*, **28**, 483.
123. Southern, E. M. (1996). *Trends Genet.*, **12**, 110.
124. Russell, W. L. (1951). *Cold Spring Harbor Symp. Quant. Biol.*, **16**, 327.
125. Rinchik, E. M. and Russell, L. B. (1990). In *Genome analysis volume 1: genetic and physical mapping* (ed. K. E. Davies and S. M. Tilghman), pp. 121–58. Cold Spring Harbor Laboratory Press, Cold Spring Harbor, NY.
126. Lyon, M. F. and Morris, T. (1966). *Genet. Res.*, **7**, 12.
127. Holdener-Kenny, B., Sharan, S. K., and Magnuson, T. (1992). *BioEssays*, **14**, 831.
128. Klebig, M. L., Russell, L. B., and Rinchik, E. M. (1992). *Proc. Natl. Acad. Sci. USA*, **89**, 1363.
129. Ruppert, S., Kelsey, G., Schedl, A., Schmid, E., Thies, E., and Schutz, G. (1992). *Genes Dev.*, **6**, 1430.
130. Grompe, M., al-Dhalimy, M., Finegold, M., Ou, C. N., Burlingame, T., Kennaway, N. G., *et al.* (1993). *Genes Dev.*, **7**, 2298.
131. Schumacher, A., Faust, C., and Magnuson, T. (1996). *Nature*, **383**, 250.
132. You, Y., Bergstrom, R., Klemm, M., Lederman, B., Nelson, H., Ticknor, C., *et al.* (1997). *Nature Genet.*, **15**, 285.
133. Thomas, J., LaMantia, C., and Magnuson, T. (1998). *Proc. Natl. Acad. Sci. USA*, **95**, 114.
134. Kushi, A., Edamura, K., Noguchi, M., Akiyama, K., Nishi, Y., and Sasai, H. (1998). *Mamm. Genome*, **9**, 269.
135. Copp, A. J. (1995). *Trends Genet.*, **11**, 87.
136. Liu, J.-P., Baker, J., Perkins, A. S., Robertson, E. J., and Efstratiadis, A. (1993). *Cell*, **75**, 59.

137. Baribault, H., Penner, J., Iozzo, R. V., and Wilson-Heiner, M. (1994). *Genes Dev.*, **8**, 2964.
138. Proetzel, G., Pawlowski, S. A., Wiles, M. V., Yin, M., Boivin, G. P., Howles, P. N., *et al.* (1995). *Nature Genet.*, **11**, 409.
139. Harvey, M., McArthur, M. J., Montgomery, C. A., Bradley, A., and Donehower, L. A. (1993). *FASEB J.*, **7**, 938.

A1

List of suppliers

Amersham

Amersham International plc., Lincoln Place, Green End, Aylesbury, Buckinghamshire HP20 2TP, UK.

Amersham Corporation, 2636 South Clearbrook Drive, Arlington Heights, IL 60005, USA.

Anderman

Anderman and Co. Ltd., 145 London Road, Kingston-Upon-Thames, Surrey KT17 7NH, UK.

Beckman Instruments

Beckman Instruments UK Ltd., Oakley Court, Kingsmead Business Park, London Road, High Wycombe, Bucks HP11 1J4, UK.

Beckman Instruments Inc., PO Box 3100, 2500 Harbor Boulevard, Fullerton, CA 92634, USA.

Becton Dickinson

Becton Dickinson and Co., Between Towns Road, Cowley, Oxford OX4 3LY, UK.

Becton Dickinson and Co., 2 Bridgewater Lane, Lincoln Park, NJ 07035, USA.

Bio

Bio 101 Inc., c/o Statech Scientific Ltd, 61–63 Dudley Street, Luton, Bedfordshire LU2 0HP, UK.

Bio 101 Inc., PO Box 2284, La Jolla, CA 92038–2284, USA.

Bio-Rad Laboratories

Bio-Rad Laboratories Ltd., Bio-Rad House, Maylands Avenue, Hemel Hempstead HP2 7TD, UK.

Bio-Rad Laboratories, Division Headquarters, 3300 Regatta Boulevard, Richmond, CA 94804, USA.

BLS

Biological Laboratory Equipments Maintenance and Service Ltd, Zsélyi Aladár M. 31. H-1165 Budapest, Hungary.

Boehringer Mannheim

Boehringer Mannheim UK (Diagnostics and Biochemicals) Ltd, Bell Lane, Lewes, East Sussex BN17 1LG, UK.

Boehringer Mannheim Corporation, Biochemical Products, 9115 Hague Road, P.O. Box 504 Indianapolis, IN 46250–0414, USA.

Boehringer Mannheim Biochemica, GmbH, Sandhofer Str. 116, Postfach 310120 D-6800 Ma 31, Germany.

British Drug Houses (BDH) Ltd, Poole, Dorset, UK.

Difco Laboratories

Difco Laboratories Ltd., P.O. Box 14B, Central Avenue, West Molesey, Surrey KT8 2SE, UK.

Difco Laboratories, P.O. Box 331058, Detroit, MI 48232–7058, USA.

Du Pont

Dupont (UK) Ltd., Industrial Products Division, Wedgwood Way, Stevenage, Herts, SG1 4Q, UK.

Du Pont Co. (Biotechnology Systems Division), P.O. Box 80024, Wilmington, DE 19880–002, USA.

European Collection of Animal Cell Culture, Division of Biologics, PHLS Centre for Applied Microbiology and Research, Porton Down, Salisbury, Wilts SP4 0JG, UK.

Falcon (Falcon is a registered trademark of Becton Dickinson and Co.).

Fisher Scientific Co., 711 Forbest Avenue, Pittsburgh, PA 15219–4785, USA.

Flow Laboratories, Woodcock Hill, Harefield Road, Rickmansworth, Herts. WD3 1PQ, UK.

Fluka

Fluka-Chemie AG, CH-9470, Buchs, Switzerland.

Fluka Chemicals Ltd., The Old Brickyard, New Road, Gillingham, Dorset SP8 4JL, UK.

Gibco BRL

Gibco BRL (Life Technologies Ltd.), Trident House, Renfrew Road, Paisley PA3 4EF, UK.

Gibco BRL (Life Technologies Inc.), 3175 Staler Road, Grand Island, NY 14072–0068, USA.

Arnold R. Horwell, 73 Maygrove Road, West Hampstead, London NW6 2BP, UK.

Hybaid

Hybaid Ltd., 111–113 Waldegrave Road, Teddington, Middlesex TW11 8LL, UK.

Hybaid, National Labnet Corporation, P.O. Box 841, Woodbridge, NJ. 07095, USA.

HyClone Laboratories 1725 South HyClone Road, Logan, UT 84321, USA.

International Biotechnologies Inc., 25 Science Park, New Haven, Connecticut 06535, USA.

Invitrogen Corporation

Invitrogen Corporation 3985 B Sorrenton Valley Building, San Diego, CA. 92121, USA.

Invitrogen Corporation c/o British Biotechnology Products Ltd., 4–10 The Quadrant, Barton Lane, Abingdon, OX14 3YS, UK.

Kodak: Eastman Fine Chemicals 343 State Street, Rochester, NY, USA.

Life Technologies Inc., 8451 Helgerman Court, Gaithersburg, MN 20877, USA.

Merck

Merck Industries Inc., 5 Skyline Drive, Nawthorne, NY 10532, USA.

Merck, Frankfurter Strasse, 250, Postfach 4119, D-64293, Germany.

Millipore

Millipore (UK) Ltd., The Boulevard, Blackmoor Lane, Watford, Herts WD1 8YW, UK.

Millipore Corp./Biosearch, P.O. Box 255, 80 Ashby Road, Bedford, MA 01730, USA.

New England Biolabs (NBL)

New England Biolabs (NBL), 32 Tozer Road, Beverley, MA 01915–5510, USA.

New England Biolabs (NBL), c/o CP Labs Ltd., P.O. Box 22, Bishops Stortford, Herts CM23 3DH, UK.

Nikon Corporation, Fuji Building, 2–3 Marunouchi 3-chome, Chiyoda-ku, Tokyo, Japan.

Perkin-Elmer

Perkin-Elmer Ltd., Post Office Lane, Beaconsfield, Bucks, HP9 1QA, UK.

Perkin-Elmer-Cetus (The Perkin-Elmer Corporation), 761 Main Avenue, Norwalk, CT 0689, USA.

Pharmacia Biotech Europe Procordia EuroCentre, Rue de la Fuse-e 62, B-1130 Brussels, Belgium.

Pharmacia Biosystems

Pharmacia Biosystems Ltd. (Biotechnology Division), Davy Avenue, Knowlhill, Milton Keynes MK5 8PH, UK.

Pharmacia LKB Biotechnology AB, Björngatan 30, S-75182 Uppsala, Sweden.

Promega

Promega Ltd., Delta House, Enterprise Road, Chilworth Research Centre, Southampton, UK.

Promega Corporation, 2800 Woods Hollow Road, Madison, WI 53711–5399, USA.

Qiagen

Qiagen Inc., c/o Hybaid, 111–113 Waldegrave Road, Teddington, Middlesex, TW11 8LL, UK.

Qiagen Inc., 9259 Eton Avenue, Chatsworth, CA 91311, USA.

Schleicher and Schuell

Schleicher and Schuell Inc., Keene, NH 03431A, USA.

Schleicher and Schuell Inc., D-3354 Dassel, Germany. Schleicher and Schuell Inc., c/o Andermann and Company Ltd.

Shandon Scientific Ltd., Chadwick Road, Astmoor, Runcorn, Cheshire WA7 1PR, UK.

Sigma Chemical Company
Sigma Chemical Company (UK), Fancy Road, Poole, Dorset BH17 7NH, UK.
Sigma Chemical Company, 3050 Spruce Street, P.O. Box 14508, St. Louis, MO 63178–9916.
Sorvall DuPont Company, Biotechnology Division, P.O. Box 80022, Wilmington, DE 19880–0022, USA.
Stratagene
Stratagene Ltd., Unit 140, Cambridge Innovation Centre, Milton Road, Cambridge CB4 4FG, UK.
Strategene Inc., 11011 North Torrey Pines Road, La Jolla, CA 92037, USA.
United States Biochemical, P.O. Box 22400, Cleveland, OH 44122, USA.
Wellcome Reagents, Langley Court, Beckenham, Kent BR3 3BS, UK.

Index

Index

Index